Reducing Pa
Emissions in
Gasoline Engines

Other SAE books of interest:

Particulate Emissions from Vehicles
Author: Peter Eastwood
PRODUCT CODE: R-389

Fuel/Engine Interactions
Author: Gautam Kalghatgi
PRODUCT CODE: R-409

Emissions and Air Quality
Author: Christian Cozzarini, Hans P. Lenz
PRODUCT CODE: R-237

Automotive Fuels Reference Book
Author: Paul Richards
PRODUCT CODE: R-297

Emission Control and Fuel Economy for Port and Direct Injected SI Engines
Author: John H. Johnson
PRODUCT CODE: PT-91

Diesel Particulate Filter Technology
Author: Timothy V. Johnson
PRODUCT CODE: PT-124

Engine Emissions Measurement Handbook
Author: Hiroshi Nakamura, Masayuki Adachi
PRODUCT CODE: JPF-HOR-002

Reducing Particulate Emissions in Gasoline Engines

THORSTEN BOGER AND WILLARD CUTLER

SAE INTERNATIONAL®

Warrendale, Pennsylvania, USA

SAE INTERNATIONAL®

400 Commonwealth Drive
Warrendale, PA 15096-0001 USA
E-mail: CustomerService@sae.org
Phone: 877-606-7323 (inside USA and Canada)
724-776-4970 (outside USA)
Fax: 724-776-0790

Library of Congress Catalog Number 2018945860
SAE Order Number R-471
http://dx.doi.org/10.4271/R-471

ISBN-Print 978-0-7680-9417-6
ISBN-PDF 978-0-7680-9543-2
ISBN-ePUB 978-0-7680-9419-0
ISBN-PRC 978-0-7680-9420-6
ISBN-HTML 978-0-7680-9418-3

To purchase bulk quantities, please contact: SAE Customer Service

E-mail: CustomerService@sae.org
Phone: 877-606-7323 (inside USA and Canada)
724-776-4970 (outside USA)
Fax: 724-776-0790

Visit the SAE International Bookstore at books.sae.org

contents

Section 1

Introduction to Particulate Emissions

CHAPTER 1

Gasoline Engine Particulate Emissions Introduction

Willard A. Cutler and Thorsten Boger

> *I durst not laugh for fear of opening my lips and receiving the bad air.*
>
> **—William Shakespeare, Julius Caesar**

The editors are pleased to participate in promoting the discussion of the important topic of gasoline vehicle particulate emissions reductions. For many years, gasoline particulates were seldom-discussed, due to the rationale that "if you can't see it, it doesn't exist." This book is our attempt to present a wide range of subjects related to gasoline particulates and their control. The reader will find chapters written by knowledgeable authors, arguably experts in their fields. While there is much yet to discover on this issue, we hope this primer will aid you on your journey to cleaner, more efficient gasoline systems.

Diesel engines have long been implicated as particle producers, producing visible smoke due to high particulates. The Euro 4 regulations in Europe (EU) required a particle mass (PM) reduction. The Euro 5b regulation required further PM reduction and introduced a particle number (PN) standard. Subsequent regulations in the

United States required particulate filters on all diesel vehicles using PM standards. In 2012, diesel soot was declared a carcinogen by the World Health Organization. Now, one seldom sees visible smoke from a diesel engine in the western world, unless the owner has tampered with the system (sometimes referred to as rolling coal) or the system has a defect. Diesel vehicles equipped with particulate filters are now cleaner than gasoline vehicles, from both a PN and PM standpoint.

While gasoline engines seldom emit visible smoke, they can and do produce a lot of particles. Gasoline engine technology has improved power and fuel economy, such as in the case of gasoline direct injection (GDI) technology, but can also produce additional unseen particles. A typical gasoline engine emits many trillions of particles every km (about a billion particles every meter). These unseen particles come in a variety of sizes from ~10 nm to several hundred nm. From a health standpoint, the smaller the particle, the deeper into the body they travel [1]. Unfortunately, once emitted, the likelihood of inhaling these particles increases, particularly near the location where the car is started, as more than half of PN is produced in the first minutes of operation. This makes car-starting locations, like curbsides, home garages, and parking garages, potentially high-concentration zones. Other potential high-risk zones include areas near stoplights and traffic throughways. Chapter 2 deals with human health impacts of particulates which, in the case of gasoline particulates, is an emerging field. These gasoline particulates can range from irritants to downright hazards. Chapter 3 also touches on some of the health and social impacts of particulates.

European entities appear to be the first to recognize the importance of limiting the number of particles emitted from both diesel and gasoline engines, in addition to the mass of the particles. Europe began to regulate gasoline direct injection engines in 2014, with a PN limit of 6×10^{11} particles/km (that's 600 billion particles/km) on the new European drive cycle (NEDC) certification test cycle. For the first 3 years, manufacturers were allowed to exceed this PN value by a factor of 10. Since September 2017, the 6×10^{11} particles/km limit value has to be met for all new type approvals, tested now over the new certification cycle, the world harmonized light vehicle test cycle (WLTC). In addition to the new test cycle, the European regulators also added a second condition for real-world driving emissions (RDE), to assure that the PN won't exceed the regulatory limit (plus a measurement error, currently set at 0.5 times), on the RDE or during normal driving. The EU regulations will be fully phased in by 2020.

China today has a much bigger pollution problem. A study from the UK medical journal, *The Lancet*, describes >8% of lives lost prematurely each year in China due to the negative impact of particles on human health [2]. China has taken steps to address its diesel vehicle emission problem with regulations [3] but believes it also needs to tackle less visible gasoline particles and has decided to be more expansive than EU and will regulate particulates on all gasoline engines, not just GDI engines. China is using the same regulatory limits on the same test as Europe, but for the time being, they have a higher RDE limit and different interpretations of methodology. However, fuel standards in China are a little different, which means the same engine tested with fuels in China likely produces more PN than when tested with EU fuels.

The United States has long led the global emission standards. However, for light-duty vehicles in the United States, both Tier 3 and LEV III standards specify a 3 mg/mile PM standard and no PN standards. The California Air Resource Board has proposed dropping the PM standard from 3 to 1 mg/mile starting in 2025. This should further reduce particulates but may suggest U.S. cars are allowed to emit about two times more particles than their European or Chinese counterparts with a PN standard. Chapter 3 deals with regulations and potential technologies to address the regulatory challenge worldwide.

Particulates are most often the result of incompletely combusted fuel, with many potential mechanisms occurring simultaneously under a wide variety of engine conditions to produce particles. Particulate output and composition change as the result of a number of varied factors including, but not limited to, temperature, altitude, fuel quality, vehicle power-to-weight ratio, drive cycle, as well as engine hardware and software. In addition, high exhaust temperatures during high-speed and high-load operation can lead to PM storage and release of material from sampling system surfaces [4] and exhaust systems [5], increasing uncertainty for OEMs. Experience from Corning's own test cars, which by no means represents a statistical sampling, suggests engine-out PN increases as mileage increases, sometimes significantly (2–3X over life). Chapter 4 provides a thorough introduction to factors influencing soot formation during combustion and operation, while Chapter 5 describes the influence of fuels and their components on soot formation.

Perhaps the most elegant way to deal with the particulate problem would be to modify combustion, so that particles are not produced at all. At some cost, various hardware and software engine modifications can be made, but given the complex interrelationships involved in particulate formation, it is often difficult to eliminate particles, while maintaining acceptable fuel economy, power, gaseous emissions, noise, and drivability for all RDE conditions, particularly in cold weather. Chapter 6 describes advanced combustion and engine control technologies to mitigate particle formation.

When particles are produced, it's possible to apply a gasoline particulate filter (GPF) to capture toxic particles on its porous walls. Fortunately, the filter can't be easily bypassed without hardware modifications, making it a secure solution for the life of the vehicle. The filter typically improves its filtration efficiency with increasing mileage, suggesting the fresh filtration efficiency will likely be the most critical operating condition. The captured particles remain in the filter until exhaust conditions are appropriate to burn the particles to "clean" or "regenerate" the filter. In gasoline applications, this often happens passively, as part of the driving cycle, due to the high gasoline exhaust temperatures and oxygen present in the exhaust during a fuel cut (as the foot comes off the accelerator). However, there are likely to be a limited number of circumstances where the computer on the vehicle must actively initiate a filter regeneration. Fortunately, since the mass of the soot collected is quite low, the regeneration exotherm is also low, relative to diesel applications. Such low exotherms suggest there is not a need for high-heat-capacity materials like aluminum titanate or silicon carbide to adsorb the exothermic reaction. Rather cordierite, a high-melting-point, low-expansion, low-thermal-conductivity material, is the preferred GPF material due to its low heat capacity, low cost, and macrostructural and microstructural flexibility.

Chapter 7 describes filter designs and fundamentals, related to particulate capture, both for combustible soot particles and the non-combustible inorganic ash particles (particles that remains in the filter for the filter life).

The basic wall-flow filtration principle has been used for many years to effectively capture and eliminate particles in diesel systems. However, diesel engines produce a lot of soot from both a number and a mass perspective. This soot output, combined with their low engine-out temperatures, often results in a soot cake on the DPF, which increases the native filtration efficiency (>95% filtration efficiency), but makes it difficult to continuously regenerate. As a result, all diesel passenger car applications use "active" regeneration software to initiate and control regenerations.

Gasoline systems often produce less soot from a mass perspective, but still significant PNs. The regulations have created specific PN and PM limits. Since engine/vehicle combinations are not all created equal, OEMs typically need 40%-95% PN filtration efficiency to reach the regulatory limit. Cleaner engine/vehicle combinations (those emitting fewer particles) require ~40%-50% PN removal to hit the regulations. On the other hand, the highest emitting engine/vehicle combinations (higher particle, engine-out) require >80% PN removal to reach the regulations. The need for high filtration efficiency (>80%) is becoming more widespread as more RDE conditions become apparent, including those during cold temperatures, high altitudes, and rapid accelerations. Chapter 8 describes filter application mechanics, as well as filter survivability and durability.

In addition, some OEMs would like to add a catalytic three-way catalyst (TWC) functionality to enable gaseous pollutant reduction to the filter. To meet the wide array of needs, a portfolio of GPF products have been developed, which can be separately used or integrated with a TWC. These products can remove particles to meet the regulatory limits, while balancing other OEM demands (fuel economy, vehicle power, catalyst integration, space constraints, etc.). Chapters 9 and 10 describe the potential of TWC integration on filters.

While everyone wants to live in a world with fewer particulates, the need hierarchy for most appears to be (1) meeting the regulations at the minimum cost and risk and (2) maintaining vehicle drivability and experience. Chapters 11 and 12 provide real examples of successful OEM filter integration with bare filters and TWC integrated filters, on actual vehicles in preparation of series introductions. This work provides the basis for current introductions in Europe and China.

In order to regulate PN and PM, one must be able to accurately measure both, in the laboratory, and on the road with portable emissions measurement systems (PEMS), topics covered in Chapter 13. In addition, on-board diagnostics (OBD) are important, as they play a role in determining whether or not a GPF is damaged, or even present on the vehicle. OBD and its various hardware and software details are introduced in Chapter 14.

While gasoline particulate control is relatively new, the industry has the benefit of >40 years of gaseous emissions controls on gasoline vehicles. This long history has provided a robust experimental foundation, creating the basis for helpful engine, component, and system models. Chapter 15 provides details on wall-scale, channel-scale, and filter-scale models that cover GPF pressure drop, filtration efficiency, soot reactions, and TWC reactions. Model inputs include relevant design parameters including cell density, wall thickness, porosity, pore size, ash loading, TWC loading, etc. It is quite likely that models will provide a foundation for system improvement for both current and future generations of control systems.

Millions of lives end prematurely each year due to ambient particulate matter pollution from vehicles and other sources [2]. Particulates generated by gasoline engines are responsible for only a small fraction of all ambient particulate matter, but the technology exists, via engine hardware/software and GPFs, to reduce the amount of particulates emitted. Many experts have shared years of experience in the accompanying chapters. This book represents thousands of hours of experimentation, modeling, and writing, making this book a helpful starting place for those interested in gasoline particulate control. We anticipate that as experience builds in this area, additional knowledge will be published that will help OEMs to better design cleaner and safer gasoline vehicles. This allows them to implement low-particle-output vehicles capable of extending the life of gasoline internal combustion engines, as a category.

References

1. Oberdörster, G., Oberdörster, E., and Oberdörster, J., "Nanotoxicology: An Emerging Discipline Evolving from Studies of Ultrafine Particles," *J. Environ. Health Perspect.* 113, no. 7 (2005): 823-839.

2. Cohen, A.J., Brauer, M., Burnett, R., Anderson, H.R. et al., "Estimates and 25-Year Trends of the Global Burden of Disease Attributable to Ambient Air Pollution: An Analysis of Data from the Global Burden of Diseases Study 2015," *The Lancet* 389, no. 10082 (April 2017): 1907-1918, doi:10.1016/S0140-6736(17)30505-6.

3. Ministry of Environmental Protection, "Limits and Measurement Methods for Emissions from Light-Duty Vehicles (China 5)," GB 18352.5-2013, MEP, Beijing, China.

4. Maricq, M.M., Szente, J.J., Harwell, A.L., and Loos, M.J., "Impact of Aggressive Drive Cycles on Motor Vehicle Exhaust PM Emissions," *Journal of Aerosol Science* 113 (2017): 1-11.

5. Xue, J., Li, Y., Wang, X., Durbin, T.D. et al., "Comparison of Vehicle Exhaust Particle Size Distributions Measured by SMPS and EEPS during Steady-State Conditions," *Aerosol Science and Technology* 49, no. 10 (2015): 984–996, doi:10.1080/02786826.2015.1088146.

About the Authors

Dr. Willard A. Cutler is currently division vice president and commercial technology director, responsible for the customer-facing technology for Corning's environmental business, for both light-duty and heavy-duty vehicles. Cutler has worked for Corning for nearly three decades, in various research, product development, commercial, and leadership roles. Cutler holds a Ph.D. in materials from the University of California, Santa Barbara, and a B.S. in materials science and engineering from the University of Utah.

Dr. Thorsten Boger is currently commercial technology director—light-duty vehicles, Corning Environmental Technologies—responsible for the customer-technology interface for Corning's global gasoline and diesel technologies applied to passenger cars and chassis dynamometer certified light commercial vehicles. He also serves as regional technology director for Europe. Boger joined Corning in 1997, working in different roles focused on new technologies for mobile emission control as well as industrial applications in the chemical and refining industry. Boger graduated with a degree in process engineering (Dipl.-Ing.) and received his doctorate (Dr.-Ing.) from the University of Stuttgart.

CHAPTER 2

Health Impact of Particulates from Gasoline Engines

Christoph Bisig, Alke Petri-Fink, and Barbara Rothen-Rutishauser
Adolphe Merkle Institute, University of Fribourg, Switzerland

Loretta Müller
University Children's Hospital Basel, Switzerland
Pediatric Respiratory Medicine, Department of Pediatrics, Inselspital, Bern University Hospital, University of Bern, Switzerland

> *Air pollution is the presence in the air of one or more substances at a concentration or for a duration above their natural levels, with the potential to produce an adverse effect.*
>
> **—Seinfeld and Pandis [1]**

2.1 Air Pollution: A Brief Historical Survey

Automobile technologies and developments have made such a tremendous impact on the quality of modern life that life without automobiles is unimaginable today. However, as with every technology, this also has its disadvantages. Diesel and gasoline soot are generated by internal combustion engines. It is also well known and accepted that these particles, together with large-scale burning of coal for power generation and in industrial processes, are the most significant contributors to air pollution, contributing to climate change and global warming, in addition to clear adverse effects in humans.

The most prominent components in polluted air include photochemical oxidants (ozone, O_3), sulfur dioxide (SO_2), carbon monoxide (CO), nitrogen oxides (NO_x), hazardous air pollutants (e.g., benzene), mercury, lead, particulate matter (PM), and organic carbon. The pollutants vary widely in concentration in space and time around the globe, depending on the respective sources, the weather, and atmospheric transformations [2, 3]. Herein, we briefly describe the history of land transport, early discoveries of the health effects of anthropogenic emissions, and measures to decrease air pollution.

The first combustion engine was designed by James Watt, who introduced the first practical steam engine in 1781. Further developments in the early 1800s led to the development of the first locomotive (1821), in which coal or wood were burnt to boil water and generate power. Steam engines were not only used for locomotives but also in ships and cable cars and by the end of the nineteenth century also in trucks and buses, transporting both goods and passengers. The further development of the steam engine was interrupted by the invention of the Otto engine, an internal combustion engine using gasoline instead of boiling water to produce motion, in the 1870s by Dr. A. N. Otto in Germany. This first engine, designed for stationary purposes, was later reinvented in parallel by Gottlieb Daimler and Karl Benz in the 1880s and applied first to bicycles and later horse carriages, nowadays known as the predecessor of the modern automobile. Only a couple of years later the diesel engine was patented by Rudolf Diesel (1893). It was not until the early 1900s that Henry Ford adapted mass production to automobile production, making them affordable for the wider public. The performance of automobiles and their fuel quality has been continuously improved ever since (reviewed by [4]).

It did not take much time from the invention and use of combustion engines to the occurrence of the first smog events. In 1930 a large air pollution event caused by industry and private coal burning in the Meuse valley, Belgium, resulted in 63 deaths over only 2 days [5]. In the winter of 1952, another large smog event occurred in London, where SO_2 and black smoke levels in particular were significantly elevated (thousands of $\mu g/m^3$), leading to several thousand premature deaths. The rates of deaths in the city remained elevated several months after the event [6].

Research into air pollution and its adverse effects, as well as the first legislation by individual countries began as early as 1955 [7]. Legislation on air pollution continued in the United States with the Clean Air Act in 1970 and the implementation of the National Ambient Air Quality Standards, aiming to decrease emissions from both stationary and mobile sources.

In the 1960s to early 1970s, acute adverse effects of CO, hydrocarbons (including polycyclic aromatic hydrocarbons (PAHs)), and lead from gasoline exhaust emissions were reported, whereas research into soot particles (black smoke) from diesel engines was scarce [8, 9]. Later, adverse effects of PM on human health were described in the 1970s and 1980s, where cardiopulmonary effects were observed during events involving high PM concentrations. The effects of lower PM levels were less clear and controversially discussed at that time until the 1990s (reviewed in [10]). The most prominent publication in that time was the so-called Harvard six-city study, where daily mortality was best correlated with increased $PM_{2.5}$ levels (i.e., PM with an aerodynamic diameter smaller than 2.5 μm) [11].

It took several decades from the first observations in the 1950s that automobile exhaust emissions can have severe health effects to the official classification of automobile exhaust emissions in 1988, where diesel exhaust was classified as a group 2A carcinogen, that is, probably carcinogenic to humans. In 2012 and 2013, diesel exhaust and outdoor air pollution were classified as group 1, carcinogenic to humans, while gasoline exhaust remained in the group 2B, possibly carcinogenic to humans [4, 12].

Ambient air quality has improved in the developed world in recent decades thanks to a range of efforts such as stronger regulations as well as improved engine and aftertreatment technologies. Engine measures such as the fuel-mixture generation, combustion process, variable valve control, exhaust gas recirculation, compression ratio, combustion chamber geometry, and ignition can be optimized, while the implementation of particle filters has been shown to be very effective for the filtering of particles [13].

In the city of Fribourg, Switzerland, levels of NO_2, O_3, and PM_{10} were found to be in the range of 36, 150, and 25 μg/m^3, respectively, in the 1990s, dropping to 27, 130, and 14 μg/m^3 in 2016. PM_{10} measurements were initiated in 1999 in the city of Fribourg, while SO_2 levels were found to be 10% of the threshold value in the late 1990s and are no longer of concern [14]. A reduction in $PM_{2.5}$ and O_3 from 1990 to 2013 has also been observed in all areas in Switzerland, as well as in other countries, that is, the United States, Russia, and Indonesia [15]. This shows that constant technical improvements and stringent regulations help to improve the air we breathe, and hence also our life expectancy. However, further research on (individual) air pollutants should be performed, and further legislation should be introduced to provide (i) a better understanding of the mechanism of how air pollution components interact with the respiratory system and (ii) further improved air quality in a global sense.

2.2 Interactions of Inhaled Air Pollutants with the Respiratory System

Inhalation of air is essential for life, in order to supply us with sufficient oxygen. Air pollution thus strongly affects quality of life and health. In the following, we describe the interaction of inhaled air pollutants with the human respiratory system, with a focus on the particulate fraction. Particles emitted from gasoline engines are usually agglomerated (**Figure 2.1A**) and consist of a carbon core with various substances attached to their surface (**Figure 2.1B**)

FIGURE 2.1 The structure of gasoline particles. (A) Transmission electron microscopy image of a gasoline exhaust particle. (B) Schematic overview of a combustion-derived particle with its carbon core coated with various chemical species such as metals, metal oxides, and organic hydrocarbons (Reproduced from Environmental Health Perspectives [16]).

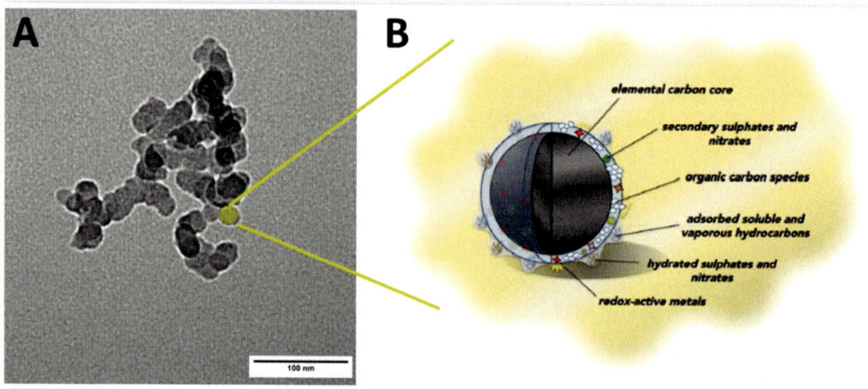

2.2.1 The Journey of an Inhaled Particle: Macro- and Microscopic Structure of the Respiratory System

With every breath, not only oxygen is inhaled, but also millions of particles enter the respiratory system. This includes gasoline exhaust particles, which are typically in the range of 10–500 nm in size. The deposition of these particles in the lung is size-dependent [17, 18]. If the particle has an aerodynamic diameter smaller than <10 μm (i.e., PM_{10}), it may enter the lungs, and if it is >100 nm or <10 nm, there is a high probability that it is deposited in the anterior nose, the nasal, pharyngeal, or laryngeal part of the airways (together with the mouth usually summarized as "nasopharyngeal region") (**Figure 2.2**) [19]. If the particles are not deposited in these regions, they can further move into the tracheobronchial region (trachea and bronchi). In this part of the airways, there is a high probability that the particle will be deposited if its aerodynamic diameter is between 1 and 100 nm. The bronchi further divide following regular dichotomous branching [20] into bronchioles and terminal bronchioles. Very small particles, that is, <1 μm, can reach the acinar airways consisting of respiratory bronchioles, alveolar ducts, and alveolar sacs [17, 18].

The interaction of the inhaled and deposited particles with the different compartments in the human lung determines their possible interactions with the tissue structures and their clearance.

The conducting airways are covered by a pseudostratified respiratory epithelium (**Figure 2.3**) composed of polarized, high epithelial cells, which either have cilia or produce mucus (called goblet cells). Additionally, there are basal cells serving as progenitor cells for other cell types and glands that also contribute to the production of mucus. The inhaled particle can be trapped in the mucus layer in the trachea and the bronchi, being cleared by mucociliary movement towards the pharynx where they will

FIGURE 2.2 Schematic overview of the respiratory system and the fractional deposition efficiencies of inhaled particles. Figure modified from [19].

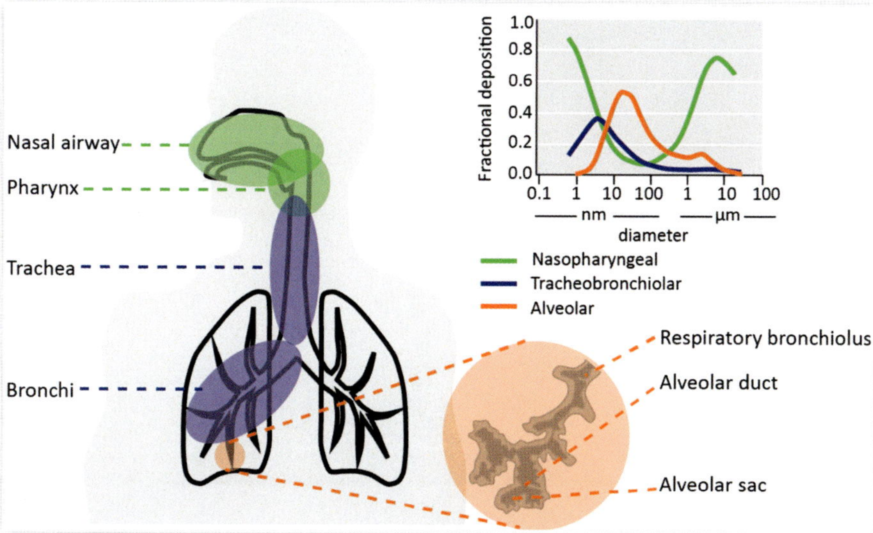

finally be swallowed. The interaction of particles with the mucus structure is size- and charge-dependent [21] and not yet fully understood for the different fractions of PM.

In the bronchial part of the respiratory tract, the epithelial cells have a cuboidal shape and the cilia finally disappear (Figure 2.3). The alveolar epithelium, which can be found deeper in the lungs, consists of squamous cells, alveolar type I epithelial cells that share a basement membrane with the endothelial cells covering the pulmonary capillaries. The alveoli also contain alveolar type II epithelial cells, which secrete lung surfactant (surface active agent) to prevent alveolar collapse [22]. Once the particle touches the surfactant layer, it is displaced into the aqueous lining layer by wetting forces coming in direct contact with airway cells [23, 24, 25]. The lung parenchyma, for example, alveolar tissue, is responsible for sufficient gas exchange. Oxygen needs to diffuse from the air compartment into the blood circulation, while carbon dioxide has to move in the opposite direction. The diffusion of oxygen and carbon dioxide is only possible if the barrier is as thin as possible and the structural barrier between air and blood is reduced to a mean arithmetic thickness of 2.2 µm [26]. The thinness of the alveolar layer might also facilitate the translocation of very small particles into the capillaries, resulting in distribution to secondary organs [27, 28, 29]. In addition to the extremely thin layer between the pulmonary lumen separating the air and the blood in the capillaries, a very large number of alveoli (approximately 300 million) offer a combined surface area of about 150 m^2 [26, 30, 31], representing optimal conditions for gas exchange [30]. While the different types of epithelial cells are the most important and structure-giving cells in the human airways, they are not the only ones. More than 40 different cell types, including endothelial cells, fibroblasts, nerve cells, lymphoid cells, gland cells, dendritic cells, and macrophages, add to the complexity of the epithelium in the lungs and perform extremely important functions, especially for innate and adaptive immune responses [3, 20].

FIGURE 2.3 Schematic overview of the airway wall structure in the different parts of the lungs. Reproduced with permission from [20].

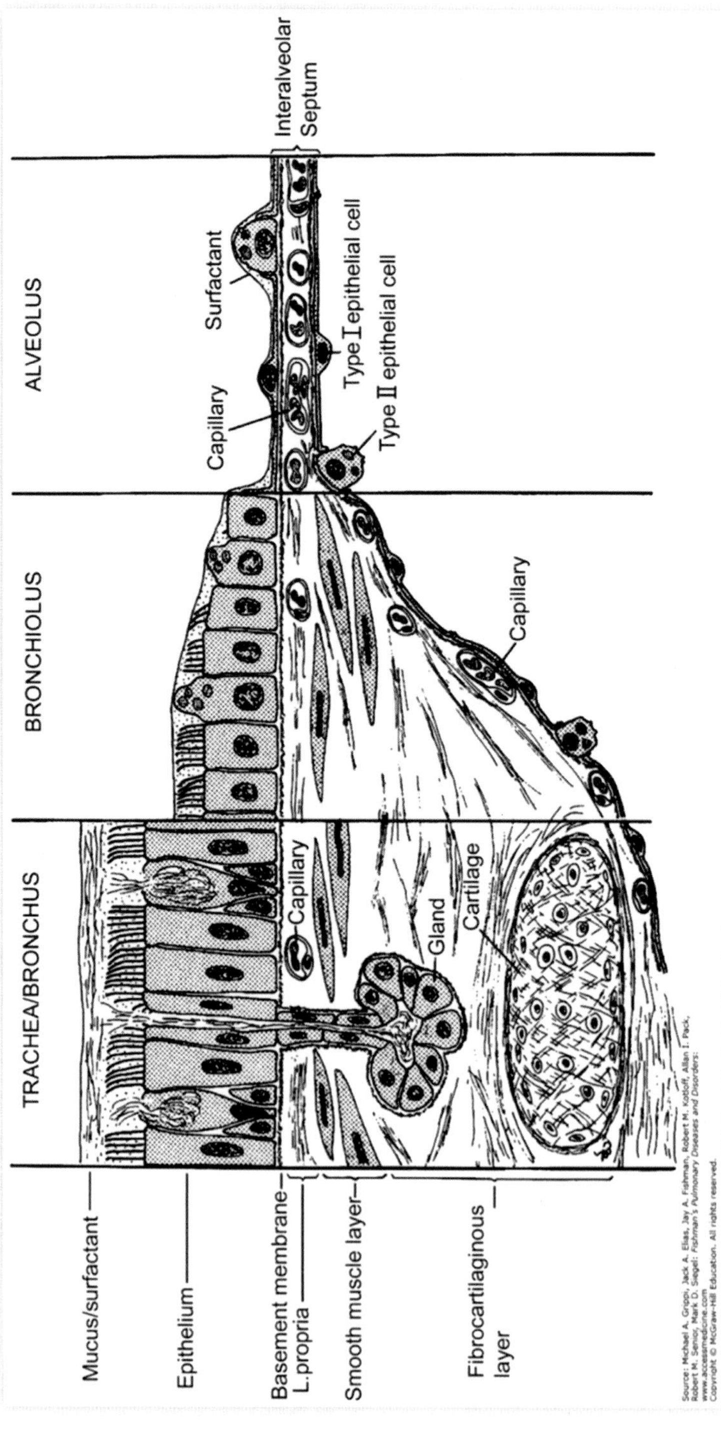

Taken together, the ultimate destination of the inhaled particle depends mostly on its aerodynamic diameter, which defines in which part of the respiratory tract the particle is deposited on the tissue wall. The location then defines what happens next with the particles. In the conducting airways, the particles are most probably transported out of the airways by clearance mechanisms, while in the lung parenchyma, the particle may interact with cells and may translocate to the bloodstream.

2.2.2 Clearance Mechanisms

Humans have different respiratory clearance mechanisms to deal with inhaled particles. While the particle size—or more specifically the aerodynamic diameter—determines the spatial location of the deposition in the different respiratory compartments, the point of deposition finally defines which clearance mechanisms will be relevant. The clearance mechanisms include two processes: (A) physical clearance of the whole particle and (B) chemical clearance processes [19, 32]. Chemical clearance includes dissolution, loss of elements from the particle matrix (leaching), and protein binding. These processes may occur in all regions of the respiratory tract, that is, in the conducting airways as well as in the gas-exchange regions. However, as gasoline exhaust particles mostly consist of carbon with different chemicals attached to their surface, complete dissolution—as well as other chemical clearance processes—is less likely to occur. Although no published studies have addressed this point, we consider physical clearance processes as more important for gasoline exhaust particles. Physical clearance mechanisms include the mucociliary escalator, phagocytosis by specialized phagocytic cells (i.e., macrophages or dendritic cells), epithelial endocytosis, translocation, lymphatic drainage, and blood circulation. Table 2.1 summarizes the physical and chemical clearance processes and specifies in which parts of the human respiratory tract they take place.

2.2.2.1 Mucociliary Clearance in the Conducting Airways

As described earlier in this chapter, the respiratory epithelium in the conducting airways includes mucus-secreting goblet cells, submucosal glands, and ciliated epithelial cells. The ciliated cells move their cilia in a directed manner, moving the mucus layer towards the pharynx. As inhaled particles may be trapped by the mucus layer, they

TABLE 2.1 Main clearance mechanisms in the human lung

Clearance mechanisms	Respiratory tract compartment		
Physical clearance processes	**Nasal**	**Tracheobronchial**	**Alveolar**
Mucociliary movement	X	X	
Epithelial endocytosis	X	X	X
Phagocytosis		X	X
Lymphatic drainage via dendritic cells		X	
Translocation into blood circulation		X	X
Chemical clearance processes	**Nasal**	**Tracheobronchial**	**Alveolar**
Dissolution	X	X	X
Leaching	X	X	X
Protein binding	X	X	X

are transported together with the mucus, which is known as mucociliary clearance. The mucus/particles eventually reach the pharynx with subsequent swallowing, finally reaching the gastrointestinal tract [33]. The subsequent fate of the particles, that is, interaction with intestine cells and/or excretion in feces, is not yet clear. In the human trachea, mucus transport velocities are between 10 and 20 mm/min [34] presupposing intact epithelial cells, ciliary structures, ciliary beat frequency, and ciliary beat pattern. The impairment of the mucociliary clearance has been attributed to several lung diseases, such as primary cilia dyskinesia, immotile cilia syndrome, bronchiectasis, cystic fibrosis, and asthma [35], and also have an impact on the clearance of particles.

2.2.2.2 Clearance Mechanisms in the Lung Periphery

Since the epithelial cells in the lung periphery do not have cilia, the peripheral clearance mechanisms depend on other mechanisms and cell types. Epithelial cells can take up particles via endocytosis, an energy-consuming process used by cells for the active uptake of molecules (e.g., nutritional substances) or particles. In this process, the cell membrane takes up the particles by invagination of the membrane (i.e., endocytosis). Additionally, the lung periphery contains phagocytic cells (i.e., macrophages) that actively take up particles by engulfing them with the cell membrane. In both cases, the particles finally end up in intracellular vesicles, that is, membrane-bound compartments. These vesicles then fuse with lysosomes to endolysosomes; it is assumed that gasoline particles cannot be digested and are kept intracellularly [36, 37]. While epithelial cells only take up small numbers of particles, macrophages are specialized for the phagocytosis of particles and are highly efficient in taking up particles. Macrophages can also pass the particle on to dendritic cells [38]. Dendritic cells also can take up the particles directly and then transport them to the draining lymph nodes, initiating an immune response. A small proportion of particles can also translocate into the bloodstream.

2.2.3 Mechanisms Underlying the Adverse Health Effects

Once inside a cell, particles may induce several biological responses. Although the exact mechanisms by which PM and traffic-derived particles cause adverse health effects are not fully understood [39], it is known that the following three main cellular effects are involved: (i) oxidative stress, (ii) inflammation, and (iii) genotoxicity (**Figure 2.4**). It has been shown that exposure to PM induces the generation of reactive oxygen species (ROS) [40], enhances the expression of pro-inflammatory cytokines [41], and causes DNA strand breakage [42].

Oxidative stress is an imbalance of oxidants and antioxidant defense proteins towards oxidants in the cells. ROS and reactive nitrogen species (RNS) are referred to as oxidants and are constantly produced intracellularly in cells or secreted by activated immune cells as signaling molecules. Due to this intrinsic ROS/RNS generation, cells also have suitable defense mechanisms (antioxidants), for example, glutathione or antioxidant enzymes. ROS/RNS can also be produced extracellularly, that is, they can be formed on redox-active particle surfaces: hydrocarbons and metals (particle bound or unbound) may catalyze the formation of ROS. Additionally, gases like O_3 and NO_2 are themselves ROS/RNS. As a result, particle deposition in the lung or

FIGURE 2.4 Overview of mechanisms causing adverse health effects of air pollution exposure. Inhalation of air pollutants leads to a deposition of particles and interaction of reactive gases in the lung, potentially leading to oxidative stress, inflammation, and genotoxicity. As a result, respiratory diseases such as chronic obstructive pulmonary disease (COPD) or lung cancer may occur, or other organs such as the cardiovascular system may be influenced, leading to systemic health effects.

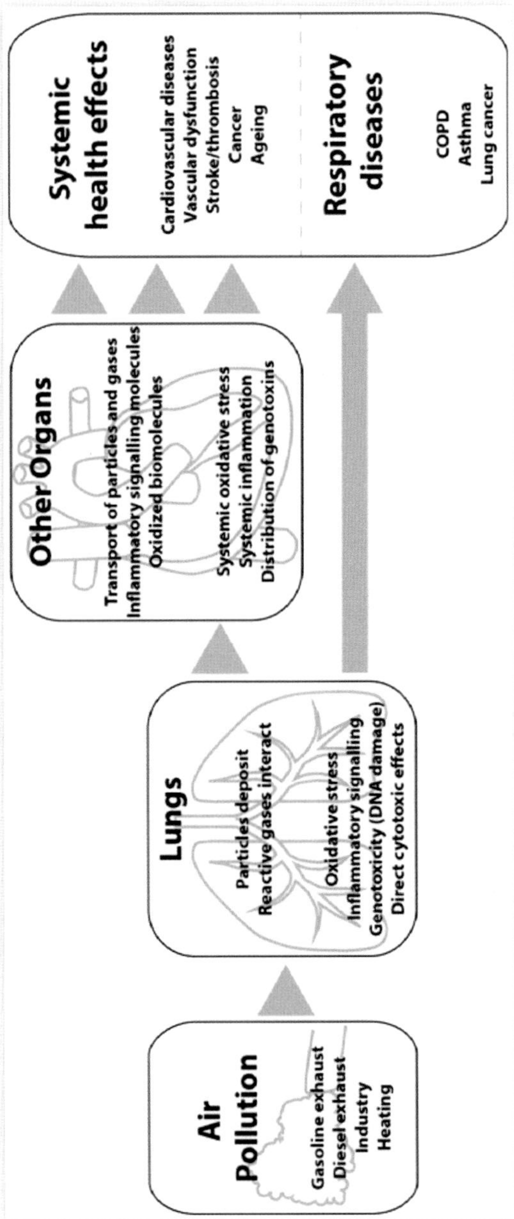

O_3 inhalation can result in increased levels of ROS. As a response to increased ROS levels, cells react by the synthesis of various defense proteins in order to prevent lipid peroxidation, DNA damage, or protein oxidation, which can be induced due to an overproduction of ROS. It should also be noted that combustion-derived particles are insoluble and therefore have a long retention time in the lungs. This can potentially lead to a depletion of antioxidative proteins and consequentially induces oxidative stress [43, 44, 45].

Inflammation on a macroscopic scale is defined as redness, swelling, heat, and pain, followed by loss of tissue/organ function. At the cellular level, the process starts with increased cytokine or chemokine release by various cell types of the human lung, for example, epithelial and endothelial cells, macrophages, and dendritic cells. This is defined as "pro-inflammation" and is often measured after particle uptake and expressed through increased levels of interleukins (IL8, IL1B, and IL6) and tumor necrosis factor alpha (TNFα). For example, IL8 release results in the invasion of neutrophils into the epithelial space. Neutrophils help the macrophages with the phagocytosis of the contaminant, but they also promote inflammation. The transcription factor nuclear factor kappa B (NFKB) seems to play a major role in the pro-inflammatory stimulus and can be activated via oxidative stress [46].

Genotoxicity is the impairment of genetic material (DNA) in the nuclei of cells; all genotoxins, that is, materials that may cause damage or changes in DNA, are mutagens, which may result in (lung) cancer. Many air pollutant constituents, such as PAHs, are known genotoxins; certain PAHs, after electrophilic activation, directly interfere with DNA and can cause false DNA replication [47]. Another mechanism is the impairment of DNA by ROS/RNS, derived either from an exogenous stimulus such as air pollutants or endogenous production through normal cellular processes. Radicals can form adducts on nucleotides or can cause DNA strand breakage [48], potentially resulting in misinterpretation of DNA polymerases and the inclusion of the wrong nucleotides [49].

These local responses to particle exposure may also spread to other organs, either via particle translocation or the induction of system oxidative stress, systemic inflammation by the release of cytokines/chemokines, or the distribution of genotoxins. As a result, particle exposure may not only lead to pulmonary diseases such as chronic obstructive pulmonary disease (COPD) [50, 51, 52] or lung cancer [53, 54, 55] but also to cardiovascular diseases, vascular dysfunction, or stroke/thrombosis [56, 57, 58, 59].

2.3 Hazard and Risk Identification of Gasoline Exhaust

2.3.1 Epidemiological Findings

Case-control studies allow for reciprocal investigations of possible relationships between an onset of a disease and exposure to gasoline exhaust emissions. Such studies have been performed frequently, but due to the nature of the study's design, that is, reciprocal, and the large number of cases required, few results are available from the past 10 years regarding modern GDI (gasoline direct injection) vehicle exhausts and their health effects in humans.

The Canadian Cancer Registries Epidemiology Research Group has published a series of studies relating workplace exposure to diesel and gasoline engine exhaust and cancer incidences [60, 61, 62]. The study design was similar, that is, cancer patients and healthy subjects were identified and asked about the workplace environment. Each workplace was then assigned to diesel and gasoline engine exhaust exposure and after adjustment for confounders such as smoking, diet, asbestos exposure, and physical activity, correlations between engine exhaust exposure and the number of cancer cases were evaluated. At first, lung cancer incidences were positively associated with the cumulative workplace burden of diesel exhaust; a dose-dependent and significant increase was found [62]. In addition, exposure to gasoline exhaust showed a trend in increased onset of lung cancer, but this trend was not significant [62]. Secondly, a risk for developing bladder cancer was shown: stronger effects were observed with diesel emissions, while effects from gasoline emissions were less consistent [61]. Finally, the authors found a significant correlation of colorectal cancer to workplace exposure to diesel emissions, while no increase was found for exposure to gasoline emissions [60].

Similar research has been performed by a Finnish group [63, 64], whereby workplace exposure of gasoline exhaust was correlated with ovarian and bladder cancer, mainly in professional drivers [63]. Kidney cancer was not associated with gasoline exhaust exposure [63]. There was only little evidence of a correlation between lung cancer and occupational exposure to engine exhaust, probably due to the small air pollution burden [64].

Epidemiological studies can also use the so-called source apportionment technique, where the total pollutant concentration is assigned to a specific primary source with different models, for example, a chemical mass balance [65]. Using this technique, Heo et al. found that no individual source is associated with $PM_{2.5}$ causing all-cause mortality. Cardiovascular mortality, however, was significantly associated with biomass burning, while respiratory mortality was correlated to gasoline and diesel emissions [66].

Gass et al. recently used a combination of four models to assess emergency hospital visits of children with asthmatic symptoms (5–18 years). Sources included, among others, biomass burning, diesel, and gasoline vehicles. They found significant correlation of emergency visits with asthma for both traffic-related events (diesel and gasoline); however, only gasoline—and not diesel—exhaust remained significant when adjusted for ozone levels [65].

2.3.2 Findings from In Vivo Studies

Tzamkiozis et al. sampled exhaust particles from five different vehicles, four diesel vehicles and one Euro3 gasoline vehicle, in an aerosol concentration enrichment system (waterborne particle sampling). Results of the physicochemical analysis showed that the gasoline engines emitted fewer particles compared to the diesel engines. However, mass analysis revealed that certain metals (sulfur, manganese, copper, and iron), ROS, and total PAHs are emitted to a higher extent by gasoline than by diesel engines. Particle suspensions with concentrations of 0.1 mg/mL (gasoline) and 0.2 mg/mL (diesel) were then instilled in two doses (50 and 100 μL) into mice. Bronchoalveolar lavage analysis revealed that the gasoline particles cause the highest accumulation of polymorphonuclear neutrophil cells in the respiratory tract, despite the lower concentrations applied, showing that gasoline exhaust particles can cause an acute inflammatory response. Diesel exhaust particles did not show this response, except those from the older Euro1 vehicle, whose response was lower than for gasoline exhaust particles. It was also shown that the lower gasoline particle dose can cause alveolar tissue injury; however, this was also observed for the diesel particles [67].

Peters et al. exposed female rats to diluted (1:20 and 1:40) gasoline engine exhaust for 12 and 24 months and found only moderate effects, such as loss of iron in bronchoalveolar lavage cells. No differences in body weight, mortality, or the other clinical endpoints were found between the air-exposed and exhaust-exposed groups [68].

A group from the National Environmental Respiratory Center (NERC) in the United States performed a series of whole exhaust exposure studies using rodents and a 1996 port fuel injection (PFI) engine (older technology). An exposure to diluted (1:10, 60 $\mu g/m^3$ PM) gasoline engine exhaust for up to 7 days resulted in a decreased clearance of bacteria in the lungs of mice [69]. Pregnant mice and their offspring were also exposed to gasoline engine exhaust, and the development of an allergic response was observed in the offspring [69]. After a 6-month exposure, the pro-inflammatory cytokine macrophage inflammatory protein (MIP)-2 was found to be elevated [69], but little lung inflammation was observed overall [70]. While no changes in lung histopathology and weight were observed, an increased cytotoxicity in the bronchoalveolar lavage fluids of female rats was demonstrated [70]. On the other hand, the vascular effects were stronger: an increase in vascular matrix metalloproteinase (MMP)-2 and -9 and ROS was observed in mice after a 7-day exposure [71]. In addition to the gasoline exhaust exposures, diesel exhaust, hardwood smoke, and simulated downwind coal emissions were applied to the rodents, as summarized in [72]. In this analysis, the vascular effects became more apparent, where six of the nine tested pro-atherosclerotic proteins were found to be upregulated after gasoline engine exhaust, but only three after diesel exhaust, and none after hardwood and simulated downwind coal emissions [72]. The authors also concluded that removal of the particles is not sufficient for exhaust detoxification (a HEPA filter was used) [70, 72].

2.3.3 Findings from Ex Vivo and In Vitro Studies

Only one study based on ex vivo tissue was found, that is, exposure of GDI exhaust to murine precision-cut lung slices [73]. A GDI engine without a three-way-catalyst (TWC) was placed in a laboratory and run under steady-state conditions (100 km/h). Emissions were diluted 100-fold and applied for 1 h at 1.5 mL/h onto the precision-cut lung slices, and filtered emissions were also included. Characterization of the unfiltered exhaust revealed 4.6×10^4 particles/cm^3, 65 ppm CO, and 20 ppm NO_x; PAHs were also analyzed. The particles were significantly reduced in the presence of the filter. Three genes were assessed, in addition to two cytochrome enzymes (cytochrome P450 (CYP)1a1 and CYP1b1) and one to gauge oxidative stress (heme oxygenase-1 (HMOX1)). The unfiltered GDI exhaust elevated all three mRNA levels significantly, while the filtered GDI exhaust elevated CYP1a1 and HMOX1, but not CYP1b1. Conclusively, due to absence of a TWC, the high emissions of volatile organic compounds were also able to induce CYP gene expression and oxidative stress in precision-cut lung slices after only 1 h of exposure, indicating that not only particles but also other components have to be eliminated more efficiently [73].

More studies have been published with in vitro cell culture experiments. Exposure of organic extracts of gasoline engine exhaust (1996 PFI engine) resulted in decreased cell proliferation in mouse macrophage cells (RAW264.7 cell line) and immunotoxic effects in rabbit alveolar macrophages [74]. The same group also pointed to possible genotoxic effects assessed by the Comet assay upon exposure to gasoline engine exhaust using human alveolar epithelial cells (A549 cell line) and the AMES test (assay with bacteria to assess the mutagenicity of chemical compounds) [75]. All these experiments were performed with extracts and were missing important details such as a complete exhaust characterization; also no information was given about the dose used.

Another approach was chosen by Lotfi et al.: whole exhaust from gasoline- and ethanol-fueled engines without TWCs were tested for carcinogenicity using the AMES test. Gasoline exhaust was determined to be a direct mutagen, while ethanol exhaust needed additional metabolic activation (indirect mutagen). Notably, the engine emitted very high CO concentrations: up to 1.75% were applied [76].

The pro-inflammatory response upon gasoline exhaust exposure was evaluated in human alveolar epithelial cells (A549 cell line) cultured at the air-liquid interface by Cheng et al. Exhaust exposure with particle number concentrations of $1–2 \times 10^6$ particles/cm^3 revealed no increased IL8 protein expression in cells; A significantly enhanced IL8 secretion was only found when cells were pretreated with TNFa to mimic an existing pro-inflammatory status. No information on engine type or characterization of the volatile fraction was given [77].

Another study exposed re-differentiated primary airway epithelial cells and the BEAS-2B airway cell line to particles of aged gasoline exhaust of a Euro5 passenger car. They found slightly increased cytotoxicity in the primary cells and reduced levels of pro-inflammatory cytokines in the cell line compared to particle-free control air [78].

Our group has studied the toxicity of gasoline exhaust in a sophisticated exposure system using freshly produced whole exhaust from different gasoline cars [79]. The exposure system includes internal controls exposed to filtered ambient air. The toxicity was evaluated using a triple-cell co-culture model of the human airways consisting of epithelial cells (16HBE14o- cell line), primary monocyte-derived macrophages and dendritic cells (i.e., cells of the innate immunity) [80], a co-culture model of epithelial cells (16HBE14o- cell line) and primary natural killer cells (important cells of the innate immunity), or monocultures of epithelial cells (also the 16HBE14o- cell line). Overall, the exposure of all applied cell models to whole gasoline exhaust over a 6 h period showed almost no toxicity, especially if compared to diesel exhaust, which induced significant oxidative stress and pro-inflammation [81, 82, 83]. In the first study, the dynamic new European driving cycle (NEDC) was driven using a 2013 GDI (Euro5b+) engine with and without an uncoated gasoline particle filter (GPF). We observed an increase in oxidative stress and aryl-hydrocarbon-receptor-related gene expressions in the triple-cell co-culture model after exposure to unfiltered GDI exhaust; these effects were diminished by addition of an uncoated GPF. On the other hand, the GPF decreased the antioxidative capacity (glutathione depletion). In the AMES test, the unfiltered GDI exhaust exhibited a higher mutagenic potential compared to filtered exhaust [82]. Overall, the GPF reduced some of the toxic effects of gasoline exhaust in our triple-cell co-culture model; however, further experiments are needed, including investigation of coated GPFs [82]. The second study investigated the effect of ethanol as a partial substitution for gasoline on the toxicity of gasoline exhaust exposure. Cell cultures as described above were exposed for 6 h to freshly produced whole exhaust of a 2012 GDI Euro5a, GDI gasoline car driving a steady-state cycle (five states, each of 20 min at 95 km/h, 61 km/h, 45 km/h, 26 km/h, and idling). Several endpoints were included: cytotoxicity, cell morphology, oxidative stress, expression of genes related to oxidative stress and pro-inflammation, and DNA damage. Again, no toxic effects were found in the triple-cell co-cultures after the acute exposure to gasoline exhaust, for either pure gasoline exhaust or of the ethanol-gasoline blend [81]. The exposure of the epithelial cell-natural killer cell co-cultures to the gasoline and ethanol-gasoline exhaust neither induced toxicity nor changes of cell activation. The studied endpoints were slightly different to those of the triple-cell co-culture models: cytotoxicity, cell morphology, oxidative stress, DNA damage, expression of surface stress receptors, and activation/inhibition of natural killer cells. While most of the endpoints were also unchanged in monocultures of epithelial cells, slightly less oxidative damage was observed after exposure to exhaust from a car driven with an ethanol-gasoline blend compared to with pure gasoline [83].

2.4 Conclusions and Future Recommendations

Epidemiological studies were able to associate gasoline exhaust with a range of adverse effects in humans. Workplace exposure to gasoline exhaust was found to be weakly associated with cancer of the lung [62, 64], bladder [61, 63], and ovaries [63]. Furthermore, using the source apportionment technique, exposure to gasoline engine exhaust was associated with higher mortality [66] and more emergency hospital visits of children with asthma [65]. These studies show that gasoline engine exhaust can have multiple effects on the human health, even though diesel exhaust was often more strongly associated with these outcomes.

Very few groups have investigated gasoline engine exhaust in vivo and in vitro since 2000. The results of the different studies are difficult to compare, as methods—and especially the gasoline exhaust composition—vary significantly, and no single method can be claimed to be optimal. For example, in mice, one group found acute inflammation [67], while another found effects on the vascular system [72].

Overall, it seems that exposure to modern gasoline exhaust can induce adverse effects in humans, but in general, more severe effects have been reported for diesel exhaust. However, due to the limited number of studies found for gasoline exhaust, especially also long-term studies, it is not yet possible to make a final assessment. We recommend that efforts to understand the consequences of exposure to gasoline exhaust emissions on possible adverse effects in humans are further extended.

Acknowledgments

This work was supported by the Adolphe Merkle Foundation, Swiss National Science Foundation (310030_159847/1), the Swiss Federal Office for Environment, the Swiss Federal Office of Energy, Schweizer Erdölvereinigung, Bernische Krebsliga, the Fondation Botnar Switzerland, and VERT Association. We thank Dr. Andreas Mayer (Technik Thermischer Maschinen, Niederrohrdorf), Dr. Norbert Heeb (EMPA Dübendorf), and Prof. Jan Czerwinski (Bern University of Applied Sciences, Biel) for the continuous support and discussions about the project.

Glossary

Airway - The passage by which air enters and leaves the lung

Alveoli - Air sacs of the lungs allowing for rapid gas exchange

AMES test - Biological assay with bacteria to assess the mutagenicity of chemical compounds

Bronchoalveolar lavage - Medical procedure involving flushing the lungs with a fluid to collect cells and proteins for analysis and to wash the lungs

Cardiopulmonary - Related to heart and lung

Conducting airways - Part of the airways by which the air enters and leaves the lung, without oxygen exchange

Cytotoxicity - Cell destruction caused by a toxic substance

Dichotomous branching - Repeated division of one branch into two

Endocytosis - The process by which cells take up particles bound to its surface

Ex vivo - Outside of the living body

Genotoxicity - Damage of the genetic material, that is, DNA

In vitro - Outside the living body and in an artificial environment, that is, cell culture experiments

In vivo - Within a living organism

Inflammatory reaction - Local response to cellular injury

Lung parenchyma - Portion of the lung involved in gas exchange (alveoli and small bronchioles)

Lysosome - Cell organelle containing degradative enzymes enclosed in a membrane

Necrosis - Cell injury caused by external factors

Oxidative stress - Imbalance between the production of reactive oxygen species (free radicals) and antioxidant defenses

Particulate Matter (PM)$_{10}$ - Fraction of particles with an aerodynamic diameter smaller than 10 μm

Phagocytosis - Engulfing of particulate matter by phagocytic cells

(Pro-) inflammation - Process capable of promoting inflammation

Surfactant - Surface active agent present throughout the respiratory system

References

1. Seinfeld, J.H. and Pandis, S.N., *Atmospheric Chemistry and Physics: From Air Pollution to Climate Change*, (New York: John Wiley & Sons, 2016).

2. IARC, "Outdoor Air Pollution," World Health Organization, 2016.

3. McClellan, R., "Particle Interactions with the Respiratory Tract," *Particle-Lung Interactions*, Edited by Gehr, P. and Heyder, J., (Boca Raton: CRC Press, 2000), 3-63.

4. Claxton, L.D., "The History, Genotoxicity, and Carcinogenicity of Carbon-Based Fuels and Their Emissions. Part 3: Diesel and Gasoline," *Mutation Research/Reviews in Mutation Research* 763 (2015): 30-85.

5. Firket, J., "Fog along the Meuse Valley," *Transactions of the Faraday Society* 32 (1936): 1192-1196.

6. Bell, M.L. and Davis, D.L., "Reassessment of the Lethal London Fog of 1952: Novel Indicators of Acute and Chronic Consequences of Acute Exposure to Air Pollution," *Environmental Health Perspectives* 109, no. Suppl. 3 (2001): 389.

7. Rogers, S.M., "Air Pollution Legislation—A Review of Current Developments," *American Journal of Public Health and the Nations Health* 50, no. 5 (1960): 642-648.

8. Stupfel, M., "Recent Advances in Investigations of Toxicity of Automotive Exhaust," *Environ Health Perspect* 17 (1976): 253-285.

9. WHO, "Urban Air Pollution with Particular Reference to Motor Vehicles: Report of a WHO Expert Committee [meeting held in Geneva from 2 to 8 July 1968]," 1969.

10. Pope III, C.A. and Dockery, D.W., "Health Effects of Fine Particulate Air Pollution: Lines That Connect," *Journal of the Air & Waste Management Association* 56, no. 6 (2006): 709-742.

11. Dockery, D.W., Pope, C.A., Xu, X., Spengler, J.D. et al., "An Association between Air Pollution and Mortality in Six US Cities," *New England Journal of Medicine* 329, no. 24 (1993): 1753-1759.

12. Loomis, D., Grosse, Y., Lauby-Secretan, B., Ghissassi, F.E. et al., "The Carcinogenicity of Outdoor Air Pollution," *The Lancet Oncology* 14, no. 13 (2013): 1262-1263.

13. Siemund, S., Stiebels, S., Bergmann, A., Scherm, P. et al., *Handbuch Verbrennungsmotor: Grundlagen, Komponenten, Systeme, Perspektiven*, Kapitel 21, (New York: Springer-Verlag, 2017).

14. Afu., accessed September 16, 2017, http://www.fr.ch/sen/de/pub/luft/feinstaub/aktuelle_situation.htm.

15. Brauer, M., Freedman, G., Frostad, J., Van Donkelaar, A. et al., "Ambient Air Pollution Exposure Estimation for the Global Burden of Disease 2013," *Environmental Science & Technology* 50, no. 1 (2015): 79-88.

16. Stone, V., Miller, M.R., Clift, M.J.D., Elder, A. et al., "Nanomaterials versus Ambient Ultrafine Particles: An Opportunity to Exchange Toxicology Knowledge," *Environ Health Perspect* 125, no. 10 (2017): 106002.

17. Heyder, J., Gebhart, J., Rudolf, G., Schiller, C.F. et al. "Deposition of Particles in the Human Respiratory-Tract in the Size Range 0.005-15-Mu-M," *Journal of Aerosol Science* 17, no. 5 (1986): 811-825.

18. Patton, J.S. and Byron, P.R., "Inhaling Medicines: Delivering Drugs to the Body through the Lungs," *Nat Rev Drug Discov* 6, no. 1 (2007): 67-74.

19. Oberdorster, G., Oberdorster, E., and Oberdorster, J., "Nanotoxicology: An Emerging Discipline Evolving from Studies of Ultrafine Particles," *Environ Health Perspect* 113, no. 7 (2005): 823-839.

20. Ochs, M. and Weibel, E.R., "Functional Design of the Human Lung for Gas Exchange," *Fishman's Pulmonary Diseases and Disorders*, Edited by Fishman, A.P., Elias, J.A., Fishman, J.A. et al., (New York: McGraw Hill, 2008), Vol. 4, 23-69.

21. Murgia, X., Pawelzyk, P., Schaefer, U.F., Wagner, C. et al., "Size-Limited Penetration of Nanoparticles into Porcine Respiratory Mucus after Aerosol Deposition," *Biomacromolecules* 17, no. 4 (2016): 1536-1542.

22. Daniels, C.B. and Orgeig, S., "Pulmonary Surfactant: The Key to the Evolution of Air Breathing," *News Physiol Sci* 18 (2003): 151-157.

23. Gehr, P.S., Schürch, S., Berthiaume, Y., Im Hof, V. et al., "Particle Retentionin Airways by Surfactant," *Journal of Aerosol Medicine* 3, no. (1) (1990): 17.

24. Gil, J., Bachofen, H., Gehr, P., and Weibel, E.R., "Alveolar Volume-Surface Area Relation in Air- and Saline-Filled Lungs Fixed by Vascular Perfusion," *J Appl Physiol* 47, no. 5 (1979): 990-1001.

25. Schurch, S., Gehr, P., Im, H.V., Geiser, M. et al., "Surfactant Displaces Particles toward the Epithelium in Airways and Alveoli," *Respir Physiol* 80, no. 1, (1990): 17-32.

26. Gehr, P., Bachofen, M., and Weibel, E.R., "The Normal Human Lung: Ultrastructure and Morphometric Estimation of Diffusion Capacity," *Respir. Physiol.* 32, no. 2 (1978): 121-140.

27. Elder, A., Gelein, R., Silva, V., Feikert, T. et al., "Translocation of Inhaled Ultrafine Manganese Oxide Particles to the Central Nervous System," *Environ Health Perspect* 114, no. 8 (2006): 1172-1178.

28. Muhlfeld, C., Gehr, P., and Rothen-Rutishauser, B., "Translocation and Cellular Entering Mechanisms of Nanoparticles in the Respiratory Tract," *Swiss Med Wkly* 138, no. 27-28 (2008): 387-391.

29. Nemmar, A., Hoet, P.H., Vanquickenborne, B., Dinsdale, D. et al., "Passage of Inhaled Particles into the Blood Circulation in Humans," *Circulation* 105, no. 4 (2002): 411-414.

30. Brain, J.D., "Inhalation, Deposition, and Fate of Insulin and Other Therapeutic Proteins," *Diabetes Technol Ther* 9, no. Suppl. 1 (2007): S4-S15.

31. Weibel, E., *Morphometry of Human Lung*, (Berlin: Springer Verlag, 1963).

32. Kreyling, W.S. and Scheuch, G., "Clearance of Particles Deposited in the Lungs," *Particle-Lung Interactions*, Edited by Gehr, P. and Heyder, J., (New York: Marcel Dekker Inc., 2000), 323-376.

33. Kilburn, K.H., "A Hypothesis for Pulmonary Clearance and Its Implications," *Am. Rev. Respir. Dis.* 98, no. 3 (1968): 449-463.

34. Morgan, L., Pearson, M., de Iongh, R., Mackey, D. et al., "Scintigraphic Measurement of Tracheal Mucus Velocity In Vivo," *Eur Respir J* 23, no. 4 (2004): 518-522.

35. Houtmeyers, E., Gosselink, R., Gayan-Ramirez, G., and Decramer, M., "Regulation of Mucociliary Clearance in Health and Disease," *Eur Respir J* 13, no. 5, (1999): 1177-1188.

36. Moller, W., Kreying, W.G., Schmid, O., Semmler-Behnke, M. et al., "Deposition, Retention and Clearance, and Translocation of Inhaled Fine and Nanosized Particles in the Respiratory Tract," *Particle-Lung Interactions*, Edited by Mühlfeld, C., Blank, F., Rothen-Rutishauser, B., Gehr, C. et al., (New York: Informa Healthcare, 2010).

37. Rothen-Rutishauser, B., Kuhn, D.A., Ali, Z., Gasser, M. et al., "Quantification of Gold Nanoparticle Cell Uptake under Controlled Biological Conditions and Adequate Resolution," *Nanomedicine (Lond)* 9, no. 5 (2014): 607-621.

38. Blank, F., Rothen-Rutishauser, B., and Gehr, P., "Dendritic Cells and Macrophages Form a Transepithelial Network against Foreign Particulate Antigens," *American Journal of Respiratory Cell and Molecular Biology* 36, no. 6 (2007): 669-677.

39. Clift, M.J., Gehr, P., and Rothen-Rutishauser, B., "Nanotoxicology: A Perspective and Discussion of Whether or Not In Vitro Testing Is a Valid Alternative," *Arch Toxicol* 85, no. 7, (2011): 723-731.

40. Gonzalez-Flecha, B., "Oxidant Mechanisms in Response to Ambient Air Particles," *Mol Aspects Med* 25, no. 1-2 (2004): 169-182.

41. Muller, J., Huaux, F., Moreau, N., Misson, P. et al., "Respiratory Toxicity of Multi-Wall Carbon Nanotubes," *Toxicol. Appl. Pharmacol.* 207, no. 3 (2005): 221-231.

42. Vinzents, P.S., Moller, P., Sorensen, M., Knudsen, L.E. et al., "Personal Exposure to Ultrafine Particles and Oxidative DNA Damage," *Environ Health Perspect* 113, no. 11 (2005): 1485-1490.

43. Park, H.S., Kim, S.R., and Lee, Y.C., "Impact of Oxidative Stress on Lung Diseases," *Respirology* 14, no. 1 (2009): 27-38.

44. Shinyashiki, M., Eiguren-Fernandez, A., Schmitz, D.A., Di Stefano, E. et al., "Electrophilic and Redox Properties of Diesel Exhaust Particles," *Environmental Research* 109, no. 3 (2009): 239-244.

45. Steiner, S., Bisig, C., Petri-Fink, A., and Rothen-Rutishauser, B., "Diesel Exhaust: Current Knowledge of Adverse Effects and Underlying Cellular Mechanisms," *Archives of Toxicology* 90, no. 7 (2016): 1541-1553.

46. Xiao, G.G., Wang, M., Li, N., Loo, J.A. et al., "Use of Proteomics to Demonstrate a Hierarchical Oxidative Stress Response to Diesel Exhaust Particle Chemicals in a Macrophage Cell Line," *J Biol Chem* 278, no. 50 (2003): 50781-50790.

47. Xue, W. and Warshawsky, D., "Metabolic Activation of Polycyclic and Heterocyclic Aromatic Hydrocarbons and DNA Damage: A Review," *Toxicology and Applied Pharmacology* 206, no. 1 (2005): 73-93.

48. Reuter, S., Gupta, S.C., Chaturvedi, M.M., and Aggarwal, B.B., "Oxidative Stress, Inflammation, and Cancer: How Are They Linked?," *Free Radical Biology and Medicine* 49, no. 11 (2010): 1603-1616.

49. Kasai, H., "Chemistry-Based Studies on Oxidative DNA Damage: Formation, Repair, and Mutagenesis," *Free Radical Biology and Medicine* 33, no. 4 (2002): 450-456.

50. Kersul, A.L., Iglesias, A., Rios, A., Noguera, A. et al., "Molecular Mechanisms of Inflammation during Exacerbations of Chronic Obstructive Pulmonary Disease," *Arch Bronconeumol* 47, no. 4 (2011): 176-183.

51. Wedzicha, J.A. and Seemungal, T.A., "COPD Exacerbations: Defining Their Cause and Prevention," *Lancet* 370, no. 9589 (2007): 786-796.

52. Yang, I.A., Relan, V., Wright, C.M., Davidson, M.R. et al., "Common Pathogenic Mechanisms and Pathways in the Development of COPD and Lung Cancer," *Expert Opin Ther Targets* 15, no. 4 (2011): 439-456.

53. Attfield, M.D., Schleiff, P.L., Lubin, J.H., Blair, A. et al., "The Diesel Exhaust in Miners Study: A Cohort Mortality Study with Emphasis on Lung Cancer," *J Natl Cancer Inst* 104, no. 11 (2012): 869-883.

54. Silverman, D.T., Lubin, J.H., Blair, A.E., Vermeulen, R. et al., "RE: The Diesel Exhaust in Miners Study (DEMS): A Nested Case-Control Study of Lung Cancer and Diesel Exhaust," *J Natl Cancer Inst* 106, no. 8 (2014).

55. Silverman, D.T., Samanic, C.M., Lubin, J.H., Blair, A.E. et al., "The Diesel Exhaust in Miners Study: A Nested Case-Control Study of Lung Cancer and Diesel Exhaust," *J Natl Cancer Inst* 104, no. 11 (2012): 855-868.

56. Brito, J.M., Belotti, L., Toledo, A.C., Antonangelo, L. et al., "Acute Cardiovascular and Inflammatory Toxicity Induced by Inhalation of Diesel and Biodiesel Exhaust Particles," *Toxicol Sci* 116, no. 1, (2010): 67-78.

57. Chin, M.T., "Basic Mechanisms for Adverse Cardiovascular Events Associated with Air Pollution," *Heart* 101, no. 4 (2015): 253-256.

58. Haberzettl, P., Bhatnagar, A. and Conklin, D., "Particulate Matter and Oxidative Stress–Pulmonary and Cardiovascular Targets and Consequences," *Systems Biology of Free Radicals and Antioxidants*, Edited by Laher, I., (Berlin/Heidelberg: Springer, 2014), 1557–1586.

59. Mills, N.L., Tornqvist, H., Gonzalez, M.C., Vink, E. et al., "Ischemic and Thrombotic Effects of Dilute Diesel-Exhaust Inhalation in Men with Coronary Heart Disease," *N Engl J Med* 357, no. 11 (2007): 1075-1082.

60. Kachuri, L., Villeneuve, P.J., Parent, M.-É., Johnson, K.C. et al., "Workplace Exposure to Diesel and Gasoline Engine Exhausts and the Risk of Colorectal Cancer in Canadian Men," *Environmental Health* 15, no. 1 (2016): 4.

61. Latifovic, L., Villeneuve, P.J., Parent, M.É., Johnson, K.C. et al., "Bladder Cancer and Occupational Exposure to Diesel and Gasoline Engine Emissions among Canadian Men," *Cancer Medicine* 4, no. 12 (2015): 1948-1962.

62. Villeneuve, P.J., Parent, M.E., Sahni, V., Johnson, K.C. et al., "Occupational Exposure to Diesel and Gasoline Emissions and Lung Cancer in Canadian Men," *Environ Res* 111, no. 5 (2011): 727-735.

63. Guo, J., Kauppinen, T., Kyyrönen, P., Heikkilä, P. et al., "Risk of Esophageal, Ovarian, Testicular, Kidney and Bladder Cancers and Leukemia among Finnish Workers Exposed to Diesel or Gasoline Engine Exhaust," *International Journal of Cancer* 111, no. 2 (2004a): 286-292.

64. Guo, J., Kauppinen, T., Kyyrönen, P., Lindbohm, M.L. et al., "Occupational Exposure to Diesel and Gasoline Engine Exhausts and Risk of Lung Cancer among Finnish Workers," *American Journal of Industrial Medicine* 45, no. 6 (2004b): 483-490.

65. Gass, K., Balachandran, S., Chang, H.H., Russell, A.G. et al., "Ensemble-Based Source Apportionment of Fine Particulate Matter and Emergency Department Visits for Pediatric Asthma," *American Journal of Epidemiology* 181, no. 7 (2015): 504-512.

66. Heo, J., Schauer, J.J., Yi, O., Paek, D. et al., "Fine Particle Air Pollution and Mortality: Importance of Specific Sources and Chemical Species," *Epidemiology* 25, no. 3 (2014): 379-388.

67. Tzamkiozis, T., Stoeger, T., Cheung, K., Ntziachristos, L. et al., "Monitoring the Inflammatory Potential of Exhaust Particles from Passenger Cars in Mice," *Inhalation Toxicology* 22, no. Suppl. 2 (2010): 59-69.

68. Peters, L., Ernst, H., Koch, W., Bartsch, W. et al., "Investigation of Chronic Toxic and Carcinogenic Effects of Gasoline Engine Exhausts Deriving from Fuel without and with Ferrocene Ddditive," *Inhalation Toxicology* 12, no. Suppl. 2 (2000): 63-82.

69. McDonald, J.D., Reed, M.D., Campen, M.J., Barrett, E.G. et al., "Health Effects of Inhaled Gasoline Engine Emissions," *Inhal Toxicol* 19, no. Suppl. 1 (2007): 107-116.

70. Reed, M.D., Barrett, E.G., Campen, M.J., Divine, K.K. et al., "Health Effects of Subchronic Inhalation Exposure to Gasoline Engine Exhaust," *Inhal Toxicol* 20, no. 13 (2008): 1125-1143.

71. Lund, A.K., Lucero, J., Lucas, S., Madden, M.C. et al., "Vehicular Emissions Induce Vascular MMP-9 Expression and Activity Associated with Endothelin-1–Mediated Pathways," *Arteriosclerosis, Thrombosis, and Vascular Biology* 29, no. 4 (2009): 511-517.

72. Mauderly, J.L., Barrett, E.G., Day, K.C., Gigliotti, A.P. et al., "The National Environmental Respiratory Center (NERC) Experiment in Multi-Pollutant Air Quality Health Research: II. Comparison of Responses to Diesel and Gasoline Engine Exhausts, Hardwood Smoke and Simulated Downwind Coal Emissions," *Inhal Toxicol* 26, no. 11 (2014): 651-667.

73. Maikawa, C.L., Zimmerman, N., Rais, K., Shah, M. et al., "Murine Precision-Cut Lung Slices Exhibit Acute Responses Following Exposure to Gasoline Direct Injection Engine Emissions," *Science of the Total Environment* 568 (2016): 1102-1109.

74. Che, W., Liu, G., Qiu, H., Zhang, H. et al., "Comparison of Immunotoxic Effects Induced by the Extracts from Methanol and Gasoline Engine Exhausts In Vitro," *Toxicology In Vitro* 24, no. 4 (2010): 1119-1125.

75. Zhang, Z., Che, W., Liang, Y., Wu, M. et al., "Comparison of Cytotoxicity and Genotoxicity Induced by the Extracts of Methanol and Gasoline Engine Exhausts," *Toxicology In Vitro* 21, no. 6 (2007): 1058-1065.

76. Lotfi, C.F., Brentani, M.M. and Bohm, G.M., "Assessment of the Mutagenic Potential of Ethanol Auto Engine Exhaust Gases by the *Salmonella typhimurium* Microsomal Mutagenesis Assay, Using a Direct Exposure Method," *Environ Res* 52, no. 2 (1990): 225-230.

77. Cheng, M.-D., Malone, B. and Storey, J.M., "Monitoring Cellular Responses of Engine-Emitted Particles by Using a Direct Air–Cell Interface Deposition Technique," *Chemosphere* 53, no. 3 (2003): 237-243.

78. Kunzi, L., Krapf, M., Daher, N., Dommen, J. et al., "Toxicity of Aged Gasoline Exhaust Particles to Normal and Diseased Airway Epithelia," *Sci Rep* 5 (2015): 11801.

79. Muller, L., Comte, P., Czerwinski, J., Kasper, M. et al., "New Exposure System to Evaluate the Toxicity of (Scooter) Exhaust Emissions in Lung Cells In Vitro," *Environ Sci Technol* 44, no. 7 (2010): 2632-2638.

80. Steiner, S., Czerwinski, J., Comte, P., Müller, L. et al., "Reduction in (Pro-) Inflammatory Responses of Lung Cells Exposed In Vitro to Diesel Exhaust Treated with a Non-Catalyzed Diesel Particle Filter," *Atmospheric Environment* 81 (2013): 117-124.

81. Bisig, C., Roth, M., Muller, L., Comte, P. et al., "Hazard Identification of Exhausts from Gasoline-Ethanol Fuel Blends Using a Multi-Cellular Human Lung Model," *Environmental Research* 151 (2016): 789-796.

82. Bisig, C., Steiner, S., Comte, P., Czerwinski, J. et al., "Biological Effects in Lung Cells In Vitro of Exhaust Aerosols from a Gasoline Passenger Car with and without Particle Filter," *Emission Control Science and Technology* 1, no. 3 (2015): 237-246.

83. Roth, M., Usemann, J., Bisig, C., Comte, P. et al., "Effects of Gasoline and Ethanol-Gasoline Exhaust Exposure on Human Bronchial Epithelial and Natural Killer Cells In Vitro," *Toxicol In Vitro* 45, no. Pt 1 (2017): 101-110.

About the Authors

Christoph Bisig studied pharmaceutical sciences at the University of Basel, Switzerland, before joining Prof. Rothen-Rutishauser's research group as a doctoral student in 2014 until the beginning of 2018. He has assessed the toxicity potential of several GDI engines with different fuels and exhaust aftertreatment systems. His work in this area has been published in a number of scientific journals. Currently he is a Postdoc at the HelmholtzZentrum München optimising an in vitro exposure protocol using a portable diesel aerosol generator and epithelial lung cells.

Prof. Alke Petri-Fink received her Ph.D. in chemistry from the University of Ulm, Germany, in 1999. After a postdoctoral stay at the University of Gainesville, Florida, she joined the Institute of Materials Science at the École Polytechnique Fédérale de Lausanne (EPFL), first as a postdoctoral researcher, then as a senior scientist. She became an Associate Swiss National Science Foundation Professor in the Department of Chemistry at the University of Fribourg in 2009, and full professor in 2011 at the Adolphe Merkle Institute, Switzerland. Her research focuses on inorganic nanoparticles, their synthesis, surfaces, and interactions with biological cells.

Prof. Dr. Barbara Rothen-Rutishauser received her Ph.D. in cell biology in 1996 from the Swiss Federal Institute of Technology (ETH) in Zurich. From 1996 to 2000 she held a postdoctoral position in Biopharmacy at the Institute of Pharmaceutical Sciences at the ETH, and in 2000 she joined Prof. Peter Gehr's research group at the University of Bern, Switzerland. Prof. Rothen-Rutishauser is an expert in the field of cell-nanoparticle interactions in the lung, with a special focus on 3D lung cell models and various microscopy techniques such as laser scanning and transmission electron microscopy. In 2011 she took up the new chair in BioNanomaterials at the Adolphe Merkle Institute, University of Fribourg, Switzerland, a position shared equally with Prof. Alke Fink. The research group's activities stretch over many fields from material synthesis and characterization to biological responses and risk assessment. Prof. Rothen-Rutishauser has published more than 200 peer-reviewed papers and is an associate editor of the journal *Particle and Fibre Toxicology.*

Dr. Loretta Müller studied environmental sciences at the Swiss Federal Institute of Technology (ETH) in Zurich and joined the group of Prof. Peter Gehr and Prof. Barbara Rothen-Rutishauser for her master's thesis and Ph.D. in 2006. For her Ph.D., she developed an exposure system to study the toxicity of exhaust emissions in cell cultures and studied the effects of scooter and diesel exhausts. After graduating in 2010, she worked for 2 years as a postdoctoral research fellow in the group of Prof. Dr. Ilona Jaspers at the University of North Carolina at Chapel Hill, USA. From 2013 to 2017 she worked as head of the wet lab of the pediatric pneumology research group of Prof. Dr. Philipp Latzin at the University Children's Hospital Basel, Switzerland. In 2018 she moved her lab to the University Children's Hospital Bern (Inselspital) to continue her work as head of wet lab of the research group of Philipp Latzin. Dr. Müller is an expert in the effects of exhaust emissions in in vitro settings and has extensive experience in working with co-culture models of the human respiratory system.

CHAPTER 3

Regulations and Environmental Technologies

Ameya Joshi
Corning Incorporated

3.1 Why the Emphasis on Particulates from Modern Gasoline Engines?

In order to limit the contributions to global warming due to burning of fossil fuels, major countries across the world have set ambitious targets on reducing vehicular fuel consumption and tailpipe CO_2 emissions (see **Figure 3.1**).

Achieving these targets would require CO_2 reductions of the order of 3%–6% per year across major markets. In response, vehicle manufacturers are rapidly advancing and implementing various technologies aimed at reducing fuel consumption and tailpipe emissions [1]. One approach that has gained popularity is that of replacing traditional port fuel injected (PFI) or multi-port injected (MPI) gasoline engines by boosted, downsized gasoline direct injected (GDI) engines [2, 3, 4, 5]. In 2016, GDI

FIGURE 3.1 Vehicular tailpipe CO_2 targets in select countries.

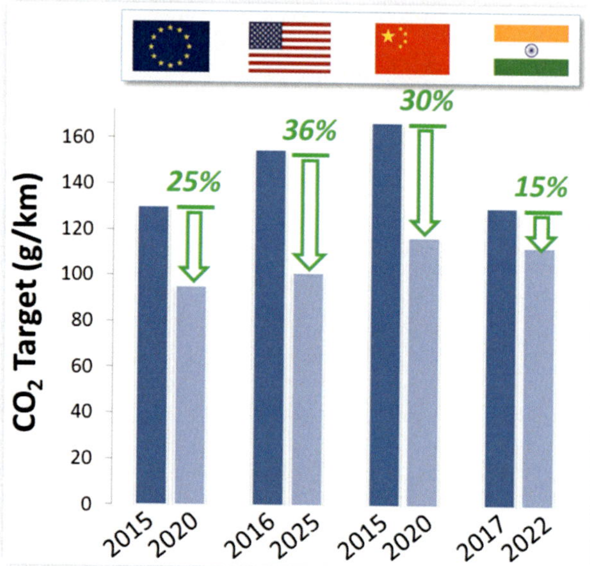

vehicles captured ~49% of the light-duty car and truck market share in the United States, within only 9 years from introduction [6]. The key difference between GDI and PFI engines lies in the fuel-air mixture preparation before combustion. In PFI engines, the fuel is injected upstream of the intake valve or in the intake manifold for mixing with air before entering the cylinder. In GDI engines, fuel is injected directly into the combustion chamber, and the associated charge cooling mitigates engine knock and enables operation at higher compression ratios. Furthermore, the amount and timing of fuel injected into the chamber can be more precisely controlled, leading to improved combustion and fuel economy gains.

However, the improved fuel economy of GDI engines is accompanied with an undesirable increase in the formation of particulate emissions. The direct injection of fuel in the combustion chamber leads to insufficient time for mixture formation and fuel impingement on the relatively cooler combustion chamber surfaces, leading to fuel-rich regions, incomplete combustion, and particulate formation. Particle emissions negatively impact human health and air quality and, for similar reasons, have been regulated in some countries for several years for diesel engines. On the vehicle side, practically all diesel engines are equipped with a diesel particulate filter (DPF) to capture these particulates. This chapter provides an overview of the particulate emissions from GDI engines, the associated high-level health risks, the changes in vehicle emission regulations, and some potential technologies to address the issue.

3.1.1 Health Risks

Exposure to ambient particulate matter has been conclusively linked with adverse impact on human health through respiratory and cardiovascular illnesses; irritation of eyes, nose, and throat; and even premature death [7]. The International Agency for Research on Cancer (IARC) has classified airborne particulate matter as carcinogenic to humans for lung cancer. A study [8] which examined lung cancer mortality over a 26-year period found that each 10 μg/m^3 increase in $PM_{2.5}$ was associated with 15%–27% increase in lung cancer mortality. $PM_{2.5}$ are defined as ambient particulates <2.5 microns ($PM_{2.5}$), which are primarily linked with combustion sources such as engines, crop burning, power generation, and domestic heating. An increase in ambient particulate concentrations is also linked with heart failure. A comprehensive review and analysis of studies done on this topic [9] showed that heart failure hospitalizations would increase by 2.1% per 10 μg/m^3 increase in ambient particulate concentrations. In the United States alone, it was estimated that reducing median daily $PM_{2.5}$ concentrations by 3.9 μg/m^3 would prevent roughly 8000 heart failure hospitalizations and save nearly a third of a billion U.S. dollars, per year. A 2015 study [10] of 79 health risks factors across 195 countries identified exposure to ambient particulate matter as the 5th largest risk factor. This was analyzed in greater detail recently [11] to arrive at the global exposures to $PM_{2.5}$ concentrations over 15 years. The analysis shows that exposure to $PM_{2.5}$ accounts for 4.2 million deaths and 103.1 million lost years of healthy life across the world. The deaths were associated with cardiovascular diseases (ischemic heart disease and cerebrovascular disease), lung cancer, lower respiratory infections, and chronic obstructive pulmonary disease.

In light of the above concerns, the World Health Organization (WHO) revised their air quality guidelines in 2005 and proposed limiting the ambient $PM_{2.5}$ concentrations to below 10 μg/m^3 for long-term (annual) and below 25 μg/m^3 for short-term (24 h) exposures [12]. These recommendations were made based on new studies at the time which provided a quantitative linkage between particulate concentrations and an increased mortality rate. Multi-city studies conducted both in the developed and developing world showed mortality effects of ~0.5% for every 10 μg/m^3 increase in short-term exposure. More than a decade since the WHO guidelines, >90% of the world's population lives in areas where ambient particulate concentrations exceed the 10 μg/m^3 recommendation [13]. Moreover, the burden of poor health is borne disproportionately by some of the most populous and rapidly growing regions of the world. China and India combined accounted for over half of the global deaths in the analyses above. It is no coincidence that these are also regions, which are now rapidly establishing stringent emissions norms.

Most of the particulates from GDI engines, the focus of this book, are less than 100 nm in diameter, at the fine end of the $PM_{2.5}$ distribution. For reference, this is orders of magnitude smaller than a human hair, which is on average 70 microns in diameter. Establishing the health effects specific to these ultra-fine particulates (UFPs) is more challenging. In one consortium study [14], emissions were measured before and after the 2008 Beijing Olympics, when aggressive air pollution control measures were implemented. Correlating various emissions with their impact on biomarkers associated with cardiorespiratory health of 125 individuals, this study showed that UFPs affect the pathophysiological pathways independent of other pollutants.

Other harmful constituents of GDI exhaust have also received attention. Polycyclic aromatic hydrocarbons (PAHs), compounds formed of multiple benzene rings, have been linked with cancer and effects on immune, reproductive, and neurologic systems [15]. GDI engines can have higher PAH emissions than PFI engines. This is not surprising, considering that soot is formed from PAH precursors. One study [16] measured particle number (PN), NOx, and hydrocarbon (HC) concentrations in the exhaust plumes of seven different 2013 Ford Focus GDI vehicles (spray-guided fuel injection) under idle, cruise, and acceleration. Compared to the Toronto fleet (light-duty (LD) and heavy-duty (HD)) average values, aromatic compounds using the BTEX (benzene, toluene, ethylbenzene, xylene) proxy were in the 62–96 percentile (higher in winter) of all vehicles on the road. A 2014 GDI pick-up truck had 14x more total PM-bound PAH emissions than a similarly powered PFI truck and ~4x more gas-phase PAH emissions [17]. The PAHs are likely associated with the soot, as can be inferred from a study on three Euro 5 GDI vehicles [18], where PAH concentrations were found to decrease by up to 90% for the larger species and 98% for benzo (α) pyrene, when fitted with a gasoline particulate filter (GPF). These molecules are relatively stable and do not break down easily once formed, leading to longer lifetimes in ambient air. Benzo (α) pyrene, a five-ring PAH, has been linked to cancer and also birth defects, and problems in liver and blood through studies with mice, and is used as a marker of PAHs in ambient air [19]. In a study [20] of six fuel blends from different refinery streams in China, 35 different PAHs were detected in the GDI engine-out emissions, of which 12 were identified on the U.S. EPA priority pollutant list [21].

From the above paragraphs, it can be concluded that inhalation of nanoparticles generated by combustion has negative implications for human health. However, it should also be mentioned that the assessment of the toxicity of particles generated by changing engine technology (such as GDI) is not straightforward. The epidemiological and in vivo studies involved are time and cost sensitive. A recent study used in vitro methods instead [22], where a model system which mimics the human epithelial airway barrier was exposed to exhaust from a GDI vehicle with and without a GPF. The study showed that while there may be beneficial effects associated with filtering of particles, it is important to ensure that secondary emissions are not being generated. Thus, adding and fine-tuning the catalytic activity to a filter can also be important for ensuring the full benefit of removing particulates. Additional health details are found in the previous chapter.

3.1.2 Global Warming Considerations

Black carbon (BC), the fraction of particulates analogous to elemental carbon, is considered a climate warming pollutant due to its ability to absorb solar radiation. The radiative forcing of BC has been reported to be as much as 55% of that due to CO_2, and greater than that due to other greenhouse gases such as CH_4, chlorofluorocarbons (CFCs), and N_2O [23]. A wide range is reported on the BC fraction, ranging from <29% [24] to over 80% [25], and even 100% during cold-start portion of the Federal Test Procedure (FTP) test cycle [26]. BC emissions are associated with both GDI and PFI vehicles, although higher particulate emissions from GDI vehicles translate to generally higher BC emissions as well. A summary of published BC emissions from 2007 to 2013 model year vehicles [27] showed that emission rates span over two orders of magnitude, from 0.1 mg/mi to upwards of 10 mg/mi. Given the global warming potential of BC noted above, there is a concern that some of the fuel efficiency benefits of GDI technology can be negated due to the BC emissions. The study noted above [15] quantified the climate impact of fuel economy/BC trade-off. It was found that replacing PFI vehicles older than 2005 with GDI ones results in a net CO_2 reduction. However, the net impact of replacing 2010 PFI vehicles or newer with a GDI actually results in *increased* net CO_2 due to the BC contribution. Considering the mean value of 1.3 mg/mi of BC emissions from GDI vehicles, a 2010 or newer GDI vehicle will have to be operated for ~20 years for the fuel efficiency improvements to offset the BC impact. GPFs can capture BC and mitigate the problem. Filtration efficiency >40% for BC was found to offset a fuel penalty of 0.5%.

3.1.3 Secondary Organic Aerosols (SOA)

A recent study [28] highlights that modern gasoline engines emit more particulates as compared to diesels fitted with diesel particulate filters (DPFs). Furthermore, the aromatics content of the gasoline soot was found to be much higher than that from the tailpipe soot from diesels with DPF. This led to increased secondary organic aerosol (SOA) formation in case of GDI vehicles. The DPFs capture any SOA precursors, while these are emitted in the case of GDI vehicles without filters, and the SOAs were found to be 6.5 times that of primary organic aerosols after atmospheric aging.

Given the growing concern with poor air quality, major countries worldwide have established limits on tailpipe particulate emissions from gasoline engines. Some of these regulatory changes are discussed in the next section.

3.2 Regulations on Vehicular Exhaust Particulate Emissions

3.2.1 United States and California

In the United States, the EPA Tier 3 and California LEV III standards are in place and regulate particulates through mass-based standards. The standards are phased in as a function of increasing percentage of vehicle sales and are fully phased in by 2021 for LEV III and by 2022 for Tier 3. For light-duty vehicles, both Tier 3 and LEV III standards specify a 3 mg/mi particulate mass (PM) limit applied to each certified vehicle, as measured over the FTP-75 (Federal Test Procedure) certification test cycle [29]. A relaxed in-use standard of 6 mg/mi applies during the phase-in period. In addition to the FTP testing, vehicles have to meet a PM limit of 6 mg/mi over the US06 test cycle, which represents aggressive driving. This limit is also phased in over a 5-year period, as a function of percentage of sales with an interim limit of 10 mg/mi.

California LEV III regulations continue to tighten further to 1 mg/mi, starting in 2025. The standards are also phased in through 2028, with increments of 25% of vehicles, which have to be certified every year. The durability requirements have been increased from 120,000 miles to 150,000 miles. The gravimetric method has been specified for PM measurements in 40 Code of Federal Regulations (CFR) Part 1065/1066 [30]. Measurement of PM down to 1 mg/mi is not trivial; there are various factors which can easily confound the accuracy of the measurement. Through extensive testing of several GDI, PFI, and diesel vehicles, staff at the California Air Resource Board (CARB) concluded that the gravimetric method is suitable for measuring PM emissions at the sub-1 mg/mi level [31]. Other studies have also shown that current gravimetric methods are capable of measuring PM down to the 1 mg/mi limit over the FTP-75 cycle [32], despite the intra- and inter-lab variability inherent to such measurements. Greater uncertainty has been reported when measuring particulates over the US06 cycle [26, 33], where the high-load operation leads to formation of nucleation mode particles in the 20–40 nm range, and which are very sensitive to engine operating conditions. A study comparing the use of gravimetric and alternate metrics has shown that the gravimetric method is relatively more sensitive to organic and semi-volatile PM [34]. Accordingly, the correlation with alternate methods was found to be cycle-dependent. While the correlation was good for FTP cycle, the gravimetric method measured consistently higher PM over the US06 cycle.

The regulation of particulate emissions through mass-based standards in the United States is different than the limit on the PN in Europe, China, and India, to be discussed next. Several studies have shown a direct correlation between solid particle number and total or black carbon mass emissions. Several studies reported a correlation of ~2×10^{12} particles/mg [35, 36, 37], although other correlations of $2.7–3.8 \times 10^{12}$ particles/mg [38] and $0.2–2.0 \times 10^{12}$ particles/mg [39] have also been reported. **Figure 3.2** shows a compilation of representative data collected by various

FIGURE 3.2 Correlation of PM and PN measurements.

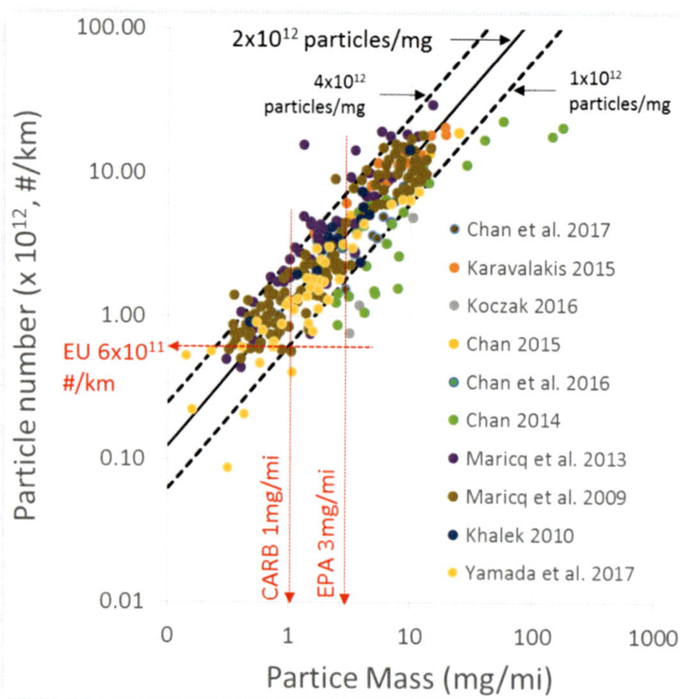

groups over several years. For reference, the PN limit of 6×10^{11} #/km translates to ~0.5 mg/mi based on the correlation of 2×10^{12} particles/mg (shown by the solid line). Also shown in the figure are the various number and mass regulations, and it is clear why the number-based standard is more challenging to meet from an engine perspective. Almost all of the data points are above the number limit threshold; a large number of data points meet the current (3 mg/mi) and even future (1 mg/mi) PM standards in the United States. This suggests that cars with a PN number standard (Europe and China) could have lower particulate amounts (mass or number) exiting the vehicles than the corresponding cars using a PM-only standard (United States and California), all else being equal.

3.2.2 Europe

3.2.2.1 Tailpipe Emission Limits and Timing

In Europe, PN limits have been in place for diesels since 2013. Starting in 2014, new GDI platforms have had to also effectively meet a PN-based limit of 6×10^{12} #/km, for solid particles larger than 23 nm in diameter. This limit tightened to 6×10^{11} #/km in September 2017 for new type approvals, and all new vehicles have to meet the limit starting September 2018. Other than the PN limits, the Euro 6 regulations also further limit tailpipe gas emissions and set PM limit at 4.5 mg/km (7.2 mg/mi). Durability is specified at 160,000 km. The emission limits and introduction timings are summarized in **Figure 3.3**, and details on the test methods are provided in the following section.

Other than the regulated limits mentioned above, the increased emphasis on public disclosure is worth highlighting. Information on real-world driving emissions (RDE) testing must be provided via free website and include maximum test values for NOx and PN [40]. Also, an "extended documentation package" is to be provided by OEMs on Auxiliary Emission Strategies.

3.2.3 Test Procedures

The Joint Research Centre (JRC) reported [41] the issue of on-road emissions far exceeding the certification values obtained under controlled lab testing. The study demonstrated the accuracy and reliability of portable emissions measurement systems (PEMS) for measurements of tailpipe emissions from on-road vehicles. Accordingly, several important changes have been recently introduced in the legislation to ensure compliance with emission norms under real-world driving conditions. Starting September 2017, the chassis-dynamometer-based certification test cycle is changed from the New European Drive Cycle (NEDC) to the World Harmonized Light Vehicle Test Procedure (WLTP) [42]. A complementary RDE test procedure is introduced and not-to-exceed (NTE) limits have been developed, expressed as Euro 6 emission limits multiplied by the "conformity factors (CF)".

The development of the RDE legislation was split in four packages. The first two outlined the basic test procedures: boundary conditions for a valid RDE test, data evaluation tools, requirement of the PEMS equipment, and specification of CFs. The third package included the evaluation of cold-start emissions, regeneration events,

FIGURE 3.3 Overview of Euro 6d regulations and implementation timeline.

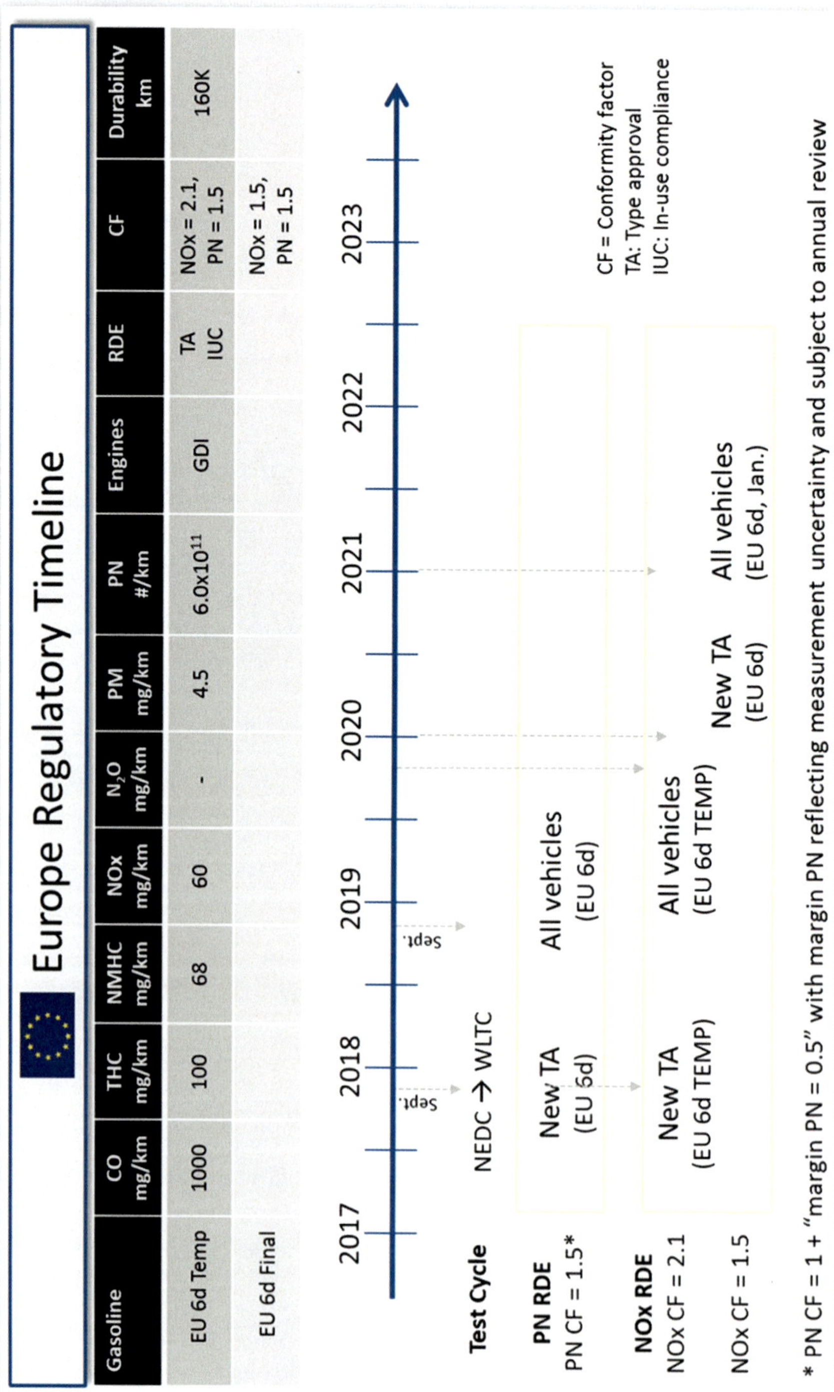

TABLE 3.1 RDE trip specifications

	Urban	Rural	Motorway
Distance	>16 km	>16 km	>16 km
Relative fraction	29%-44%	23%-43%	23%-43%
Driving speeds	15-40 km/h	60-90 km/h	>90 km/h >100 km/h for at least 5 min V_{max} < 145 km/h

and test provisions for hybrids, while a fourth and final package covers in-service compliance, surveillance tests, and provisions for light commercial vehicles. At the time of this printing, the fourth package had not been published.

The conditions for a valid RDE test have been designed to be representative of normal driving conditions on European roads. The vehicle is to be driven for 90–120 min on a mix of urban, rural, and motorway (highway) routes, starting with the urban segment. The segments are identified by the vehicle speeds, and these are noted in **Table 3.1** summarizing some of the RDE trip specifications.

The RDE emission limits along with the associated CFs must be met over the entire RDE trip and the urban part separately as well. For new type approvals, the CF is set at 1.5 for PN and temporarily at 2.1 for NOx starting September 2017, while the latter is also reduced to 1.5 in January 2020 (the timing is extended by 1 year for all vehicles in both cases). The conformity factors are expressed as "1+margin error" with the equipment margin of error being set at 0.5 for the final factors for both NOx and PN.

Other boundary conditions have to be complied with for a test to be valid:

- The altitude difference between the start and end points of the test should be less than 100 m, and the cumulative altitude gain should be limited to 1200 m/100 km.
- Moderate and extended temperature and altitude ranges are specified-emissions associated with any data points wlhich fall in the extended category are divided by 1.6:
  - The moderate conditions are 0 to 700 m altitude and 0°C to 30°C ambient temperature.
  - The extended conditions cover 700–1300 m altitude and between –7°C to 0°C and 30°C to 35°C ambient temperatures.
  - If any data point falls outside these boundary conditions, the entire trip is considered invalid.
- For valid tests, the data has to be further processed using at least one of two methods: the moving-average window method (with associated EMROAD software) and the power-binning method (with associated software called CLEAR).

Vehicle preconditioning is described: the vehicle has to be driven for at least 30 min and parked with engine turned off for 6–56 h before starting the RDE test.

In the first two packages of RDE, cold-start emissions, as defined by emissions occurring in first 5 minutes or until engine coolant reaches 70°C, were excluded. Recognizing that a significant fraction of emissions occur during cold start, the 3rd package [43] specifies that these emissions are now included in the analyses. Also included are emissions associated with regeneration events, corrected through the use of Ki-factors. If regeneration occurs during the first test, it can be voided at the request of the manufacturer, but the 2nd test is considered valid even if regeneration occurs, and the associated emissions during regeneration are included in the overall evaluation.

The 3rd RDE package also describes the calculation of emissions from hybrid vehicles. The internal combustion engine has to be working for at least 12 km under urban driving conditions, and it is recommended that the trip be started in charge-sustaining mode. The emissions are weighted according to the CO_2 emissions under charge-sustaining mode as described below. Emissions of any species (M) are calculated according to the following equation:

$$M\left(\frac{\text{mg}}{\text{km}},\frac{\#}{\text{km}}\right)=\frac{\text{Total emissions}\,(\text{mg or}\,\#)}{\text{Total CO}_2\,(\text{mg})}\times \text{CO}_{2,\text{WLTC}}\left(\frac{\text{mg}}{\text{km}}\right)$$

where $CO_{2,WLTC}$ is measured in charge-sustaining mode. Looking ahead, the accuracy of PEMS equipment is expected to improve, and with the availability and review of RDE data, a downward revision of CFs to 1 may be expected. Furthermore, there are also considerations to include ultra-fine particles below 23 nm in the regulated limits, and the emission profiles and measurement capabilities are being reviewed [44].

3.2.4 China

In December 2016, China's Ministry of Environmental Protection (MEP) issued the final China 6 light-duty vehicle regulations. The standards will be phased in over two steps. China 6a introduces WLTC as the certification test cycle, tighter gaseous emission limits, and a PN limit of 6×10^{11} #/km, starting 2020 (nationwide). China 6b further tightens the gas emission standards and enforces compliance with RDE, starting 2023 (nationwide). The definition of RDE in China is still under discussion and boundary conditions may be different compared to Europe to better reflect the local driving patterns. Also, CFs are not set, but a monitoring phase of RDE starts with 6a, and these will likely inform the CF decisions, which are slated to be finalized in 2022. China is moving very aggressively to curb its pollution problems, and there are discussions of moving the timing earlier for gaseous emissions, especially in some of the densely populated provinces.

While the China 6 regulations (see **Figure 3.4**) are following a similar framework to Europe's, there are some important differences:

- The regulations apply to all vehicles, in contrast to Europe, where PN regulations apply to GDI vehicles only. Particulate formation during combustion is not limited to GDI engines, and it is known that PFI engines also

FIGURE 3.4 Overview of China 6 regulations and implementation timeline.

China Regulatory Timeline

	CO mg/km	THC mg/km	NMHC mg/km	NOx mg/km	N_2O mg/km	PM mg/km	PN #/km	Engines	RDE	CF	Durability km
CN 6a	700	100	68	60	20	4.5	Before 2020/7: 6.0×10^{12} After 2020/7: 6.0×10^{11}	All	TT COP IUC(1)	2020/7: Monitor 2023/7: 2.1, to be adjusted(2) in 2022	160K
CN 6b	500	50	35	35	20	3.0		All			200K

TT: type test; COP: compliance of production; IUC: in-use compliance
(1) OEMs need to run RDE at TT & IUC. Regulator can check RDE for COP & IUC. (2) Likely adjusted downward. CF for PN & NOx only. Necessary to measure & record CO during RDE test.

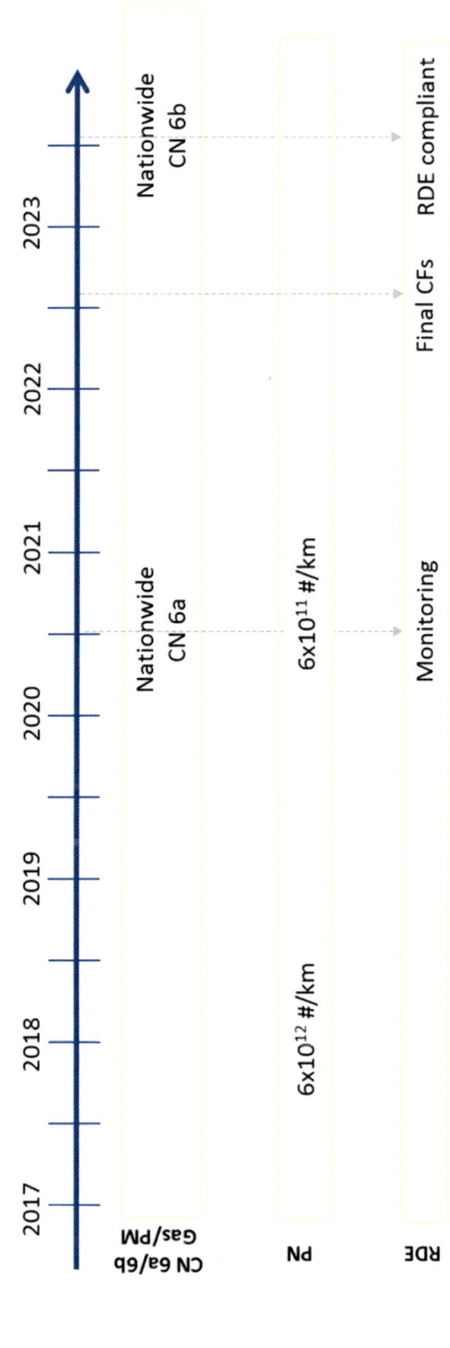

CHAPTER 3

emit particulates, although these are orders of magnitude lower under warm engine operating conditions. On the other hand, a recent study [45] shows that under sub-zero ambient temperatures, not uncommon in many parts of the world during winter, the emissions from PFI engines can be of the same magnitude as GDI, during cold start.

- When completely implemented (6b), the gaseous emission limits are tighter than Europe by a factor of 2 for CO and HC and ~40% for NOx.
- Unlike in Europe, N_2O is also regulated, and meeting this is not trivial under real-world driving conditions.
- For RDE, the altitude boundary condition is extended to 2,400 m (compared with 1,300 m in Europe). The emissions of RDE tests at extended high altitude (1,300–2,400 m) will be divided by a factor of 1.8.
- Per the current regulatory text, cold-start emissions are not included in the RDE analysis, but since this was also only recently adopted in Europe after the 3rd package, there is an expectation that cold start will be given more attention prior to China 6b.
- The durability is increased to 240,000 km starting with China 6b.

Any new regulations are only as good as its implementation strategy, including enforcement. A summary by ICCT [46] provides a good summary of the China 6 regulations and on the compliance program, describing it as "the most comprehensive ever in China." The compliance program includes three elements: (1) type testing, to be done by the OEM at certified labs. Measurements include gas and particle emission testing on WLTC after cold start at both normal and low ambient temperatures. (2) Conformity of production (COP) tests. The OEM is also responsible for preparing and submitting a plan to guarantee COP to authorities. (3) In-service compliance tests to be done by the OEM on a subset of the type tests at low (10,000–60,000 km), medium (60,000–110,000 km), and high (110,000–160,000 km) mileages. The regulatory agency can perform verification tests via random testing of in-use vehicles up to 160,000 km of driving.

3.2.5 India

India announced its plans to skip one stage of emission regulations, moving directly from Bharat Stage (BS) 4 to BS 6 regulations [47], the latter being comparable to Euro 6. This is the only major market which has resorted to "leapfrogging" one stage of regulations and is again a testimony to the urgency that is being felt in cleaning up the air in some of the most populated parts of the world. The BS 6 norms are applicable starting April 2020, although full implementation to include RDE will occur starting April 2023. The regulatory framework in India follows Europe, with some differences. At this point, there is no decision made to move away from NEDC to WLTP testing. Also, work is being done to develop an RDE test that is better reflective of the local driving conditions, which are very different from those in Europe. Similar to other markets, the move to stringent regulations is accompanied by an introduction of higher-quality fuel. In an effort to combat the worsening air quality in New Delhi, the government announced plans to introduce 10 ppm sulfur fuel by

April 2018, 2 years ahead of the introduction of BS 6 regulations. For the first time, the regulations are also introduced nationwide, as opposed to the previous approach of introducing first in major cities followed by the rest of the country with a time lag.

3.3 Technology Pathways

Various strategies are being pursued to minimize particulates [48, 49]. A detailed description of all strategies will be discussed in subsequent chapters. This section provides a brief overview of technical considerations and pathways for reduced particulate formation and tailpipe emissions.

3.3.1 In-Cylinder Methods

A lot of work is being done to minimize the in-cylinder formation of particulates and thereby reduce the emissions at the source. Primary consideration is given to achieving mixture preparation and avoiding fuel-rich regions. Piston bowl, engine, and injection system design is aimed at reducing impingement of fuel on the cool piston and cylinder surfaces, known sources for increased particulates. The combustion system is broadly classified into spray-guided and wall-guided systems, which differ primarily in terms of the strategy used for charge stratification. In spray-guided systems, the injector and spark plug are in close proximity and some of the fuel is directed towards the spark plug. Wall-guided systems rely on the fuel impingement on piston bowl surfaces and the subsequent transport of fuel towards the spark plug by the charge motion and flow field. Both systems have their challenges relevant to particulate formation: in spray-guided systems, the close proximity of injector and spark results in shorter time for fuel evaporation, whereas in wall-guided systems, fuel impingement with cold cylinder and piston surfaces leads to higher particle emissions.

3.3.1.1 Injection Systems

Proper design of the injection system is critical to achieve the desired mixture preparation within a short time between the opening of inlet valves and spark timing. The rapid, fine atomization of fuel requires the use of higher injection pressures as compared to PFI engines. Injectors are probably the most important element in the system. These have to be designed for accurate, rapid metering of fuel, at correct timing and small droplet size to ensure quick evaporation. Several factors of injector design and calibration are considered for achieving high fuel economy, low emissions, and low particulates [50, 51], including (a) the number of injection holes, spray pattern, duration and number of injections, fuel rail pressure, etc., and (b) higher fuel rail pressures (>250 bar), to help with better atomization and penetration of fuel spray and lower particulates. In one study [52], optimization of injection timing, and increased injection pressure from 200 to 500 bar was shown to reduce the particulate emissions by half. Coking of injectors should be minimized, as this is a key contributor to increased particulate emissions over the lifetime of a vehicle [53, 54, 55, 56, 57]. Coking of injectors leads to increased spray penetration and risk of fuel impingement on walls, reduced fuel atomization quality, increased fuel leakage during injector closing, and diffusion flame due to fuel absorbed by porous deposits. Injector prototypes developed recently [58] show that a significant reduction in PN emissions can be realized through the use of injectors which protrude into the combustion chamber, thus exposing them to higher temperatures which minimize deposit formation. Particulate formation is sensitive not only to the hardware but to the injection timing as well [59]. High emissions occur with early injection at intake stroke, due to impingement of fuel on piston surfaces. On the other hand, high emissions can also occur with highly retarded injection into the compression stroke, as this leads to insufficient time for mixture formation before onset of combustion. A split injection strategy is commonly used where the first injection introduces a fraction of the fuel early in the cycle to form a well-mixed lean charge, followed by a second injection done late in the compression stroke for increased heat release and early catalyst light-off. A combination of port fuel injection and direct injection can be used to reduce particulates [60]. Also multiple (double or triple) injections are carried out at low temperatures during warm-up or under high loads for improved fuel economy, knock mitigation, and lower emissions.

3.3.1.2 Fuel Efficiency Improvements

In order to meet the CO_2 targets, a wide variety of advanced gasoline engine technologies are being pursued which have implications for tailpipe emissions [61]. Some of these engine improvements have an added benefit in terms of reduced particulates.

3.3.1.3 Variable Compression Ratio (VCR)

VCR has been developed over several years and primarily offers improved fuel economy through the optimal selection of compression ratio (CR) per the load requirement. During cold start, low CR results in increased distance between piston and injector, and therefore reducing propensity of particulate formation due to fuel impingement. Also, the improved combustion overall is expected to lower emissions in general. Despite its increased complexity, the benefits have led to commercialization of this system [62].

3.3.1.4 Exhaust Gas Recirculation (EGR)

EGR is increasingly being used to reduce NOx emissions and also offer improved fuel economy. A benchmarking study done by EPA [63] showed a 7.5% improvement in fuel economy possible with the use of cooled EGR, starting with an Atkinson engine with a 14:1 compression ratio engine. Cooled EGR is also effective in reducing particulate emissions. In one study [64], EGR ratios (fraction of exhaust gas reintroduced in the engine) of up to 25% were studied. At part loads, brake-specific soot mass and solid particle number were reduced by more than 50% and 38%, respectively. At high-speed, high-load conditions, fuel enrichment can be used to mitigate engine knock and to limit the high temperatures, which are detrimental to downstream components, such as turbochargers. Fuel enrichment, however, leads to high particulate and HC emissions and also leads to a fuel penalty, so it is generally avoided. EGR is also effective in reducing the gas temperatures and therefore eliminate the need for fuel enrichment. Thus, particulate emissions reduction was observed even for the high-load conditions. In a more recent study [65], up to 75% reduction in PN was observed through the use of 15% EGR when fuel enrichment at high loads was avoided. The use of a GPF to clean the recirculated exhaust before entering the engine was found to yield a further reduction in particulates.

3.3.1.5 Lean and Low-Temperature Combustion

GDI engines can be operated in both stoichiometric and lean-burn combustion modes. Lean-burn operation is attractive due to its potential to reduce fuel consumption by 5%-15% compared to stoichiometric GDI engines. By definition, lean combustion avoids the fuel-rich combustion regime and is inherently suited for low particulate emissions. In practice though, particulate emissions were found to be high, in both the lean stratified and lean homogenous combustion modes. The modulated rich periods required to make ammonia for non-urea-based NOx control for lean-burn operation are associated with transient particulate emissions [66], and the use of GPFs was shown to be effective in capturing particulates with >95% filtration efficiency.

Much progress is also being made on achieving low-temperature combustion via compression ignition engines. Oak Ridge National Laboratory has provided an excellent summary on the various strategies being pursued [67], differing in the level of fuel stratification. The overall goal is to achieve (a) sufficient premixing of fuel/air to avoid soot formation associated with fuel-rich combustion and (b) reduce peak combustion temperatures via dilution with air or EGR to avoid NOx formation. The key challenge for this technology is that by definition, exhaust temperatures are low, and this poses a challenge to keep the catalyst above light-off temperatures. An example is the development of the gasoline direct injection compression ignition (GDCI) engine targeted to meet Tier3-Bin30 limits [68]. A BSFC (brake-specific fuel consumption) of 211-214 g/kWh was demonstrated over a wide load range, utilizing a high CR of 16:1 and injection pressure of 350 bar. The difficulty of achieving the very high conversion efficiencies (>99%) needed for HC has translated into a rather complex aftertreatment system, which at the latest stage includes a pre-turbo catalyst, HC trap, SCR, and a passive GPF for off-cycle particulates. An announcement of the commercialization of a spark-assisted HCCI engine was recently made [69] and promises to cut fuel consumption and likely emissions as well.

3.3.1.6 Electrification

Electrification of the powertrain is an attractive approach for reducing tailpipe emissions. Increasing on-board power on hybrid architectures—ranging from 48 V mild hybrids to full and plug-in hybrid electric vehicles (PHEVs)—can be used for propulsion of the vehicle and thereby reducing fuel consumption and associated emissions. However, there is increasing evidence that particulate emissions for hybrids can be equivalent or even higher than vehicles without any electrification. In a study in Japan [70], particulate emissions were measured on the JC08 test cycle from several gasoline and diesel vehicles, including one GDI and one MPI hybrid. Most of the particulate emissions for GDI and MPI engines were associated with cold start, with almost no emissions after the engine warmed up. However, in the case of hybrids, emissions were found to occur even after engine warm-up, most likely due to transient engine turn-on events. Overall, emissions from hybrid MPI vehicles were found to exceed those from diesels fitted with DPFs. In another study, a GDI engine was used to simulate HEV start-stop operation on an NEDC test [71]. Although the engine only operated 28% of the time during the cycle, the PN emissions were 4.5X higher than when the engine was run in conventional mode. In the case of PHEVs, attention is being given to particulate emissions associated with "high-powered cold starts" [72]. This is the scenario when the engine turns on for the first time when the vehicle is already moving at a high speed or under high load and results in cold-start emissions which are larger than would be under normal cold starts. A recent study focused on this issue and measured real-world emissions from a 1.5L GDI PHEV certified to Euro 6b standards and driven with various levels of charge on the battery [73]. As expected, the PN emissions in urban conditions were eliminated when the vehicle was propelled by battery alone. When the battery was depleted though, the demand on the IC engine was high and PN emissions were the highest. A peak in particulate emissions was observed when the engine turn-on was coupled with a high power demand. Hybrids and their power management strategies are evolving, and clearly there is room to improve on some of the above shortcomings. In the meantime, robust aftertreatment solutions are seen to be necessary to cover the operation of hybrids under widely varying real-world conditions.

3.3.2 Fuel Considerations

Concurrent with engine improvements, high fuel quality is a prerequisite when moving to the next level of regulations, both for its impact on combustion and aftertreatment systems. Major markets noted above have committed to low-S (<10 ppm) fuel. The chemical composition and physical properties of the fuel affect the injection quality and combustion characteristics, and there are several studies which link these properties with particulate formation [74, 75, 76, 77,78, 79, 80]. Unfortunately, in many markets, fuel quality varies widely, even when fuels are within the legal specifications. The topic of fuel and its influence on particle formation is also a topic covered in more detail in another chapter of this book.

A PM index (PMI) model has been proposed [81], which quantifies the relationship between fuel composition and the particulate emissions. The model accounts for the double bond equivalents, vapor pressure, and weight fraction of each component in the fuel. A strong correlation has been confirmed between the PMI index and particulate emissions. Aromatics are clearly linked with increased particulates, both through their role in the formation of PAHs and through their lower volatility which leads to slower mixture formation. One study [79] found that compared with a domestic Korean fuel with high oxygen content, the use of Euro 5 fuel with higher aromatics led to a 77% increase in PN emissions over the NEDC test cycle. In a study of the emissions from various GDI and PFI engines on the LA92 test cycle [82], Tier2-Bin2 and SULEV vehicles showed an increase in cumulative PM by 148% when the aromatics content was increased from 15% to 25%, and further increase by 73% when the aromatics content was increased to 35%. In light of these studies, the availability of fuel with low aromatics content will be an important consideration to achieving low in-use emissions. The aromatics content in gasoline has been lowered over the years. In the United States, the Tier 3 gasoline fuel specifications cap the aromatics content at 25%. Other components such as oxygenates which improve combustion or volatility can be expected to lower particulates. One such oxygenate which is already added mandated in some regions is ethanol.

In the United States, the Renewable Fuel Standard requires the use of renewable fuel to be blended with gasoline, and ethanol is primarily used. Ethanol is considered potentially beneficial for lowering particulates due to the increased oxygen content and the displacement of aromatics. There are conflicting reports, which also found an increase in emissions when using ethanol [37], and the differences in findings likely stem from the sensitivity of particulate formation to other properties such as its impact on physical properties and the engine operating conditions. Other studies confirm the benefit of using ethanol-blended fuels during the cold-start phase of the FTP test cycle [83] and on the WLTC [84]. Other oxygenates have also been studied. The impact of M15 methanol/gasoline blends [85] on spray behavior and particulate emissions showed that the blend did not consistently lead to low particulate emissions. Higher oxygenates, such as butanol, also show promise. In one study [86] testing a GDI vehicle on FTP75 and US06 cycles, both PN and PM emissions were found to reduce, compared to reference E0 fuel when it was splash blended with 10% ethanol (E10) and 16% iso-butanol (iB16). The blend with butanol gave higher reductions. On the US06 test cycle, PM emissions reduced from 2 mg/mi for E0 to 1.9 mg/mi for E10 and to 0.4 mg/mi for iB16. In another study, addition of 33% butanol to E5 gasoline (5% ethanol) was shown to reduce particulates while keeping a similar particle size distribution [87].

3.3.3 Aftertreatment System

It is clear that there are various pathways to mitigate particulate emissions via engine design and fuel improvements. Nevertheless, there is potential for deterioration of emissions with vehicle mileage. As mentioned previously, coking of injectors is known to increase emissions and could potentially lead to an increase in particulates over the lifetime of the vehicle [88, 89, 90]. Coking of injectors leads to increased spray penetration and increased risk of fuel impingement on walls, reduced fuel atomization quality leading to modified droplet size and velocity, increased fuel leakage during injector closing and diffusion flame due to fuel absorbed by porous deposits. There is limited data on changes in emissions with vehicle mileage. A study [91] on two 2010 GDI vehicles found that particulate emissions on the FTP test cycle improved from 3–4 mg/mi for a relatively new vehicle to ~1.5 mg/mi at 30–60K mileage, and then followed by an increase again to 2–3 mg/mi for 90–150K miles.

The final option for eliminating particles before leaving tailpipe is via aftertreatment systems. Existing aftertreatment of gasoline vehicles consists of three-way catalysts (TWCs), which primarily convert gaseous emissions. Studies have shown that TWCs can help eliminate some of the smaller particles (<20 nm) especially during cold start, when the catalyst is below light-off temperature [92, 93]. This is likely due to condensation of nucleation mode particles formed from heavy HC on the catalyst, which are either oxidized or released later, when the temperature is above light-off.

The use of a GPF is seen as a reliable way to ensure that the emissions are controlled well below the regulated limits, over a broad range of driving conditions.

FIGURE 3.5 Potential product choices and gasoline aftertreatment system layouts including GPFs.

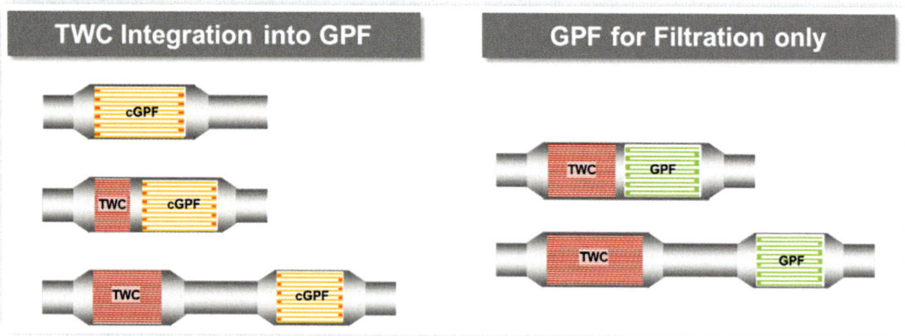

Particulate filter technology is not new. All modern diesel vehicles in advanced markets are equipped with DPFs, and these have shown robust performance over millions of miles in accumulated driving history. There are some key differences in the design of the filters for GPFs due to the higher operating temperatures for gasoline (up to 900°C compared to <400°C for diesel), lower PM (1–3 mg/km compared to 10 mg/km for diesel), lower PNs (~10^{12} #/km compared to ~10^{13} #/km for diesel), and lower O_2 concentrations.

Potential aftertreatment system layouts include use of either bare (uncoated) or TWC-coated GPFs (see **Figure 3.5**). In the former, the main functionality is filtration only, while the latter provides efficient space utilization via adding functionality of gas emissions conversion. The filters can be added in the close-coupled (CC) or in the under-floor (UF) position, as dictated by the emissions and space constraints.

The key requirements are high filtration efficiency and also low pressure drop, the latter to minimize the impact on fuel economy and driving characteristics. In addition, the filters must have adequate porosity to host the catalyst, sufficient channel volume for soot and ash accumulation, and thermomechanical robustness to withstand the stresses during canning and exotherms associated with soot regeneration. These often conflicting requirements are met through selection of the appropriate material, optimizing the overall size and cell design, and engineering the pore structure. Subsequent chapters will delve into details of the design and performance aspects. It is only noted here that several studies now confirm that the replacement of a TWC with a coated GPF can be done without impacting the fuel economy. The performance of GPFs–filtration, pressure drop, and gas conversion–has been studied extensively in the past few years, and there is ample evidence that this is a robust technology which helps meet the current regulations over diverse, real-world driving conditions. For the future, as was mentioned in the previous section, particles below 23 nm may be regulated. Studies are emerging which show that GPFs are very effective at capturing very fine, sub-23 nm particles. This is probably not surprising, given that the increased Brownian motion of smaller particles lends to increased filtration efficiency. In a study [94] involving 5 GDI vehicles, high PN emissions (~6 × 10^{12} #/km) were recorded when sub-23 nm particles were included. GPFs were found to be effective in near total capture of these small particles. In another study [95] of a 1.4 liter Euro 6b GDI vehicle under real-world driving conditions, an increase in engine-out PN by >50% was recorded when particles below 23 nm were included, while the tailpipe emissions after the GPF increased only by ~20%. The GPF filtration efficiency increased by ~10% when filtration of the smaller particles was included, and all tailpipe emissions comfortably met the regulated PN limit of 6×10^{11} #/km.

3.4 Summary

Regulations for vehicular emissions continue to tighten globally to near zero tailpipe levels, with an emphasis on achieving these not only on certification testing but under real-world driving conditions. With the increasing adoption of GDI technology, particulate emissions are a focus, and both PN and PM standards are driving the rapid adoption of advanced technologies to mitigate these emissions. The confluence of new engine and vehicle technologies to help reduce fuel consumption and the demand for lower tailpipe particulate emissions requires a concerted approach for engine-based and aftertreatment solutions to mitigate particulates. The next round of regulations will likely see one more round of tightening, to include finer particulates, with conformity factors revised downwards, and with further reduced tailpipe CO_2 requirements. This justifies the ongoing research on advanced gasoline vehicle technologies, fuel blends and additives, low-temperature combustion, powertrain electrification, and advanced aftertreatment systems. The addition of GPFs is a major change to the gasoline aftertreatment system, with filters likely to be installed for most GDI vehicles and many PFI vehicles, beginning in Europe and China.

References

1. Johnson, T. and Joshi, A., "Review of Vehicle Engine Efficiency and Emissions," SAE Technical Paper 2018-01-0329, 2018, doi:10.4271/2018-01-0329.

2. Zhao, F., Harrington, D.L., and Lai, M.C., *Automotive Gasoline Direct-Injection Engines*, (Warrendale: SAE International, 2002).

3. Bandel, W., Fraidl, G., Kapus, P., Sikinger, H. et al., "The Turbocharged GDI Engine: Boosted Synergies for High Fuel Economy Plus Ultra-low Emission," SAE Technical Paper 2006-01-1266, 2006, doi:10.4271/2006-01-1266.

4. Woldring, D., Landenfeld, T., and Christie, M., "DI Boost: Application of a High Performance Gasoline Direct Injection Concept," SAE Technical Paper 2007-01-1410, 2007, doi:10.4271/2007-01-1410.

5. Davis, R., Mandrusiak, G., and Landenfeld, T., "Development of the Combustion System for General Motors' 3.6L DOHC 4V V6 Engine with Direct Injection," *SAE Int. J. Engines* 1, no. 1 (2008): 85-100, doi:10.4271/2008-01-0132.

6. Davis, S.C., Williams, S.E., Boundy, R.G., and Moore, S., "2016 Vehicle Technologies Market Report," Oak Ridge National Lab, ORNL, 2016.

7. United States Environmental Protection Agency, "2009 Final Report: Integrated Science Assessment for Particulate Matter," 2009, http://cfpub.epa.gov/ncea/risk/recordisplay.cfm?deid=216546.

8. Turner, M.C., Krewski, D., Pope III, C.A., Chen, Y. et al., "Long-Term Ambient Fine Particulate Matter Air Pollution and Lung Cancer in a Large Cohort of Never-Smokers," *Am J Respir Crit Care Med* 184 (2011): 1374–1381.

9. Shah, A.S.V., Langrish, J.P., Nair, H., McAllister, D.A. et al., "Global Association of Air Pollution and Heart Failure: A Systematic Review and Meta-Analysis," *Lancet* 382 (2013): 1039–1048.

10. GBD 2015 Risk Factors Collaborators, "Global, Regional, and National Comparative Risk Assessment of 79 Behavioural, Environmental and Occupational, and Metabolic Risks or Clusters of Risks, 1990–2015: A Systematic Analysis for the Global Burden of Disease Study 2015," *Lancet* 388 (2016): 1659–1724.

11. Cohen, A.J., Brauer, M., Burnett, R., Anderson, H.R. et al., "Estimates and 25-Year Trends of the Global Burden of Disease Attributable to Ambient Air Pollution: An Analysis of Data from the Global Burden of Diseases Study 2015," *Lancet* 389 (2017): 1907-1918.

12. WHO, "WHO Air Quality Guidelines for Particulate Matter, Ozone, Nitrogen Dioxide and Sulfur Dioxide, Global Update 2005, Summary of Risk Assessment," 2005, whqlibdoc.who.int/hq/2006/WHO_SDE_PHE_OEH_06.02_eng.pdf.

13. "State of Global Air 2017," Special Report, Boston, MA: Health Effects Institute.

14. Gong, J., Zhu, T., Kipen, H., Wang, G. et al., "Comparisons of Ultrafine and Fine Particles in Their Associations with Biomarkers Reflecting Physiological Pathways," *Environ. Sci. Technol.* 48 (2014): 5264–5273.

15. Agency for Toxic Substances and Disease Registry, https://www.atsdr.cdc.gov/csem/csem.asp?csem=13&po=11.

16. Zimmerman, N., Wang, J.M., Jeong, C.-H., Ramos, M. et al., "Field Measurements of Gasoline Direct Injection Emission Factors: Spatial and Seasonal Variability," *Environ. Sci. Technol.* 50, no. 4 (2016): 2035-2043, doi:10.1021/acs.est.5b04444.

17. Karavalakis, G., Short, D., Vu, D., Yanf, J. et al., "Monoaromatic and Polycyclical Aromatic Emissions from GDI and PFI Vehicles on Ethanol and

Iso-Butanol Blends," presentation at *the SAE Powertrain, Fuels, and Lubes Conference*, Baltimore, October 2016.

18. Muñoz Fernandez, M. and Heeb, N., "PAH and Nitro-PAH Emissions from GDI Vehicles," presentation at *the 19th ETH-Conference on Combustion Generated Nanoparticles*, Zurich, June 28-July 1, 2015.

19. Directive 2004/107/EC of the European Parliament and of the Council of 15 December 2004 Relating to Arsenic, Cadmium, Mercury, Nickel and Polycyclic Aromatic Hydrocarbons in Ambient Air, http://eur-lex.europa.eu/LexUriServ/LexUriServ.do?uri=CONSLEG:2004L0107:20090420:EN:PDF.

20. Yinhui, W., Rong, Z., Yanhong, Q., Jianfei, P. et al., "The Impact of Fuel Compositions on the Particulate Emissions of Direct Injection Gasoline Engine," *Fuel* 166 (2016): 543-552.

21. https://www.epa.gov/sites/production/files/2015-09/.../priority-pollutant-list-epa.pdf.

22. Bisig, C., Steiner, S., Comte, P., Czerwinski, J. et al., "Biological Effects in Lung Cells In Vitro of Exhaust Aerosols from a Gasoline Passenger Car with and without Particle Filter," *Emiss. Control Sci. Technol.* 1 (2015): 237–246

23. Ramanathan, V. and Carmichael, G., "Global and Regional Climate Changes due to Black Carbon," *Nature Geoscience* 1 (2008): 221-227.

24. Price, P., Stone, R., OudeNijeweme, D., and Chen, X., "Cold Start Particulate Emissions from a Second Generation DI Gasoline Engine," SAE Technical Paper 2007-01-1931, 2007, doi:10.4271/2007-01-1931.

25. Chan, T.W., Meloche, E., Kubsh, J., and Brezny, R., "Black Carbon Emissions in Gasoline Exhaust and a Reduction Alternative with a Gasoline Particulate Filter," *Environ. Sci. Technol.* 48 (2014): 6027–6060.

26. Maricq, M.M., Szente, J.J., Harwell, A.L., and Loos, M.J., "Impact of Aggressive Drive Cycles on Motor Vehicle Exhaust PM Emissions," *J. Aerosol Science* 113 (2017): 1-11.

27. Zimmerman, N., Wang, J.M., Jeong, C.-H., Wallace, J.S. et al., "Assessing the Climate Trade-Offs of Gasoline Direct-Injection Engines", *Environ. Sci. Technol.* 50, no. 15 (2016): 8385–8392

28. Platt, S.M., El Haddad, I., Pieber, S.M., Zardini, A.A. et al., "Gasoline Cars Produce More Carbonaceous Particulate Matter Than Modern Filter-Equipped Diesel Cars," *Scientific Reports* 7 (2017): 4926.

29. U.S. Environmental Protection Agency, "40 CFR Parts 79, 80, 85, et al., Control of Air Pollution from Motor Vehicles: Tier 3 Motor Vehicle Emission and Fuel

Standards; Final Rule," April 28, 2014, http://www.gpo.gov/fdsys/pkg/FR-2014-04-28/pdf/2014-06954.pdf.

30. LEV III PM Technical Support Document Appendix P, "Development of Particulate Matter Mass Standards for Future Light-Duty Vehicles," 2012, http://www.arb.ca.gov/regact/2012/leviiighg2012/leviiighg2012.htm.

31. ARB, "An Update on the Measurement of PM Emissions at LEV III Levels," 2015, https://www.arb.ca.gov/msprog/levprog/leviii/lev_iii_pm_measurement_feasibility_tsd_20151008.pdf.

32. Hu, S., Zhang, S., Sardar, S., Chen, S. et al., "Evaluation of Gravimetric Method to Measure Light-Duty Vehicle Particulate Matter Emissions at Levels below One Milligram per Mile (1 mg/mile)," SAE Technical Paper 2014-01-1571, 2014, doi:10.4271/2014-01-1571.

33. Shields, E., "Particulate Matter (PM) Emissions from Low Greenhouse Gas Engine Technologies," presented at *the Advanced Clean Cars Symposium*, September 2016.

34. Xue, J., Li, Y., Quiros, D., Hu, S. et al., "Investigation of Alternative Metrics to Quantify PM Mass Emissions from Light Duty Vehicles," *Journal of Aerosol Science* 113 (2017): 85–94.

35. Khalek, I.A., Bougher, T., and Jetter, J.J., "Particle Emissions from a 2009 Gasoline Direction Injection Engine Using Different Commercially Available Fuels," *SAE Int. J. Fuel Lubr.* 3, no. 2 (2010): 623–637, doi:10.4271/2010-01-2117.

36. Maricq, M.M., Szente, J., Loos, M., and Vogt, R., "Motor Vehicle PM Emissions Measurement at LEV III Levels," *SAE Int. J. Engines* 4, no. 1 (2011): 597-609, doi:10.4271/2011-01-0623.

37. Yamada, H., Inomata, S., and Tanimoto, H., "Particle and VOC Emissions from Stoichiometric Gasoline Direct Injection Vehicles and Correlation between Particle Number and Mass Emissions," *Emiss. Control Sci. Technol.* 3 (2017): 135–141.

38. Maricq, M.M., Szente, J., Adams, J., Tennison, P. et al., "Influence of Mileage Accumulation on the Particle Mass and Number Emissions of Two Gasoline Direct Injection Vehicles," *Environ. Sci. Technol.* 47 (2013): 11890–11896.

39. Chan, W., Meloche, E., Kubsh, J., and Brezny, R., "Black Carbon Emissions in Gasoline Exhaust and a Reduction Alternative with a Gasoline Particulate Filter," *Environ. Sci. Technol.* 48 (2014): 6027–6034.

40. http://www.acea.be/publications/article/access-to-euro-6-rde-monitoring-data.

41. https://ec.europa.eu/clima/sites/clima/files/transport/vehicles/docs/2011_pems_jrc_62639_en.pdf.

42. Commission Regulation (EU) 2017/1151, June 1, 2017, http://publications.europa.eu/resource/cellar/7d1c640d-62d8-11e7-b2f2-01aa75ed71a1.0006.01/DOC_1.

43. Commission Regulation (EU) 2017/1154, June 7, 2017, http://eur-lex.europa.eu/legal-content/EN/TXT/PDF/?uri=CELEX:32017R1154&from=EN.

44. Joint Research Center, "PMP IWG Progress Report," Informal Document GRPE-75-17, 2016, http://www.unece.org/trans/main/wp29/wp29wgs/wp29grpe/grpeinf75.html.

45. Badshah, H., Kittelson, D., and Northrop, W., "Particle Emissions from Light-Duty Vehicles during Cold-Cold Start," *SAE Int. J. Engines*, 9, no. 3 (2016): 1775-1785, doi:10.4271/2016-01-0997.

46. He, H. and Yang, L., "China's Stage 6 Emission Standard for New Light-Duty Vehicles (Final Rule)," ICCT Policy Update, March 2017.

47. Gazette of India, "BS VI Notification," http://egazette.nic.in/WriteReadData/2016/168300.pdf.

48. Piock, W., Hoffmann, G., Berndorfer, A., Salemi, P. et al., "Strategies Towards Meeting Future Particulate Matter Emission Requirements in Homogeneous Gasoline Direct Injection Engines," *SAE Int. J. Engines* 4, no. 1 (2011): 1455-1468, doi:10.4271/2011-01-1212.

49. Pritchard, J. and Cheng, W., "Effects of Secondary Air on the Exhaust Oxidation of Particulate Matters," *SAE Int. J. Engines* 8, no. 3 (2015): 1088-1097, doi:10.4271/2015-01-0886.

50. Whitaker, P., Kapus, P., Ogris, M., and Hollerer, P., "Measures to Reduce Particulate Emissions from Gasoline DI engines," *SAE Int. J. Engines* 4, no. 1 (2011): 1498-1512, doi:10.4271/2011-01-1219.

51. McNeil, S., Adamovicz, P., and Lieder, F. "Bosch Motronic MED9.6.1 EMS Applied on a 3.6L DOHC 4V V6 Direct Injection Engine," SAE Technical Paper 2008-01-0133, 2008, doi:10.4271/2008-01-0133.

52. Peer, J., Backes, F., Sauerland, H., Härtl, M. et al., "Development of a High Turbulence, Low Particle Number, High Injection Pressure Gasoline Direct Injection Combustion System," *SAE Int. J. Engines* 9, no. 4 (2016): 2301-2311, doi:10.4271/2016-01-9046.

53. Smith, S. and Imoehl, W., "Measurement and Control of Fuel Injector Deposits in Direct Injection Gasoline Vehicles," SAE Technical Paper 2013-01-2616, 2013, doi:10.4271/2013-01-2616.

54. Aradi, A., Imoehl, B., Avery, N., Wells, P. et al., "The Effect of Fuel Composition and Engine Operating Parameters on Injector Deposits in a High-Pressure Direct Injection Gasoline (DIG) Research Engine," SAE Technical Paper 1999-01-3690, 1999, doi:10.4271/1999-01-3690.

55. Wen, Y., Wang, Y., Fu, C., Deng, W. et al., "The Impact of Injector Deposits on Spray and Particulate Emission of Advanced Gasoline Direct Injection Vehicle," SAE Technical Paper 2016-01-2284, 2016, doi:10.4271/2016-01-2284.

56. Wang, C., Xu, H., Herreros, J.M., Wang, J. et al., "Impact of Fuel and Injection System on PM Emissions from a DISI Engine," *Appl. Energy* 132 (2014): 178e91.

57. Prakash, A., Nelson, E., Jones, A., Macias, J. et al., "Particulate Mass Reduction and Clean-up of DISI Injector Deposits via Novel Fuels Additive Technology," SAE Technical Paper 2014-01-2847, 2014, doi:10.4271/2014-01-2847.

58. Prakash, A., Aradi, A., Imoehl, W., and Armitage, P., "Impact of Injector Design and Fuel Composition on Particulate Number Generation," SAE Technical Paper 2017-01-2395, 2017, doi:10.4271/2017-01-2395.

59. Ketterer, J.E. and Cheng, W.K., "On the Nature of Particulate Emissions from DISI Engines at Cold-Fast-Idle," *SAE Int. J. Engines* 7, no. 2 (2014): 986-994, doi:10.4271/2014-01-1368.

60. Heiduk, T., Kuhn, M., Stichlmeir, M., and Unselt, F., "The New 1.8L TFSI Engine from Audi Part 2: Mixture Formation, Combustion Method and Turbocharging," *MTZ* 72, no. 7-8 (2011): 58-64, doi:10.1365/s38313-011-0078-1.

61. Joshi, A., "Progress and Outlook on Gasoline Vehicle Aftertreatment Systems," *Johnson Matthey Technol. Rev.* 61, no. 4 (2017): 311.

62. Fujimoto, Y., "Introduction of Variable Compression Turbo Engine," *Advanced Clean Cars Symposium: The Road Ahead*, September 27-28, 2016.

63. Schenk, C. and Dekraker, P., "Potential Fuel Economy Improvements from the Implementation of cEGR and CDA on an Atkinson Cycle Engine," SAE Technical Paper 2017-01-1016, 2017, doi:10.4271/2017-01-1016.

64. Hedge, M., Weber, P., Gingrich, J., Alger, T. et al., "Effect of EGR on Particle Emissions from a GDI Engine," *SAE Int. J. Engines* 4, no. 1 (2011): 650-666, doi:10.4271/2011-01-0636.

65. Fischer, M., Kreutziger, P., Sun, Y., and Kotrba, A., "Clean EGR for Gasoline Engines – Innovative Approach to Efficiency Improvement and Emissions Reduction Simultaneously," SAE Technical Paper 2017-01-0683, 2017, doi:10.4271/2017-01-0683.

66. Parks, J.E., Storey, J.M.E., Prikhodko, V.Y., Debusk, M.M. et al., "Filter-based control of particulate matter from a lean gasoline direct injection engine," SAE Technical Paper 2016-01-0937, 2016, doi:10.4271/2016-01-0937.

67. Dempsey, A.B., Curran, S.J., and Wagner, R.M. "A Perspective on the Range of Gasoline Compression Ignition Combustion Strategies for High Engine Efficiency and Low NOx and Soot Emissions: Effects of In-Cylinder Fuel Stratification," *Int. J. Eng. Res.* 2016: 1-21.

68. Sellnau, M., "Aftertreatment for Low-Temperature Combustion and US Tier3-Bin 30 Emissions," presented at *the SAE 2017 Light Duty Emissions Control Symposium*, Washington DC, January 2017

69. http://www2.mazda.com/en/publicity/release/2017/201708/170808a.html.

70. Yamada, H., Inomata, S., and Tanimoto, H., "Particle and VOC Emissions from Stoichiometric Gasoline Direct Injection Vehicles and Correlation between Particle Number and Mass Emissions," *Emiss. Control Sci. Technol.* 3 (2017): 135–141.

71. Zinola, S., Raux, S., and Leblanc, M., "Persistent Particle Number Emissions Sources at the Tailpipe of Combustion Engines," SAE Technical Paper 2016-01-2283, 2016, doi:10.4271/2016-01-2283.

72. Nicholas, M., Tal, G., and Turrentine, T., *Advanced Clean Cars Symposium*, https://www.arb.ca.gov/msprog/consumer_info/advanced_clean_cars/pev_data_from_uc_davis_household_study_first_year_michael_nicholas.pdf, 2016.

73. AECC, "Real-Driving Emissions Test Programme Results from a Plug-In Hybrid Electric Vehicle (PHEV)," *13th Integer Emissions Summit*, Dresden, June 2017.

74. Leach, F., Stone, R., and Richardson, D., "The Influence of Fuel Properties on Particulate Number Emissions from a Direct Injection Spark Ignition Engine," SAE Technical Paper 2013-01-1558, 2013, doi:10.4271/2013-01-1558.

75. Khalek, I., Bougher, T., and Jetter, J., "Particle Emissions from a 2009 Gasoline Direct Injection Engine Using Different Commercially Available Fuels," *SAE Int. J. Fuels Lubr.* 3, no. 2 (2010): 623-637, doi:10.4271/2010-01-2117.

76. Butler, A., Sobotowski, R., Hoffman, G., and Machiele, P., "Influence of Fuel PM Index and Ethanol Content on Particulate Emissions from Light-Duty Gasoline Vehicles," SAE Technical Paper 2015-01-1072, 2015, doi:10.4271/2015-01-1072.

77. Qin, J., Li, X., and Pei, Y., "Effects of Combustion Parameters and Lubricating Oil on Particulate Matter Emissions from a Turbo-Charged GDI Engine Fueled with Methanol/Gasoline Blends," SAE Technical Paper 2014-01-2841, 2014, doi:10.4271/2014-01-2841.

78. Tanaka, D., Uchida, R., Noda, T., Kolbeck, A. et al., "Effects of Fuel Properties Associated with In-Cylinder Behavior on Particulate Number from a

Direct Injection Gasoline Engine," SAE Technical Paper 2017-01-1002, 2017, doi:10.4271/2017-01-1002.

79. Kim, Y., Kim, Y., Kang, J., Jun, S. et al., "Fuel Effect on Particle Emissions of a Direct Injection Engine," SAE Technical Paper 2013-01-1559, 2013, doi:10.4271/2013-01-1559.

80. Storey, J., Lewis, S., Szybist, J., Thomas, J. et al., "Novel Characterization of GDI Engine Exhaust for Gasoline and Mid-Level Gasoline-Alcohol Blends," *SAE Int. J. Fuels Lubr.* 7, no. 2 (2014): 571-579, doi:10.4271/2014-01-1606.

81. Aikawa, K., Sakurai, T., and Jetter, J. J., "Development of a Predictive Model for Gasoline Vehicle Particulate Matter Emissions," SAE Technical Paper 2010-01-2115, 2010, doi:10.4271/2010-01-2115.

82. Karavalakis, G., Short, D., Vu, D., Russell, R. et al., "Evaluating the Effects of Aromatics Content in Gasoline on Gaseous and Particulate Matter Emissions from SI-PFI and SIDI Vehicles," *Environ. Sci. Technol.* 49 (2015): 7021–7031.

83. Chan, T.W., Saffaripour, M., Liu, F., Hendren, J. et al., "Characterization of Real-Time Particle Emissions from a Gasoline Direct Injection Vehicle Equipped with a Catalyzed Gasoline Particulate Filter during Filter Regeneration," *Emiss. Control Sci. Technol.* 2, no. 2 (2016): 75-88.

84. Zhu, R., Hu, J., Bao, X., He, L. et al., "Tailpipe Emissions from Gasoline Direct Injection (GDI) and Port Fuel Injection (PFI) Vehicles at Both Low and High Ambient Temperatures," *Environmental Pollution* 216 (2016): 223-234.

85. Murad, S., Camm, J., Davy, M., and Stone, R., "Spray Behaviour and Particulate Matter Emissions with M15 Methanol/Gasoline Blends in a GDI Engine," SAE Technical Paper 2016-01-0991, 2016, doi:10.4271/2016-01-0991.

86. Chan, T., "The Impact of Isobutanol and Ethanol on Gasoline Fuel Properties and Black Carbon Emissions from Two Light-Duty Gasoline Vehicles," SAE Technical Paper 2015-01-1076, 2015, doi:10.4271/2015-01-1076.

87. Hergueta, C., Bogarra, M., Tsolakis, A., Essa, K. et al., "Butanol-Gasoline Blend and Exhaust Gas Recirculation, Impact on GDI Engine Emissions," *Fuel* 208 (2017): 662–672.

88. Smith, S. and Imoehl, W., "Measurement and Control of Fuel Injector Deposits in Direct Injection Gasoline Vehicles," SAE Technical Paper 2013-01-2616, 2013, doi:10.4271/2013-01-2616.

89. Wen, Y., Wang, Y., Fu, C., Deng, W. et al., "The Impact of Injector Deposits on Spray and Particulate Emission of Advanced Gasoline Direct Injection Vehicle," SAE Technical Paper 2016-01-2284, 2016, doi:10.4271/2016-01-2284.

90. Wang, C., Xu, H., Herreros, J.M., Wang, J. et al., Impact of Fuel and Injection System on PM Emissions from a DISI Engine," *Appl. Energy* 132 (2014): 178e91.

91. Maricq, M.M., Szente, J., Adams, J., Tennison, P. et al., "Influence of Mileage Accumulation on the Particle Mass and Number Emissions of Two Gasoline Direct Injection Vehicles," *Environ. Sci. Technol.* 47 (2013): 11890–11896.

92. Samuel, S., Hassaneen, A., and Morrey, D., "Particulate Matter Emissions and the Role of Catalytic Converter during Cold Start of GDI Engine," SAE Technical Paper 2010-01-2122, 2010, doi:10.4271/2010-01-2122.

93. Bogarra, M., Herreros, J.M., Hergueta, C., Tsolakis, A. et al., "Influence of Three-Way Catalyst on Gaseous and Particulate Matter Emissions during Gasoline Direct Injection Engine Cold-Start," *Johnson Matthey Technol. Rev.* 61, no. 4 (2017): 329.

94. Czerwinski, J., Comte, P., Heeb, N., Mayer, A. et al., "Nanoparticle Emissions of DI Gasoline Cars with/without GPF," SAE Technical Paper 2017-01-1004, 2017, doi:10.4271/2017-01-1004.

95. Andersson, J., Demuynck, J., and Hamje, H., "AECC/Concawe2016 GPF RDE PN Test Programme: PN Measurement above and below 23 nm," *21st ETH-Conference on Combustion Generated Nanoparticles*, Zurich, 2017.

About the Author

Dr Ameya Joshi is director of Emerging Technologies and Regulations at Corning Incorporated, USA. He holds a PhD in mechanical engineering from the University of Delaware and completed his postdoctoral research in chemical engineering at the Colorado School of Mines. His technical background is in reaction engineering as applied to combustion and emissions aftertreatment. His current role is to follow advances in engine technologies and regulations pertinent to vehicular emissions. Previously he has been a regional technology manager responsible for introducing innovative products to customers in Japan and Korea, prior to which he was research manager of Modeling & Simulation, in the Reaction Engineering group where he was responsible for developing a suite of modeling tools to predict the in-use performance of cellular ceramic products.

Section 2

Fundamentals of Particulate Emissions

CHAPTER 4

Soot Formation in Combustion

Thomas Koch, Jürgen Pfeil, Markus Bertsch, Christian Disch, Alexander Heinz, and Denis Notheis
Institut für Kolbenmaschinen, Karlsruher Institut für Technologie

4.1 Combustion-Generated Particles

The particles emitted by combustion engines typically consist of three distinct types, labeled “Nucleation Mode,” “Accumulation Mode,” and “Course Mode.” Coarse mode particles are of varying nature, such as rust from the exhaust system. They are not emitted directly but formed from the other two modes. The predecessors stored within the exhaust system become attached to each other and re-enter the exhaust flow as larger particles. Because of these storage and release processes, the coarse mode is an inconsistent emission. However, it is suggested that they consist of a solid core and an outer layer of volatile material. As the coarse mode is of unsettled state, artificial nature, and comparative rarity, these particles have been little studied [13]. Historically, nucleation mode particles have been less studied, because they are at the limit of detection for many instruments. The consistency of these particles is still a topic of research. A literature survey by Giechaskiel,

Manfredi, and Martini gives an overview about the current state of research on nucleation mode particles [16, 17]. They state:

> *The structure of primary particles is sometimes different (more amorphous) and unburned hydrocarbons or volatile organics can be found. This means that differences in the thermal pre-treatment (temperature, residence time of PN systems) might lead to different results. A lot of studies have found a solid core mode with older and modern diesel engines, both at low and high loads. Solid core mode is often observed at gasoline engines with port fuel injection (G-PFI) and it is assumed to originate from the metals of the lube oil or from fuel additives. At GDIs a shoulder at 10-20 nm appears quite often. For mopeds very often the size distribution after thermal pre-treatment peaks at or below 20 nm. It should be mentioned that in many studies it was recognized that the 'solid' core mode was re-nucleation artifact of the PMP (Particle Measurement Programme) method and the dilution factors employed.*

Accumulation mode particles consist of a collection of much smaller "primary" particles. The size of the primary particles ranges from about 20 to 50 nm. It should be mentioned that accumulation mode particles vary in size because they contain both greater or fewer numbers of primary particles, not because the primary particles vary in size. The number of primary particles to form an agglomerate is not a fixed number and varies from tens through hundreds to thousands. The morphology of agglomerates is also of a diverse nature, as the surface is coated by a layer of volatile or semi-volatile material. The chemical composition of particles can be defined according to the four-layered conceptual model by Eastwood [13], depicted in **Figure 4.1**. The five main components of the particles are subsequently presented according to the discussion by Eastwood [13]. As the sulfate bound in the fuel mainly contributes to the sulfate fraction, it is important to mention that this model was based upon diesel particles.

FIGURE 4.1 Conceptual composition of particle composition according to Eastwood [13].

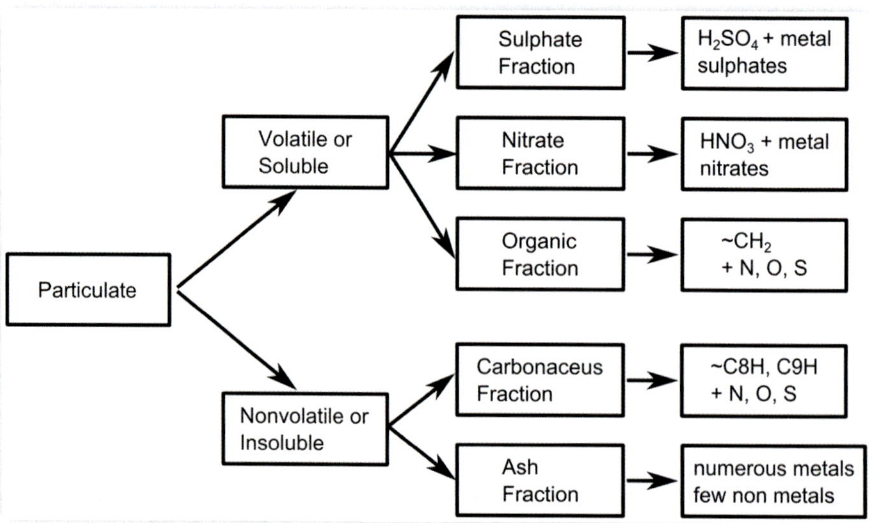

4.1.1 Organic Fraction

Based on the assay method, the organic fraction is also known as soluble organic fraction, or volatile organic fraction. However, the mass of soluble organic fraction is usually quite close to the mass of volatile organic fraction, even though the separation processes are based on different properties that do not necessarily generate the same end result. The organic fraction consists of several hundred compounds, such as alkenes, alkanes, alcohols, esters, ketones, acids, and aromatics. Even lighter C_4 to C_8 compounds were detected, which should be in the gaseous state, thus supporting the assumption that surface interactions are strong.

4.1.2 Sulfate Fraction

The sulfate fraction mainly consists of water-soluble sulfates, or the SO_4^{2-} ion, with the chief component sulfuric acid, H_2SO_4. There is a connection between the amount of sulfuric acid and the amount of water on the filter, thus dependent on the humidity in the filter's immediate environment. Therefore, prior to gravimetric measurements, the filters must be conditioned for a certain period of time in a closely defined environment.

4.1.3 Nitrate Fraction

The nitrate fraction consists of water-soluble nitrates, like the NO_3^- ion, with the chief compound nitric acid, HNO_3. Compared to the sulfate fraction, the amount of the nitrate fraction is usually lower. Additionally, nitric acid has a lower boiling point as compared to sulfuric acid, thus showing a greater volatility.

4.1.4 Carbonaceous Fraction

The title "carbonaceous" implies that this fraction is predominantly but not exclusively carbon. In the research community, this fraction is also titled as "Soot," "Graphite Carbon," "Elemental Carbon," or "Black Carbon."

4.1.5 Ash Fraction

The ash fraction covers inorganic compounds or elements, such as metals and a few nonmetals, and is a mixture of highly variable composition. A method to characterize the ash fraction is to burn all the particles and thereafter analyzing the remains, which is the "Incombustible Ash." The ash fraction is mostly caused by fuel additives and oil as well as the elemental composition of the working fluid itself.

The ratio for each of these fractions on the total particle emissions depends upon the composition of the working fluids, the engine operating parameters, the wear performance of the engine, as well as influences of the exhaust gas aftertreatment devices. Besides the chemical composition, particles are classified according to their diameters. As smaller particles do not contribute strongly to the particle mass, the legislative authority saw it necessary to limit the particle number and thus the emission of fine particles. Because of their small diameters, these particles are respirable and represent a considerable health risk. A typical distribution of the particle emissions of a diesel engine can be found in works by Kittelson et al. [24] and Dageforde [7]. Thereby, particle size distributions are plotted in three ways: according to number, area, and mass. Fundamentally this is the domination of different ranges by number and mass. Concerning particle numbers, most of the particles reside in the nucleation mode, while for the particle mass, the maximum is found in the accumulation mode. Generally, the particles are classified by their diameter as follows:

- PM10 for particles with a diameter not exceeding 10 μm.
- PM2.5 for particles with a diameter not exceeding 2.5 μm ("fine particles").
- PM0.1 for particles with a diameter not exceeding 100 nm ("ultra-fine particles").
- Particles with a diameter not exceeding 50 nm are known as "nanoparticles."

Even though the term "diameter" implicates particles to be of spherical shape, particles are more likely to show branched or reticular structures. Therefore, the measured diameter of particles by the actual measurement system is a comparative value and gives the reference diameter of an ideal round particle, which would behave comparatively to the measured particle. According to Hinds, different definitions for the reference diameter are common [19]:

- *Hydrodynamic Diameter*—diameter of a reference particle showing equal diffusion properties
- *Aerodynamic Diameter*—diameter of a reference particle with the density of 1 g/cm^3 and equal descent rate
- *Electrical Mobility Diameter*—diameter of a reference particle with a well-defined charge that shows equal mobility in an electrical field

The different diameters can be converted to each other as exemplarily shown by Jimenez et al. [21] and McMurray et al. [28]. However, the existence of different definitions with different values for the diameter supports the assumption that shape, mass, and surface of particles with equal diameters can show significant differences. Furthermore, the nucleation mode particles, in particular, can consist of a large number of volatile particles. As the nucleation mode particles are mostly dominated by volatile particles, the sampling position and the thermal pretreatment of the exhaust gas can influence the measured size distributions.

4.2 Soot Formation

From the above-mentioned particle fractions, ash and soot are generated within the engine's combustion process, while the other components form later in the exhaust system or even not before entering the surrounding air [13]. Even though today the formation process of soot is not determined in every detail, it is known that polycyclic aromatic hydrocarbons (PAH) play a major role in the process. This mechanism proposed by Bockhorn et al. [3] is schematically shown on the left-hand side of **Figure 4.2**. The formation process of PAH is dominated by acetylene (C_2H_2), which

FIGURE 4.2 Soot formation process according to Bockhorn [3] with pictures of soot structure created with HRTEM (left figure) and soot output dependency of temperature and air-fuel ratio by Böhm et al. [4] (right figure).

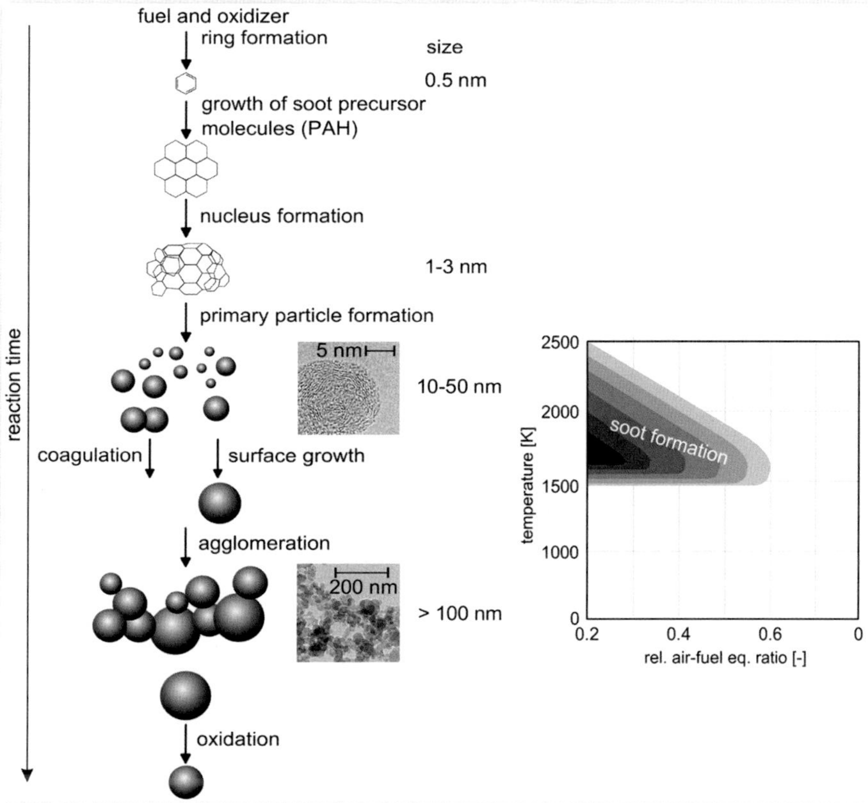

is present in a greater amount in fuel-rich flames. By reactions with CH and CH_2, C_3H_3 is created, which tends to build up a benzene ring (aromatic *ring formation*) by recombination and rearrangement. By abstraction of hydrogen and acetylene addition (HACA mechanism) [15], spatial structures with a graphite-like structure are formed from the plain PAH molecules which are *soot precursor molecules*. The three-dimensional structures resulting from the initial *nucleus formation* are known as *primary soot particles*. Due to *surface growth and coagulation*, which means a conglomeration of particles, the soot formation continues and extends. The growth of singular particles is limited, even though the reasons are not yet fully understood. This process is followed by the *agglomeration* of colliding particles, creating loosely structured agglomerates. The soot output is thereby dependent on pressure, carbon-to-oxygen ratio and temperature, as shown in Figure 4.2 on the right-hand side by Böhm et al. [4]. A maximum output of soot is thereby visible for rich air-fuel equivalence ratios and temperatures from 1600 to 1700 K. For lower temperatures, the formation of radical precursors like C_3H_3 is inhibited. For higher temperatures, these precursors are pyrolyzed and oxidized. Therefore, the soot formation is limited to temperatures from about 1000 to 2000 K [40]. Soot oxidation can occur at the precursor, nuclei, and particle stages of the soot formation process. Heywood states that in general, the rate of heterogeneous reactions such as the oxidation of soot depends on the diffusion of reactants to and products from the surface as well as the kinetics of the reaction [18]. There are many species in or near the flame that could oxidize soot, such as O_2, O, OH, CO_2, and H_2O. For the homogeneous, stoichiometric combustion in gasoline engines, the soot oxidation process is assumed to be dominated by OH, as the concentration of O and O_2 is at a low level [18]. However, in compression ignition engines, a large fraction of the soot formed is oxidized within the cylinder due to the excess of oxygen [37]. There are a number of models available which describe the dependency of the soot oxidation process on oxygen partial pressure and temperature [13, 23]. As it has been proven difficult to follow the oxidation of soot aerosols in flames, studies of bulk samples of pyrographite can be used to understand the soot oxidation process. The semi-empirical formula of Nagle and Strickland-Constable [29] has shown to correlate pyrographite oxidation with oxygen partial pressures below 0.1 MPa and temperatures in the range of 1100 to 2500 K and is therefore often used to describe soot oxidation mechanisms [18].

4.3 Influencing Factors on Soot Formation in Gasoline Engines

The influencing factors on soot formations in gasoline direct injection (GDI) engines are shown in **Figure 4.3**. Spray-wall interactions and the resulting impingements of liquid fuel on piston and liner result in insufficient mixture preparation. If the time for evaporation of the liquid fuel films is too short, a diffusive combustion with increased soot emissions is the consequence. However, at steady-state operation, these emissions can be prevented by a proper dimensioning of spray targeting, injection pressure, and injection timing. Concerning multi-hole injectors, the deposit formation at the injector tip is a major source of particle number emissions. At the end of the injection process, the impulse of the injection is low and the liquid fuel of the nozzle sac and the nozzle hole volumes can wet the injector tip. Due to insufficient evaporation of this liquid fuel, deposits are formed in a diffusive combustion at the injector tip. Furthermore, the built-up deposits further on act as a reservoir for liquid fuel in a self-enforcing process. Due to this effect, the particle number emissions can be increased by more than one order of magnitude without influencing any other emission characteristic, such as THC or NO_x.

Also, a dominant reason for particle emissions is inhomogeneities caused by insufficient mixture preparation. Possible sources for inhomogeneities are, for instance, spray-wall interactions, spray-valve interactions, or late injection timings and thus not enough time for mixture preparation.

4.3.1 Inhomogeneity

Soot formation generally depends on the local temperature and on the local air-fuel ratio [3]. Theoretically, fuel-rich zones should not exist for a homogeneous, stoichiometric operation of a GDI engine. Investigations show that spray-wall interaction is one of the main causes for soot formation in GDI engines [20, 25, 27, 39].

Concerning soot formation, this spray-wall interaction becomes critical, only if the time before ignition is too short to evaporate and sufficiently homogenize the

FIGURE 4.3 Influences on particle formation according to Bertsch et al. [5].

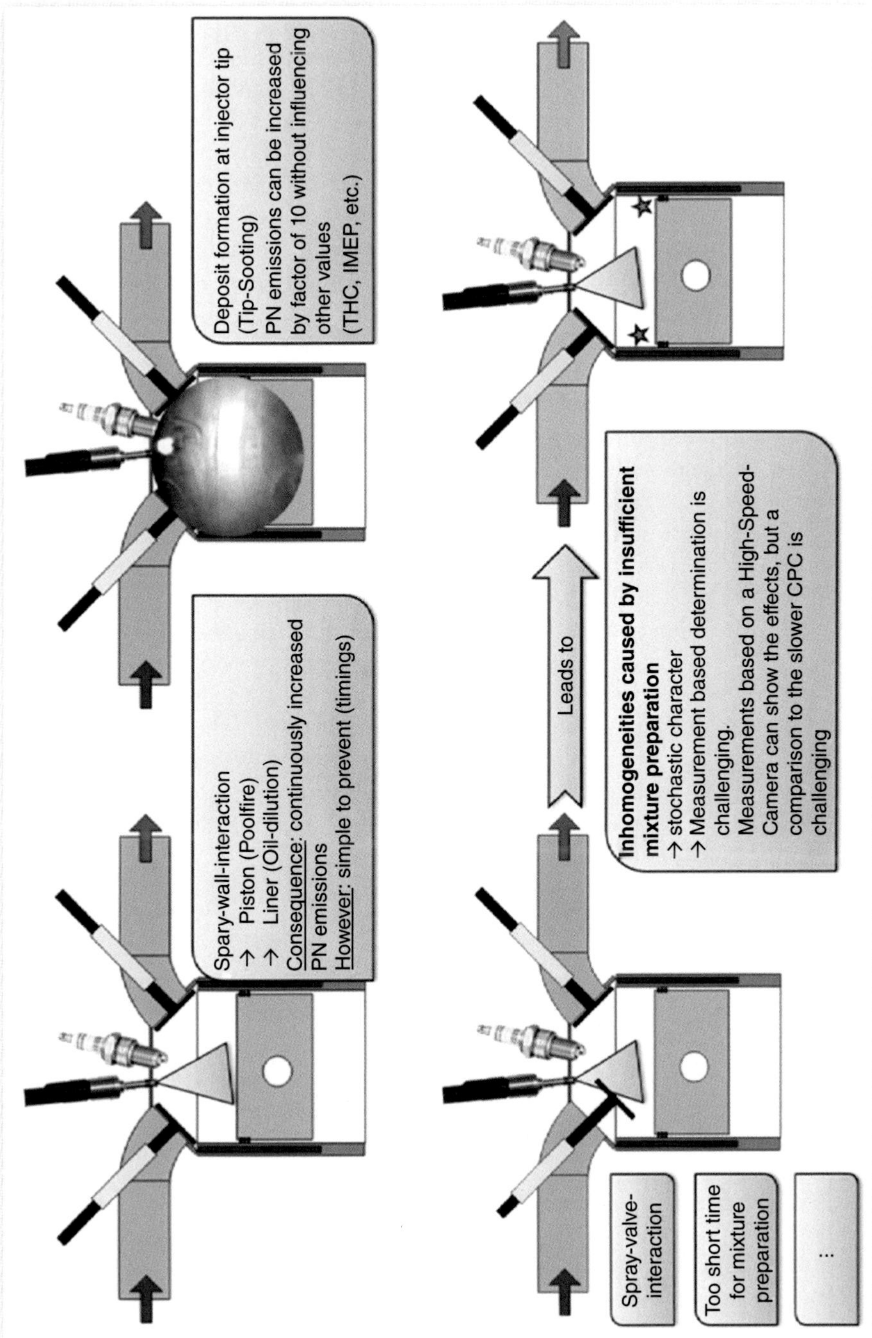

liquid fuel. During the following combustion process, most of the injected fuel is oxidized. After regular combustion, the temperature in the combustion chamber is at a high level. In combination with the low oxygen content, due to the stoichiometric operation, ideal conditions for soot formation are present. Thus, even small amounts of fuel stored in the wall films can significantly lead to increased particle number emissions. Contrary to diesel engines, soot cannot be oxidized after the combustion because of the low oxygen content [26].

Schulz et al. have shown that in addition to the initial temperature, the injection pressure, and the distance from injector to the hot plate, the angle of the injector to the plate plays an important role, regarding the wall film area and the wall film mass [34]. Bertsch [6] shows that with the correct injection timings, piston film can be prevented even if high amounts of fuel have to be injected. They also show that hydraulic flow and large-scale motion have a significant effect on the wall impingement. In the studies of Dageförde et al. [7] and Notheis et al. [30], the influence of large-scale motion on the spray targeting could be shown and explain the different effects on the particle number concentration resulting from a poor interaction of the in-cylinder charge motion and the spray.

This is also one reason for inhomogeneities caused by an insufficient mixture preparation. Fuel-rich zones with oxygen deficiency are formed, which are a potential source of soot emissions. In contrast to global fuel-rich conditions, a post-oxidation of the soot built up in the fuel-rich zones is possible. An insufficient mixture preparation caused by valve impingement or cold engine conditions can also lead to remaining droplets in the gas phase and therefore to diffusive flames.

Steimle et al. used high-speed imaging to locate different forms of inhomogeneity in the combustion chamber [36]. The different effects are shown in **Figure 4.4**. Largely, impingement on different locations leads to an increase in particle formation. Under cold engine conditions, inhomogeneities in the mixture formation can lead to diffusive combustion in large areas of the combustion chamber and the occurrence of remaining fuel droplets as an additional soot source.

4.3.2 Tip Sooting

Using multi-hole injectors, another possible reason for soot emissions is a deposit formation at the tip of the injector, known as "Tip Sooting." These deposits are formed as a result of a liquid fuel film on the tip. Using a clean injector, the fuel evaporates almost completely before the flame front reaches the injector. However, if deposits are formed at the tip, fuel is stored within the porous structure of the deposits. The evaporation process is delayed and leads to an insufficient mixture preparation of the stored fuel. The fuel is oxidized after the initial combustion under oxygen deficiency and leads to increased soot formation. Measurements by Berndorfer et al. [2] and Wiese et al. [42] have shown that this effect can increase the emitted particle number concentrations at steady-state operation by more than one order of magnitude, while neither the hydraulic flow nor the spray pattern are influenced by the deposits. Piock et al. have shown that an increased injection pressure is capable to reduce the deposit formation [31]. And even when deposits are built up at the injector tip, increased injection pressure of up to 40 MPa showed a reduction in particle number emissions.

FIGURE 4.4 Some of the most important sources of soot emissions in GDI engines according to Steimle et al. [36].

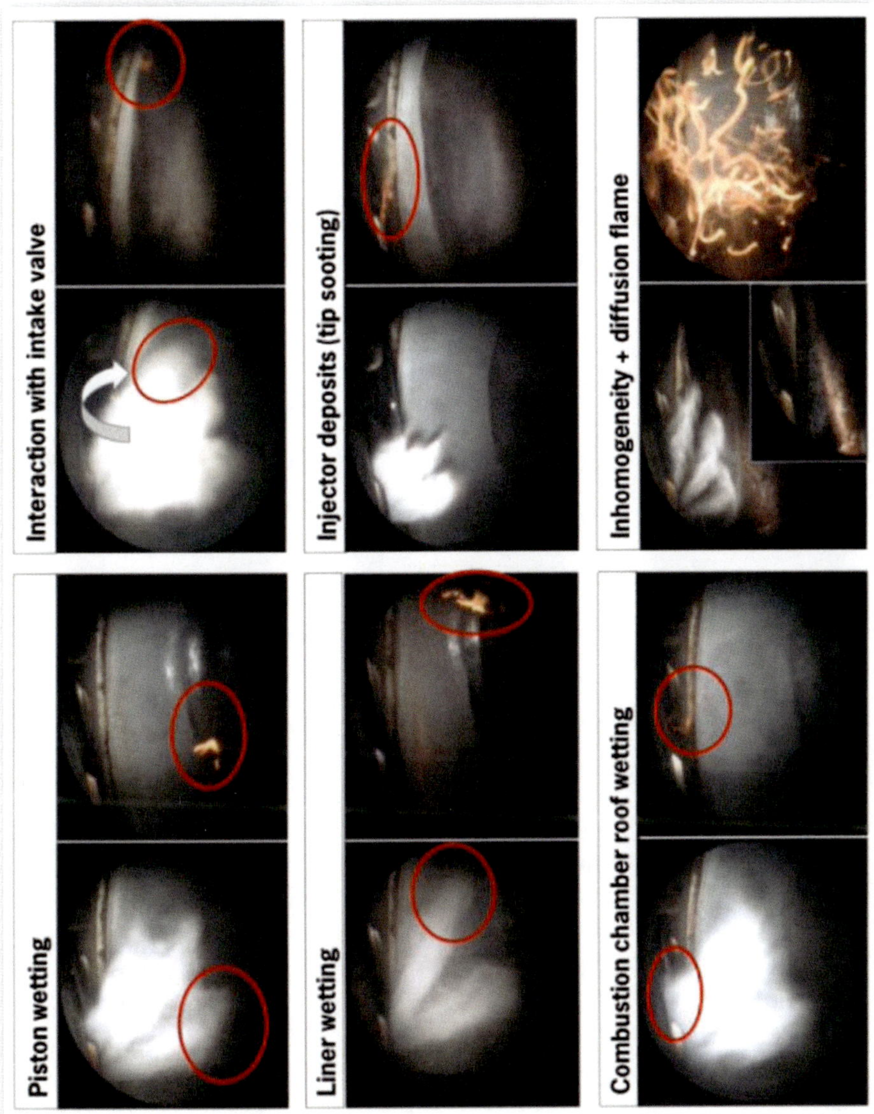

FIGURE 4.5 Deposit formation on the injector tip measured by Dageförde [9].

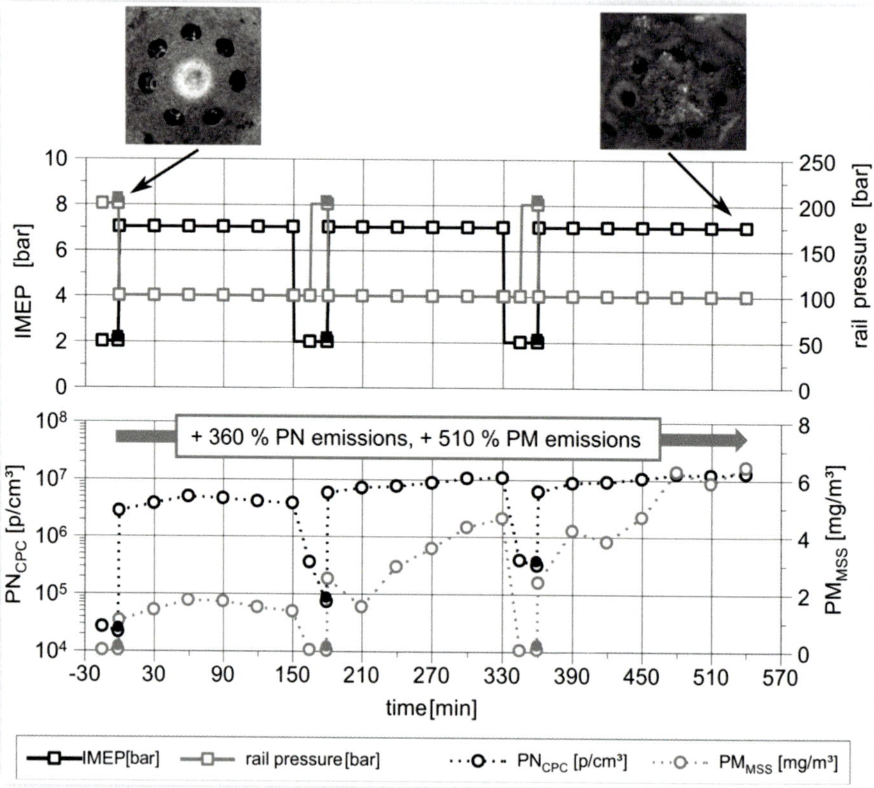

Figure 4.5 shows the result of tests on deposit formation using a multi-hole injector by Dageförde [9]. The picture on the left shows the injector tip before the start of the test. After running the engine about 500 minutes at 10 MPa injection pressure, the particle number emissions increased by about 360%, while the particle mass emissions increased by about 510%. After this test, the injector had deposits formed on the injector tip, as shown in the picture on the right.

4.4 Soot Formation in Different Operating Points

4.4.1 Higher Specific Loads and Low Engine Speeds

As downsizing and downspeeding are the most promising solutions to reduce the fuel consumption of modern engines, the engines operate more frequently at high engine loads and low engine speeds compared to engines with a larger displacement. For the

FIGURE 4.6 Influencing factors on PN formation in GDI engines at high specific loads according to Bertsch [6].

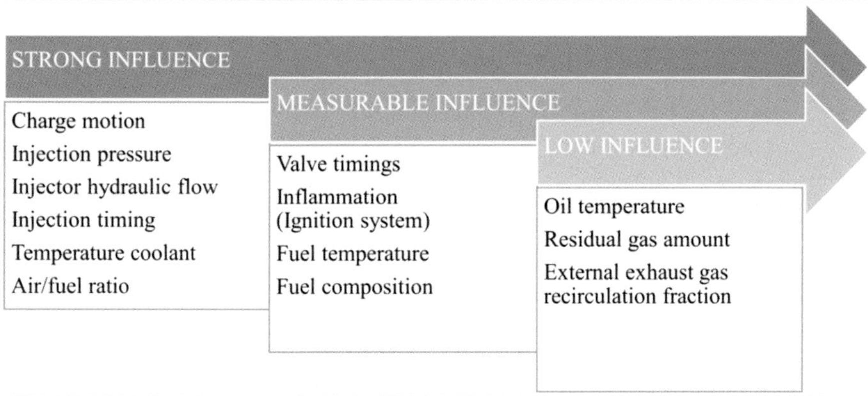

higher engine load, a larger amount of fuel needs to be injected. In combination with the low in-cylinder charge motion caused by the low engine speeds, the mixture preparation is challenging. Bertsch and Notheis investigated different engine strategies and their influence on particle number emissions experimentally [5] and in comparison to CFD simulations [30]. The investigations using a single-cylinder research engine were used to classify the different influencing factors according to their particle number reduction potential. The three groups "Strong Influence," "Measurable Influence," and "Low Influence" are illustrated in **Figure 4.6**.

One of the parameters with the strongest influence on particle number formation investigated in this project was the in-cylinder charge motion. Increased in-cylinder charge motion (tumble or swirl) leads to enhanced mixture formation and thus to reduced particle number formation, if the level of in-cylinder charge motion is matched to the engine. If not adjusted correctly, negative effects of the implemented charge motion are possible. Numerical studies on the same engine have shown that different charge motion strategies change the spray targeting. This could lead to a higher amount of cylinder wall impingement which increases the particle number emissions. With the relative cold temperatures of the cylinder walls, the evaporation of the liquid film is reduced. The fuel on the cylinder wall can get in the top land and burn diffusively as soon as the piston moves downwards.

Another promising solution to reduce particle number emissions is a combination of a reduced hydraulic flow of the injector with increased injection pressure. The average droplet size is reduced and the evaporation process is enhanced due to the increased surface-to-volume ratio. By using an increased injection pressure, the necessary amount of fuel can be injected within a shorter time with lower average droplet sizes.

It's likely that the different parameters leading to particle number emissions reduction potential are not additive and most of the effects interact with each other. It can also be stated that a "clean" combustion process, emitting particle number emissions slightly above the level of ambient air, can be realized in a gasoline direct injection engine at high engine load under steady-state, hot engine conditions.

Also, it could be recommended to operate gasoline direct injection engines with low particle number emissions at high engine load [6]:

- A short time for engine heat-up is required for low particle number emission engine concepts.
- The spray targeting, injection pressure, and injection timing need to be optimized synchronously and adjusted to the engine operation point.
- Operation with air/fuel ratio below $\lambda = 0.9$ should be prevented.
- Large-scale charge motion enhances the mixture formation process but needs to be applied to the engine geometry and operation settings.
- The ignition system can stabilize the ignition process and thus reduce particle number emissions.
- The fuel composition influences the evaporation process, soot formation, and soot oxidation. Therefore, future gasoline fuel for low particle number emissions should contain low amounts of aromatics and a high vapor pressure.

4.4.2 Engine in Transient Operation Points

Understanding the emission behavior of engines in transient operation points is one of the great challenges in order to optimize engine application for Real-Driving-Emission (RDE) standards regarding particle emissions. Transient conditions refer to changing engine conditions, which can be characterized in four major classifications as shown in Figure 4.7 proposed by Disch et al. [11]. The first-order transient is the reciprocating engine process itself. The cylinder pressure changes in a short timescale and processes have a high impact on engine and emission performance. Even in stationary engine operation, these processes are highly transient. Second-order transient describes cycle-to-cycle variations. These are governed by driver requests such as changes in engine load and speed. The engine control unit (ECU) control strategies have a great effect on these timescales as well. Third-order transients describe changes in engine temperatures such as oil, coolant, and combustion chamber surface temperatures. These processes

FIGURE 4.7 Criteria for interpretation and evaluation of transient engine operation [11].

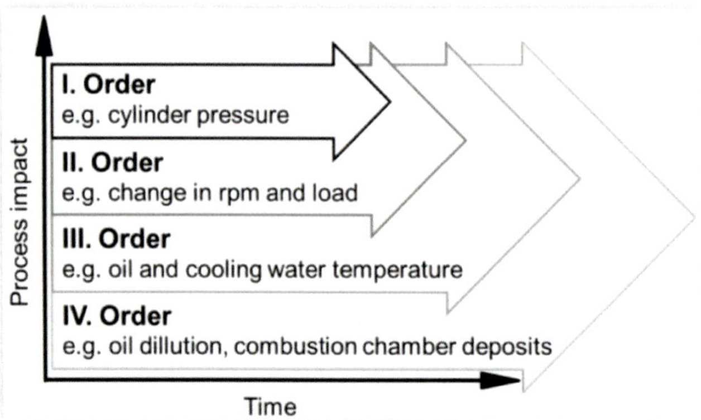

are slower and the impact on the emissions is high, but not as significant as first- and second-order processes. Cold start emissions in particular are influenced by the engine heat-up behavior. The fourth order describes engine processes over a longer timeframe such as wear, combustion chamber deposit build-up, and oil dilution [11].

4.4.2.1 Effects in Transient Operation during Load Step

The ECU calibration during load step conditions has a significant influence on mixture formation and especially the lambda value. Studies by Disch et al. have shown that during load steps, especially after coasting conditions, the lambda value control during the first combustion cycles is emission critical [10, 12]. To ensure a reliable engine torque build-up, the calibration of the initial combustion cycles, before the lambda control becomes active, has to be adjusted so that lean misfire is prevented. For this reason, a precise stoichiometric mixture with a high degree of homogenization is indispensable. An under-stoichiometric ECU calibration and/or poor mixture formation, for the initial combustion cycles, causes peak values for particle number concentrations. In addition, the injection strategy during engine load changes has a superimposed impact.

The in-cylinder conditions at start of injection (SOI) change significantly during the first few combustion cycles. Pool flames, as a result of piston wetting caused by early SOI applications can be detected. The cooling of the piston and wall surface during coasting requires a temporary SOI shift to later SOI timings. The temporary shift is an effective opportunity to reduce peak values of particle number concentrations. However, at later SOI timings, higher fluctuations of local lambda values close to the spark plug were observed. This leads to increased particle number emissions due to inhomogeneity and higher cycle-to-cycle variations.

Figure 4.8 shows an example of pool flame occurrence during a fast tip-in and the corresponding particle number concentration and size distribution. Pool flames can be observed in early phases of the tip-in. In later phases, only a few droplets can be identified as sources of particle emissions, even with constant SOI timings. This is due to the heat-up of the piston surface during the firing operation.

Tan et al. also showed a significant influence of spark timing and lambda value development during load changes. They state that in order to improve the transient behavior, extra development effort has to be invested in order to control the lambda value and its influence on transient emission behavior [38].

4.4.2.2 Catalyst Heating Operation

The catalyst heating operation is a special case of transient engine operation. To effectively reduce exhaust gas emissions after engine start-up, the catalyst needs to be heated up above the light-off temperature. To facilitate a quick heat-up of the catalyst, a high exhaust enthalpy flow is needed in combination with low gaseous emissions and low cycle-to-cycle variations. During this operation mode with cold engine conditions, more than 50% of the cumulative particle number emissions are produced in the first 100 seconds of the NEDC test cycle [33, 41].

Therefore, the quality of the mixture formation process mainly influences the particle number emissions. For GDI engines with a centrally mounted injector, a multi-pulse injection strategy with a short injection pulse just before ignition timing can lead to low particle number emissions [32]. The split-up of the injection process reduces the penetration and thus can help avoid wall wetting. Additionally, the short

FIGURE 4.8 Comparison of particle emission and high-speed imaging during tip-in.

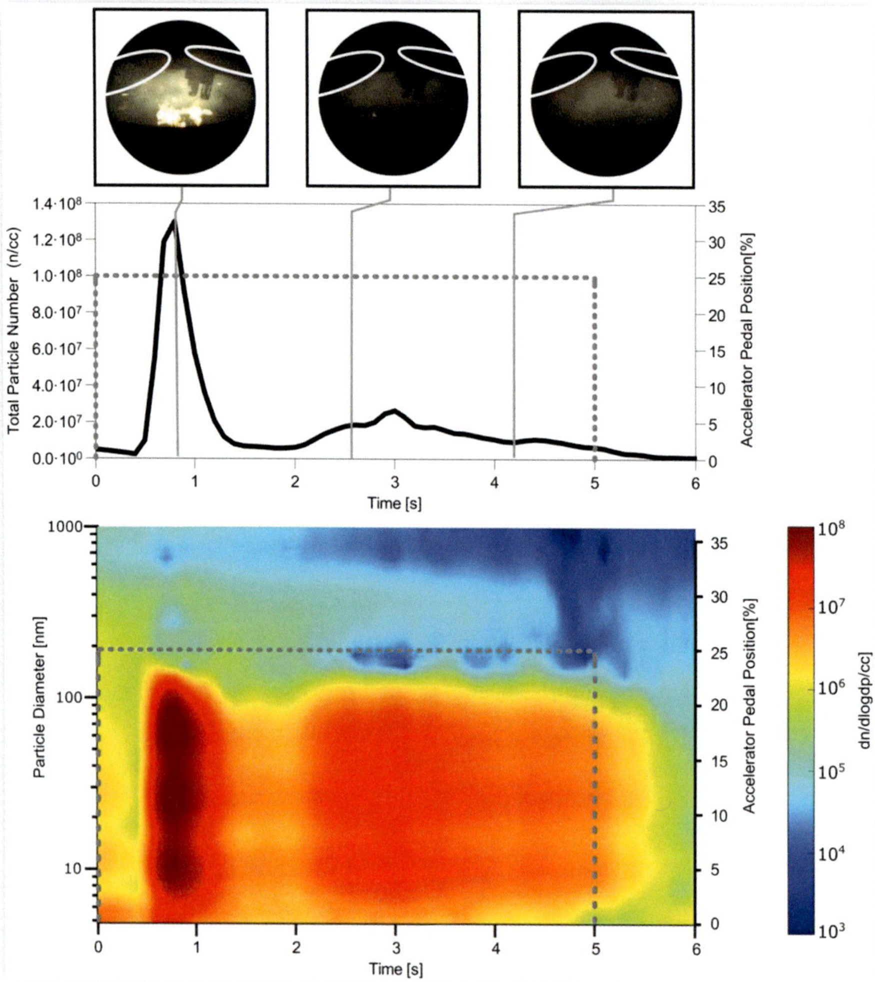

injection pulse just before ignition timing leads to a local fuel enrichment and increased turbulence within the spark plug area [1]. While the turbulence supports the stabilization of the inflammation and combustion process, the local fuel enrichment and the very short duration between the last injection pulse and ignition timing can lead to a substantial increase of the particle number emissions [8]. Therefore, the timing and injected fuel mass must be carefully and precisely chosen. Besides the optimization of the injector and the spray targeting by new laser drilling methods [22, 26], a further increase in the injection pressure is a promising solution to reduce particle number emissions. Schumann demonstrated with a single-cylinder research engine that an injection pressure of up to 80 MPa enables stratified operation in the catalyst heating mode with low particle number and NOx and HC emissions [35].

Besides the optimization of the fuel supply system, increased in-cylinder charge motion supports the mixture formation process. Dageförde therefore reduced the maximum intake valve lift of a single-cylinder research engine and combined the increased in-cylinder charge motion with a split injection strategy and the use of different fuels [7]. Results showed that the increased in-cylinder charge motion and the short injection pulse, right before ignition timing, leads to a stabilization of the combustion process and lower particle number emissions. De Francqueville and Pilla performed detailed investigations on particle formation in catalyst heating operation, using an optical accessible engine with a transparent sapphire ring [14]. They used Laser-Induced Incandescence (LII) to localize soot in the combustion chamber and Laser-Induced Fluorescence (LIF) to measure the fuel distribution. It was possible to determine five mechanisms as causes for the soot formation: ballistic impacts on the piston or on the lower part of the liner, non-direct impacts on the upper part of the liner due to fuel transportation in the combustion chamber or rebounds, and the last mechanism identified was mixture heterogeneities in the bulk gases. In summary, they state that fuel impacts on the piston, and to a lesser extent on the liner, were identified as the main mechanisms for the soot formation in GDI engines.

References

1. Altenschmidt, F., Gildein, H., Sauter, W., and Schaupp, U., "The Spray-Guided Mercedes-Benz Combustion System - Developed not only for Stratified Mode," *SIA International Conference and Exhibition, the Spark Ignition Engine of the Future*, Strasbourg, 2011, 1–9.
2. Berndorfer, A., Breuer, S., Piock, W., and Bacho, P.V., "Diffusion Combustion Phenomena in GDi Engines caused by Injection Process," SAE Technical Paper 2013-01-0261, 2013, doi:10.4271/2013-01-0261.
3. Bockhorn, H., *Soot Formation in Combustion: Mechanisms and Models* (Springer Series in Chemical Physics), Vol. 59, (Berlin/New York: Springer-Verlag, 1994), ISBN:3-540-58398-X.
4. Böhm, D., Hesse, D., Jander, H., Luers, B. et al., "The Influence of Pressure and Temperature on Soot Formation in Premixed Flames," *Symposium (International) on Combustion* 22, no. 1 (1988): 403–412, doi:10.1016/S0082-0784(89)80047-5.
5. Bertsch, M., "Investigations to Reduce Particle Number Emissions from GDI-Engines II," FVV-Abschlussbericht, Project Nr. 1144 and 1188, 2016.
6. Bertsch, M., "Experimental Investigations on Particle Number Emissions from GDI Engines," Dissertation, Karlsruhe Institute of Technology (KIT), Karlsruhe, 2017, doi:10.5445/IR/1000063994.
7. Dageförde, H., "Untersuchung innermotorischer Einflussgrößen auf die Partikelemission eines Ottomotors mit Direkteinspritzung," Dissertation, Karlsruhe Institute of Technology (KIT), Karlsruhe, 2015, ISBN:978-3-8325-4054-8.
8. Dageförde, H., Koch, T., Beck, K.W., and Spicher, U., "Influence of Fuel Composition on Exhaust Emissions of a DISI Engine during

CHAPTER 4

Catalyst Heating Operation," SAE Technical Paper 2013-01-2571, 2013, doi:10.4271/2013-01-2571.

9. Dageförde, H., Koch, T., and Spicher, U., "Untersuchung von Maßnahmen zur Reduktion der Partikel-Anzahlemissionen bei Otto-DI-Motoren," Abschlussbericht zu FVV-Projekt Nr. 1046 (in German), 2014.

10. Disch, C., Koch, T., Spicher, U., and Donn, C., "Engine-in-the-Loop as a Development Tool for Emissions Optimisation in the Hybrid Context," *MTZ Worldwide* 75 (2014): 40, doi:10.1007/s38313-014-0234-5.

11. Disch, C., Kubach, H., Pfeil, J., Koch, T. et al., "Cycle-Resolved Combustion Diagnostics of a Direct Injection Gasoline Engine in Transient Operation," *11th International Symposium on Combustion Diagnostics*, Baden-Baden, July 2014.

12. Disch, C., Pfeil, J., Kubach, H., Koch, T. et al., "Experimental Investigations of a DISI Engine in Transient Operation with Regard to Particle and Gaseous Engine-out Emissions," *SAE Int. J. Engines* 9, no. 1 (2016): 262-278, doi:10.4271/2015-01-1990.

13. Eastwood, P., *Particulate Emissions from Vehicles* (Wiley-PE Publishing Series), (Warrendale/Chichester: Published on behalf of SAE International/ John Wiley, 2008), ISBN:978-0-7680-2060-1.

14. Francqueville, L.d. and Pilla, G.L., "Investigation of Particle Formation Processes in a GDI Engine in Catalyst Heating Operation," *International Symposium on Combustion Diagnostics*, At Strasbourg, France, 2012, 129–137, ISBN:9783000326691.

15. Frenklach, M. and Wang, H., "Detailed Modeling of Soot Particle Nucleation and Growth," *Symposium (International) on Combustion* 23, no. 1 (1991): 1559–1566, doi:10.1016/S0082-0784(06)80426-1.

16. Giechaskiel, B., Manfredi, U., and Martini, G., "Engine Exhaust Solid Sub-23 nm Particles: I. Literature Survey," *SAE Int. J. Fuels Lubr.* 7, no. 3 (2014): 950–964, doi:10.4271/2014-01-2834.

17. Giechaskiel, B. and Martini, G., "Engine Exhaust Solid Sub-23 nm Particles: II. Feasibility Study for Particle Number Measurement Systems," *SAE Int. J. Fuels Lubr.* 7, no. 3 (2014): 935–949, doi:10.4271/2014-01-2832.

18. Heywood, J.B., *Internal Combustion Engine Fundamentals* (McGraw-Hill Series in Mechanical Engineering), (New York: McGraw-Hill, 1988), ISBN:0-07-100499-8.

19. Hinds, W.C., *Aerosol Technology: Properties, Behavior, and Measurement of Airborne Particles*, 2nd ed., (New York: Wiley, 1999), ISBN:978-0-471-19410-1.

20. Jiao, Q. and Reitz, R.D., "Modeling Soot Emissions from Wall Films in a Direct-Injection Spark-Ignition Engine," *International Journal of Engine Research* 16, no. 8, (2015): 994-1013, doi:10.1177/1468087414562008.

21. Jimenez, J.L., Bahreini, R., Cocker, D.R.I., Zhuang, H. et al., "New Particle Formation from Photooxidation of Diiodomethane (CH2I2)," *Journal of Geophysical Research* 108, no. D10 (2003): 4318, doi:10.1029/2002JD002452.

22. Kazour, J., Befrui, B., Husted, H., Raney, M. et al., "Innovative Sprays and Particulate Reduction with GDi Injectors," SAE Technical Paper 2014-01-1441, 2014, doi:10.4271/2014-01-1441.

23. Kennedy, I.M., "Models of Soot Formation and Oxidation," *Progress in Energy and Combustion Science* 23, no. 2 (1997): 95–132, doi:10.1016/S0360-1285(97)00007-5.

24. Kittelson, D., Patwardhan, U., Zarling, D., Gladis, D. et al., "Issues Associated with Measuring Nothing or Almost Nothing: Real-Time Measurements of Metallic Ash Emissions from Engines," *Cambridge Particle Meeting*, 2013, http://www.cambridgeparticlemeeting.org/sites/default/files/Presentations/2013/DKittelson%28UofMinnesota%29_2013_Issues%20asssociated%20with%20measuring%20nothing.pdf.

25. Köpple, F., Seboldt, D., Jochmann, P., Hettinger, A. et al., "Experimental Investigation of Fuel Impingement and Spray-Cooling on the Piston of a GDI Engine via Instantaneous Surface Temperature Measurements," *SAE Int. J. Engines* 7, no. 3 (2014): 1178–1194, doi:10.4271/2014-01-1447.

26. Kufferath, A., Berns, S., Hammer, J., Busch, R. et al., "The EU 6 Challenge at GDI – Assessment of Feasible System Solutions," 2012, accessed November 18, 2015, http://www.övk.at/veranstaltungen_/symposien/2012/nachlese_en.pdf.

27. Kufferath, A., Köpple, F., Bargende, M., and Jochmann, P., "Investigation of the Parameters Influencing the Spray-Wall Interaction in a GDI Engine – Prerequisite for the Prediction of Particulate Emissions by Numerical Simulation," *SAE Int. J. Engines* 6, no. 2 (2013): 911–925, doi:10.4271/2013-01-1089.

28. McMurry, P.H., Wang, X., Park, K., and Ehara, K., "The Relationship between Mass and Mobility for Atmospheric Particles: A New Technique for Measuring Particle Density," *Aerosol Science and Technology* 36, no. 2 (2002): 227–238, doi:10.1080/027868202753504083.

29. Nagle, J. and Strickland-Constable, R.F., "Oxidation of Carbon between 1000–2000°C," *Proceedings of the Fifth Conference on Carbon*, University Park, 1962, 154–164, doi:10.1016/B978-0-08-009707-7.50026-1.

30. Denis, N. and Bertsch, M., "Investigations to Reduce Particle Number Emissions from GDI-Engines III FVV-Abschlussbericht," Project Nr. 1188 and 1193, 2017.

31. Piock, W.F., Befrui, B., Berndorfer, A., and Hoffmann, G., "Fuel Pressure and Charge Motion Effects on GDi Engine Particulate Emissions," *SAE Int. J. Engines* 8, no. 2 (2015): 464–473, doi:10.4271/2015-01-0746.

32. Piock, W.F., Berndorfer, A., Hoffmann, G., Salemi, P. et al., "Strategies Towards Meeting Future Particulate Matter Emission Requirements in Homogeneous Gasoline Direct Injection Engines," *SAE Int. J. Engines* 4, no. (1) 2011: 1455–1468, doi:10.4271/2011-01-1212.

33. Sabathil, D., Koenigstein, A., Schaffner, P., Fritzsche, J. et al., "The Influence of DISI Engine Operating Parameters on Particle Number Emissions," SAE Technical Paper 2011-01-0143, 2011, doi:10.4271/2011-01-0143.

34. Schulz, F., Schmidt, J., Kufferath, A., and Samenfink, W., "Gasoline Wall Films and Spray/Wall Interaction Analyzed by Infrared Thermography," *SAE Int. J. Engines* 7, no. 3 (2014): 1165–1177, doi:10.4271/2014-01-1446.

35. Schumann, F., "Experimentelle Grundlagenuntersuchungen zum Katalysatorheizbetrieb mit strahlgeführter Benzin-Direkteinspritzung und Einspritzdrücken bis 800 bar," Dissertation, Karlsruhe Institute of Technology, Karlsruhe, 2014, http://nbn-resolving.org/urn:nbn:de:swb: 90-431354.

36. Steimle, F., Kulzer, A., Richter, H., Schwarzenthal, D. et al., "Systematic Analysis and Particle Emission Reduction of Homogeneous Direct Injection SI Engines," SAE Technical Paper 2013-01-0248, 2013, doi:10.4271/2013-01-0248.

37. Stiesch, G., *Modeling Engine Spray and Combustion Processes* (Heat and Mass Transfer), (Berlin/New York: Springer, 2003), ISBN:3-540-00682-6.

38. Tan, C., Xu, H., Ma, H., and Ghafourian, A., "Investigation of VVT and spark timing on combustion and particle emission from a GDI Engine during transient operation," SAE Technical Paper 2014-01-1370, 2014, doi:10.4271/2014-01-1370.

39. Velji, A., Yeom, K., Wagner, U., Spicher, U. et al., "Investigations of the Formation and Oxidation of Soot Inside a Direct Injection Spark Ignition Engine Using Advanced Laser-Techniques," SAE Technical Paper 2010-01-0352, 2010, doi:10.4271/2010-01-0352.

40. Warnatz, J., Maas, U., and Dibble, R.W., *Combustion: Physical and Chemical Fundamentals, Modeling and Simulation, Experiments, Pollutant Formation*, 4th ed., (Berlin/Heidelberg [u.a.]: Springer, 2006), ISBN:3-540-25992-9.

41. Whitaker, P., Kapus, P., Ogris, M., and Hollerer, P., "Measures to Reduce Particulate Emissions from Gasoline DI engines," *SAE Int. J. Engines* 4, no. 1 (2011): 1498–1512, doi:10.4271/2011-01-1219.

42. Wiese, W., Kufferath, A., Storch, A., and Rogler, P., "Requirements for Multi-Hole Injectors to Meet Future Emission Standards in Direct-Injection Gasoline Engines [in German]," *2nd International Engine Congress*, 2015, 63–80, ISBN:978-3-658-08861-3.

About the Authors

Prof. Dr. sc. techn. Thomas Koch has been heading the Institute of Internal Combustion Engines at Karlsruhe Institute of Technology (KIT) since 2013 and has been responsible for combustion engine issues in the areas of research, teaching, and innovation. For 10 years prior to this, Professor Koch, as a graduate mechanical engineer, held various positions in the development of commercial vehicle engines at Daimler AG. Already in his doctoral thesis and as postdoctoral researcher at ETH Zürich, Professor Koch has dealt with thermodynamic issues relating to fuel combustion.

Dipl.-Ing. Jürgen Pfeil joined the Institute of Internal Combustion Engines (IFKM) of Karlsruhe Institute of Technology (KIT) in 2001. Today, he works as group leader in the research group "Optics and Oil Circuit" of the Institute. Jürgen Pfeil and his group are applying and improving optical measurement tools for diagnostic tasks in the application fields combustion, sprays, fluid mechanics, and exhaust aftertreatment systems.

Dr.-Ing. Markus Bertsch studied mechanical engineering at Karlsruhe Institute of Technology. He was research associate at the Institute of Internal Combustion Engines (IfKM) at Karlsruhe Institute of Technology (KIT) and worked on several research projects concerning combustion process analysis, optical diagnostics, and exhaust gas analysis on GDI engines with special focus on particle number emissions.

His dissertation titled "Experimental Investigations on Particle Number Emissions from GDI Engines" was published in December 2016.

Dipl.-Ing. Christian Disch received his diploma and doctorate at the Karlsruhe Institute of Technology (KIT). Much of his experimental work has involved the application of sophisticated optical diagnostics to analyze mixture formation and combustion processes of both single-cylinder and multi-cylinder GDI engines. His publication on the optical investigation of transient engine operation presented at the JSAE/SAE Powertrains, Fuels & Lubricants Meeting 2015 (Kyoto, Japan) was honored with the Best Paper Award.

Alexander Heinz (M.Sc.) studied mechanical engineering at Karlsruhe Institute of Technology (KIT). Since 2015, he has been working as a researcher at the Institute of Internal Combustion Engines. His work includes the investigation of particle emission with a focus on transient and real driving engine behavior of gasoline engines as well as the use of special investigation methods in order to characterize this behavior.

Denis Notheis (M.Sc.) has been working as a researcher at the Institute of Internal Combustion Engines at Karlsruhe Institute of Technology (KIT) since 2015. He investigated particle formation in gasoline engines with direct injection for the research project "Investigations to reduce particle number emissions from GDI-engines III" of the Forschungsvereinigung Verbrennungskraftmaschinen (FVV). In his work, he focused on numerical studies of the spray/flow field interactions and their influences on the particle number emission.

CHAPTER 5

Fuel Impact on Particle Formation

Georgios Karavalakis
University of California, Riverside

Gasoline direct injection (GDI) engines are becoming the standard for gasoline light-duty vehicles in both the United States and Europe, replacing the traditional port fuel injection (PFI) engines, due to their improved thermal efficiency and lower carbon dioxide (CO_2) emissions [1, 2]. Especially in the United States, GDI platforms have been introduced by automakers as a pathway to reduce net greenhouse gas (GHG) emissions from the transportation sector and meet the federal Corporate Average Fuel Economy (CAFE) standard. However, GDI engines produce higher particulate matter (PM) emissions compared to their PFI counterparts. GDI involves the direct spray of fuel into the combustion chamber. Late evaporation of this fuel can lead to localized poor air-fuel mixing or diffusion-governed combustion that favors PM formation, especially during cold-start conditions [3, 4].

The use of biofuels in the United States and Europe has also been promoted for the past decades in an effort to reduce GHG emissions from the transportation sector due to environmental protection concerns. Biomass-derived ethanol is the most popular biofuel in the United States, where all the gasoline sold contains up to 10% ethanol by volume (E10). The ethanol utilization is on the rise in the United States, with the U.S. Environmental Protection Agency (EPA) allowing 15% of ethanol by volume (E15) to be sold in the market [5]. The positive environment for ethanol growth was also favored by mandates put in place by the Energy Independence and Security Act of 2007 (EISA) and the Renewable Fuel Standard. In addition to lower concentrations of gasoline-ethanol blends, gasoline is allowed to contain as much as 83% of ethanol by volume and as little as 51% of ethanol by volume [6]. Higher levels of ethanol can be used in flexible fuel vehicles (FFVs), which are designed for this purpose and are certified for emissions compliance by testing with E0 and E85 [7, 8]. Analogous to ethanol, higher alcohols have been the subject of increased interest as potential fuels in spark ignition (SI) engines [9, 10, 11]. Particular emphasis has been given to butanol isomers, which combines the advantages of an energy density, closer to that of gasoline, with the oxygen content and renewability of ethanol [12, 13].

PM formation in SI direct injection engines is well documented [14, 15, 16, 17]. Polycyclic aromatic hydrocarbons (PAHs) are known species to play a major role in the process of soot formation in GDI combustion [18, 19]. The formation process of PAH is dominated by acetylene (C_2H_2), which is present in a larger amount in fuel-rich flames. By reactions with CH and CH_2, C_3H_3 is formed, which tends to form a benzene ring (aromatic ring) by recombination and rearrangement. By abstraction of hydrogen and acetylene addition (HACA mechanism), these PAH molecules grow planar. Spatial structures with a graphite-like structure are formed from the plain PAH molecules and grow further by the HACA mechanism. These built-up three-dimensional structures are known as primary soot particles. Due to surface growth and coagulation, the soot formation continues and grows. This process is followed by the agglomeration of colliding particles, building up loosely structured agglomerates. Soot oxidation can occur at the precursor, nuclei, and particle stages of the soot formation process. There are many species in or near the flame that could oxidize soot, such as O_2, O, OH, CO_2, and H_2O. For the premixed, stoichiometric combustion in gasoline engines, the soot oxidation process is assumed to be dominated by OH, as the concentration of O and O_2 is on a low level [20, 21, 22, 23].

Fuel composition is critical to the formation of PM in GDI engines [24, 25, 26]. It should be expected that different fuels have different concentrations of PM emissions [27, 28]. So when comparisons are made between tests using different fuels, it would be useful to have a particulate matter index (PMI) that enables a correction to be made based on the key fuel properties. Aikawa et al. [29] developed a model to correlate PM emissions with the vapor pressure and the double bond equivalent (DBE) of the fuel components. The DBE is a measure of how unsaturated a hydrocarbon is and can be defined from:

$$\mathrm{DBE} = \frac{2\mathrm{C} - \mathrm{H} + 2}{2}$$

where C, H, and N are the number of carbon, hydrogen, and nitrogen atoms, respectively, in an organic compound. The PMI links the PM emissions with the vapor pressure and the DBE of the components in the fuel weighted by mass fraction (Wt) as follows:

$$\text{PM Index} = \sum_{i=1}^{n} I_{(443\,\mathrm{K})} = \sum_{i=1}^{n} \frac{\mathrm{DBE}_i + 1}{V \cdot P_{(443\,\mathrm{K})}} \cdot Wt_i$$

As above mentioned, the DBE is a measure of the number of double bonds and rings in the fuel. Particulate emissions can correlate with DBE values and higher boiling points for different chemical species in the fuel. Differences in molecular structure and DBE between paraffins and aromatic hydrocarbons contribute to greater particulate emissions for aromatics [30, 31]. For example, the DBE values of paraffins is 0, whereas the DBE values of aromatic hydrocarbons, depending on the molecular structure, are approximately 4 to 7. In addition, components with higher boiling points and lower corresponding vapor pressures evaporate more slowly, resulting in a greater tendency for diffusion combustion instead of premixed combustion. This results in more imperfect fuel mixing and diffusion-controlled burning of thin films of liquid fuel on the piston, leading to higher particulate emissions [32]. The work undertaken by Aikawa et al. [29] used a PFI engine and had no independent control of the fuel vapor pressure or DBE, as commercial gasolines were used, along with a base fuel, to which different components were added. Indolene was used as a base fuel, to which were added 10% by mass of components such as 2,2,4-trimethylpentane, dodecane, ethylbenzene, and 1,2,4-trimethylbenzene. The PMI range of the fuels tested was 1.01–3.86. The index range of over 1,400 worldwide fuels available was calculated with a mean PMI of 2.12. Furthermore, Aikawa et al. [29] evaluated the vapor pressure at a range of temperatures and found the best correlation between PMI and particle emissions for the vapor pressure at a temperature of 443 K.

Leach at al. [33] further investigated the model of Aikawa et al. [29] using a spray-guided GDI engine with designed fuels where the DBE and the vapor pressure were varied independently. In contrast to Aikawa et al. [29], who calculated the PMI by volume fractions, Leach et al. [33] calculated a particle number index (PNI), using mass fractions. However, they stated that in the majority of cases the relative difference in calculation of PMI and PNI is less than 15%. It should also be noted that in their work, the engine was operated at low engine speed (1500 rpm), low load (0.18 MPa IMEP), and with a rich mixture ($\lambda = 0.9$). Dageförde et al. [34] investigated different fuels and fuel blends at catalyst heating operation, and their results supported the findings of Aikawa et al. [29] and Leach et al. [33], as the particle number emissions could be reduced by operating the engine with oxygenated fuel blends as well as with alkylate fuel. The ethanol and n-butanol blends (E40 and B40) have a reduced amount of aromatic compounds and thus a lower DBE. However, using toluene as an aromatic compound and thus increasing DBE did not increase the PN emissions. A study by Khalek et al. [35] indicated that reductions in high boiling point aromatics and the DBE value played important roles in reducing both the PM mass and PN emissions. Karavalakis et al. [32] showed that reducing the aromatic content could effectively decrease PM mass, PN, and black carbon emissions from a fleet of PFI and GDI vehicles. The same study showed a tight correlation between the PMI and particulate emissions, as well as a strong contribution of aromatic species with high boiling points to PM formation [32]. Similar results were observed in a study conducted by Kim et al. [36] when they tested a GDI vehicle using three fuels with different aromatic contents over the New European Driving Cycle (NEDC) and the Federal Test Procedure (FTP) test cycles. The authors found that higher aromatic contents and less volatile fuel led to increases in PN emissions. Zhu et al. [37] tested two identical GDI vehicles over the WLTC test cycle using seven fuels with varying levels of aromatics and olefins contents, as well as T50 (50% vol distillation temperature) and T90 distillation parameters. The results of this study showed decreases in PM and PN emissions when the aromatic content and T90 decreased. Chan et al. [38] evaluated the particulate emissions from two vehicles when operating on two fuels with different PM indices over the FTP and the US06 Supplemental Federal Test Procedure (US06) driving cycles. They found that replacing naphthylenic aromatic hydrocarbons in the fuel (i.e., high PMI fuel) with monoaromatics and naphthenes (i.e., low PMI fuel) reduced PN and black carbon emissions. They also found that the low PMI fuel led to formation of smaller particles with diameters of about 50 nm. Jiao et al. [39] predicted the soot emissions of GDI engines using a computational fluid dynamics (CFD) model. The study confirmed that the light components of gasoline can be more easily evaporated, which affects the ignition and combustion, whereas the heavy components have a large impact on the engine emissions, especially in the presence of the fuel film. Fushimi et al. [40] tested four GDI vehicles and concluded that the high boiling point components may cause the increase in PM emissions.

A major study conducted by the U.S. EPA has shown that PM emissions are greatly influenced by aromatic content and T90, followed by ethanol, T50, and RVP (Reid vapor pressure) [41]. This study, the EPAct/V2/E-89 (hereafter "EPAct") program, was the first large parametric study of gasoline fuel effects to include PM measurements for all tests and was conducted on a fleet of 15 PFI vehicles. The results of this study showed positive coefficients for total aromatics, T90, and ethanol in PM fuel effect models, meaning that relative increases in these fuel properties were associated with higher emissions. Further analysis was conducted on the EPAct data, which replaced total aromatics and T90 fuel parameters in the original design with the PMI parameter [42]. The results showed a large coefficient for PMI, confirming its high correlation with PM emissions. This study also showed that ethanol exacerbates the propensity of low-volatility fuel components to form PM [42]. This phenomenon was further confirmed by a follow-up study from the same group of authors where they showed that the presence of ethanol was found to have a reinforcing interaction with PMI resulting in augmented PM emissions. The authors suggested that ethanol's high heat of vaporization hinders evaporation of the higher molecular weight components of the fuels [43]. Karavalakis et al. [44] conducted a comprehensive literature review and statistical analysis for the identification of possible correlations between fuel properties, such as the distillation parameters, aromatics, and octane indices, and PM mass and total and solid PN emissions. In their review, the authors included some of the major fuel studies available in the literature such as the EPAct study, the European "Particulates" project, the Coordinating Research Council (CRC) E-94 study (phases 1 and 2), studies conducted by the California Air Resources Board (CARB), and others. The authors reported that for GDI vehicles, aromatics and distillation temperatures had the strongest positive correlations with PM mass and total PN emissions. Distillation end point was statistically significant for both PM mass and PN emissions, while T70 and T90 showed a consistent positive correlation only for total PN emissions. Other properties, such as octane indices AKI, research octane number (RON), and motor octane number (MON), and T10, generally showed more negative correlation with PM and PN emissions for GDI vehicles. The authors also presented cases where statistically significant interactions showed trends in the data related to particular combinations of engine types/model years, number of engine cylinders, and drive cycles. For example, they showed that T10 had a more significant impact on PM mass emissions (positive correlation) for larger engines for GDI vehicles, whereas end point had stronger statistically significant effects on PM mass emissions (positive correlation) for vehicles equipped with wall-guided GDI naturally aspirated engines.

It is still not conclusive as to whether the use of ethanol causes an increase or decrease in particulate emissions from current SI engines. Storch et al. [45] reported measurements of two-dimensional soot volume fraction in GDI ethanol-blended spray flames ignited by a spark plug inside a spray injection chamber. They found reduced evaporation for pure ethanol and E85 blends compared to isooctane and E20, which they were ascribed to the high enthalpy of evaporation of ethanol delaying the evaporation process. They also showed higher sooting tendency of isooctane-E20 compared to pure isooctane, as well as that ethanol addition to a toluene-isooctane mixture and gasoline can increase the frequency of soot volume fractions. The authors claim that is due to the fact that ethanol changes the evaporation characteristics of the mixture due to non-ideal mixing behavior and high enthalpy of evaporation, which leads to an increased soot formation tendency. Chen et al. [46] investigated the effects of gasoline/ethanol blends in different blending proportions for PN and particle mass emissions using a single-cylinder optical access direct injection engine under cold and warm conditions at a stoichiometric condition (1500 rpm, 0.5 bar manifold absolute pressure). They found increases in both PN and PM emissions with increasing ethanol content, with these increases being more profound under cold conditions than under warm conditions (**Figure 5.1**).

The increases in particulate emissions were attributed to the deleterious effect of ethanol on spray break-up and the evaporation efficiency as a result of its high vaporization enthalpy and low energy density. Similar findings have been reported by the same group of authors where higher ethanol concentrations led to increased PM mass and PN emissions during engine cold-start conditions [47]. These results were explained by mixture inhomogeneities caused by poor evaporation at low temperature due to the high enthalpy of evaporation of ethanol. Di Iorio et al. [48] studied the PM emissions of ethanol/gasoline blends (E0, E50, E85, and E100) in homogeneous and stratified conditions in a supercharged GDI engine. They found that the accumulation mode particles increased in the stratified mode. When the fuel was injected at the early stage of the intake stroke, there was sufficient time for evaporation and mixing with the fresh air, and improved the air/fuel mixture homogeneity. When the fuel was injected at a later stage of the intake stroke, the evaporation and mixing time was short even though the in-cylinder higher temperature was not enough to promote the fuel evaporation, and this resulted to a certain extent in stratification and uneven mixing. Once the flame front arrived at the wet surface, the diffusion combustion that occurred on the oil film contributed highly to PM formation. Similar trends have been seen by Catapano et al. [49] when they studied the effects of engine speed on PM emissions of ethanol blends. They observed a decrease in particle size distribution and PN emissions as the ethanol content in the blend increased for the cases with no significant fuel impingement at 2000 rpm. But as the engine speed was increased to 4000 rpm, the longer injection durations for ethanol blends resulted in more impingement with the piston, and the time available for vaporization was decreased, which increased the PN and sizes for the ethanol blends, as shown in **Figure 5.2**.

Analogous increases in PN emissions were observed in a study by He et al. [50] using certification gasoline and an E20 blend at four steady-state engine operating conditions when testing a wall-guided GDI engine. It was found that fuel injection timing was the dominant factor impacting PN emissions. Chan et al. [51] showed

FIGURE 5.1 Total particulate number (top) and total particulate mass (bottom) emission for different gasoline/ethanol blends in a cold (20°C) and a warm (80°C) engine. Adapted from Chen et al. [46].

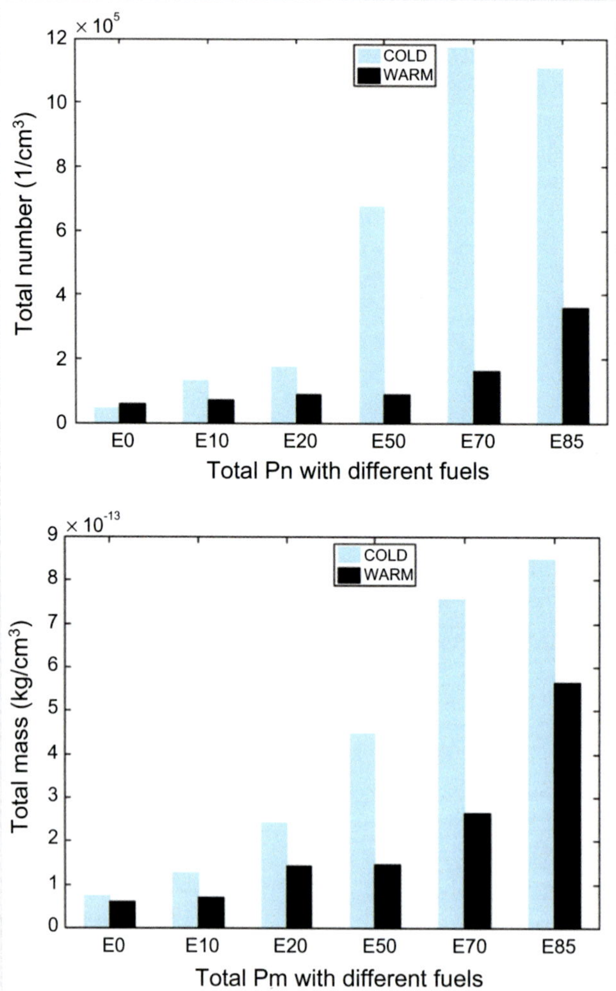

that the use of E10 fuel reduced the particle emissions from a PFI vehicle and showed moderate particle reductions over the FTP cycle at standard ambient test temperature (22°C), but significantly increased particle emissions compared to neat gasoline at other ambient test temperatures (–7°C and –18°C).

The majority of the published literature have shown the beneficial effects of ethanol fueling on PM formation from GDI engines and vehicles. Esarte et al. [52] found that the presence of alcohols reduced the formation of soot and that a higher oxygen/carbon (O/C) ratio in the reacting mixture resulted in a higher soot reduction due to the enhancement of oxidation reactions by the presence of oxygen in the mixture. Thus, the reaction rate of the particle surface growth by large hydrocarbons for the

FIGURE 5.2 Particle size distributions for the gasoline fuel (E0) and the ethanol blends at 2000 rpm (top) and 4000 rpm (bottom) partial load. Adapted from Catapano et al. [49].

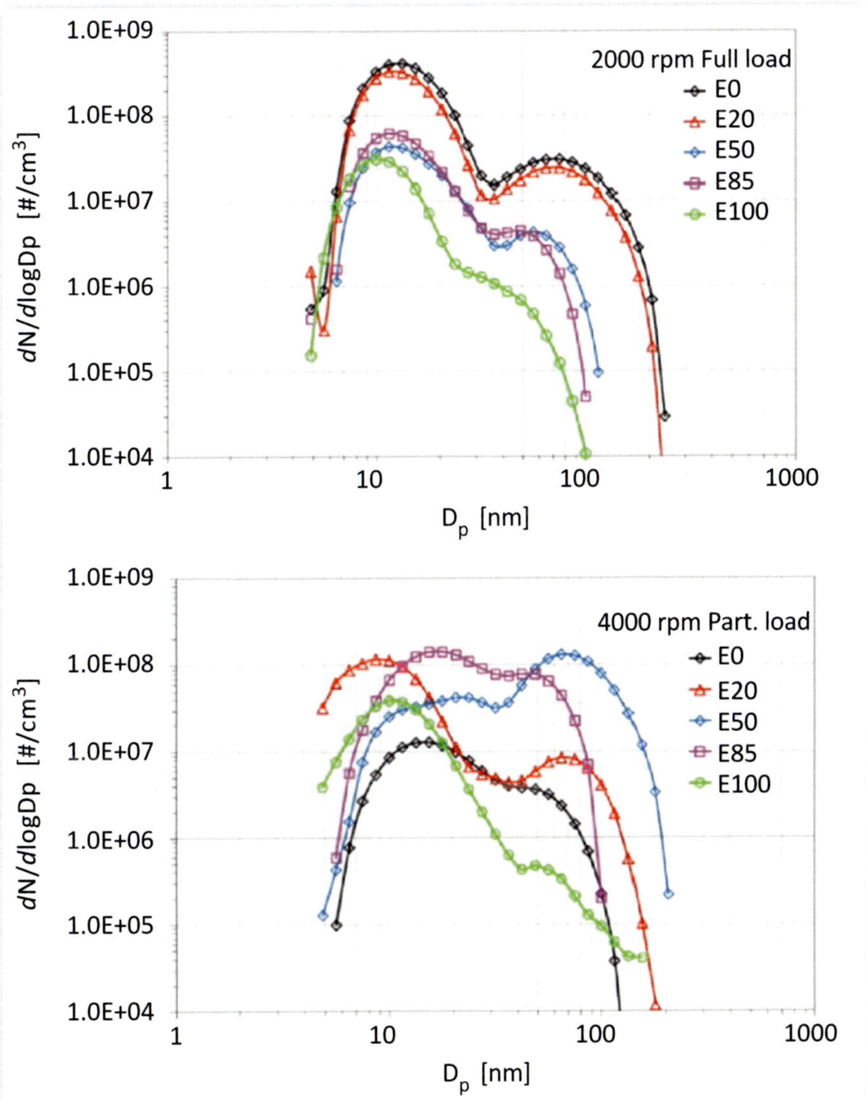

formation of the accumulation mode particles decreases. Maricq et al. [53] examined the impact ethanol/gasoline blends from a turbocharged GDI vehicle with two engine calibrations over the FTP cycle. They showed statistically significant reduction in PM mass and PN emissions when the ratio of ethanol was higher than 30% by volume for both engine calibrations. However, as the ethanol level ranged from 0% to 20%, a small benefit in PM mass and PN emissions was observed, which was within the test variability. The same study showed that the PM composition was determined to be primarily elemental carbon (EC) in nature. Storey et al. [54] compared the PM and PN emissions of pure gasoline, E10, and E20 fuels under the FTP and US06 driving cycle on a GDI vehicle. They found that the addition of ethanol could simultaneously reduce the PM mass and PN emissions in the transient phase and transition phase, which was mainly due to the PM reduction under the fuel-rich condition, whereas the PM emissions in the rich-fuel conditions were much higher than those in the stoichiometric conditions. Under steady-state conditions, the PN concentrations decreased with an increase in ethanol level. Similar to Maricq et al. [53], this work showed that most of the organic carbon (OC) in PM was effectively removed by the three-way catalyst (TWC) leaving a larger fraction of EC. In a study by Fatouraie et al. [55], they conducted experiments in a single-cylinder GDI engine at the same load conditions and at a fixed engine speed (1500 rpm) using E0/indolene and E100. They found over an order of magnitude less soot with ethanol under all operating conditions compared to indolene.

Sakai and Rothamer [56] studied different ethanol blends (E10, E20, E30, E40, and E50) on a single-cylinder GDI engine to evaluate the relative sooting propensity of ethanol fueling under completely premixed and prevaporized operation. By operating the engine in this mode, the presence of liquid fuel in-cylinder was eliminated, allowing the examination of soot production from gas phase sources. The results showed that higher ethanol concentrations decreased particulate emissions. As ethanol levels increased, the particle size distribution changed from agglomeration to nucleation mode dominated profiles. At higher equivalence ratios, this shift also corresponds to the change from an accumulation mode dominated distribution for low-level ethanol blends to a nucleation mode dominated distribution at higher blends. Karavalakis et al. [57] studied the particulate emissions of a GDI FFV with a wall-guided injection system fueled with E10, E51, E83, and an isobutanol blend at a proportion of 55% by volume over the FTP and LA92 driving cycles. They found that the addition of higher ethanol blends and the isobutanol blend resulted in large reductions in PM mass, black carbon, and total and solid particle emissions, with specific fuel properties having an obvious effect on particulate emissions, such as the oxygen content and aromatic compounds. Mamakos et al. [58] also showed important reductions in total and solid PN emissions when they tested GDI FFVs on high ethanol volumetric fractions (75–85%) over the NEDC and the motorway portion of the Artemis cycle. Jin et al. [59] examined the impact of different low-, mid-, and high-ethanol blends (E0, E10, E30, E50, and E85) on a GDI vehicle over the FTP cycle. Their results showed strong PN and PM mass reductions for the high-ethanol blends relative to E0 and E10 fuels. As the ethanol content was increased, the size-resolved PN concentration showed a decrease in the accumulation mode. Finally, Zhang et al. [9] showed that with an increase in the ethanol ratio, the accumulation mode particles were significantly reduced, with the nucleation mode particles being the majority of the total particle concentration. This finding was attributed to the lower combustion temperature with the increase of the ethanol ratio, whereas fewer primary carbon particles were generated by the pyrolysis of fuel vapor and dehydrogenation reaction decreased due to the lower aromatic content and the higher oxygen content in the fuel.

According to the above statements, the main cause of the gasoline engine particles is the non-premixed combustion of the incompletely evaporated and mixed fuel. Thus, if the fuel molecules contain oxygen to improve the combustion of the local zone, then it will help to inhibit the generation of soot particles. Based on this hypothesis, Oh et al. [60] studied the effects of oxygen content on the PM emissions of a GDI engine equipped with a centrally mounted injector, a GDI engine equipped with a side-mounted injector, and a PFI engine. They found that increasing the oxygen content in gasoline helped reduce the PM emissions of the GDI engine, especially during the start-up phase. Wang et al. [61] compared the PM emissions of gasoline and ethanol in a single-cylinder spray-guided GDI engine under the conditions of 3.5 bar to 8.5 bar IMEP and stoichiometric air/fuel ratio. It was found that the ethanol would produce less PM emissions but more PN emissions. Vuk and Vander Griend [62] researched the particulate emissions from a GDI vehicle that was fueled with various ethanol/gasoline blends and operated over the FTP and US06 driving cycles. They found major reductions in both PM mass and PN emissions with ethanol concentrations as low as 10%, suggesting that ethanol promoted faster evaporation of wetted walls in situations of fuel impingement, while also providing additional oxygen in fuel-rich diffusion flames. On a different study conducted on a spray-guided GDI engine, it was found that the accumulation mode particles and the average diameter decreased with an increase in the oxygen concentration in the rich air/fuel mixture [63]. For the stoichiometric mixture, the particle concentration of E85 was significantly lower than that of the other fuels, whereas the particle concentrations of E30, M30 (30% by volume of methanol), and gasoline were equivalent. The effect of air/fuel ratio on the PM of the E85 was less pronounced compared to gasoline and E30, and the accumulation mode particles were very low, especially under the rich mixture condition. From these results, it can be inferred that the oxygen content of the fuel molecules can reduce the concentration of the intermediate substance of the soot formation precursor.

Another possible biofuel that is receiving widespread attention for SI engine applications is butanol [64, 65], whose isomers have research octane numbers ranging from 96 for straight n-butanol to 105 for the branched tert-butanol [66]. A study by Li et al. [67] conducted in a 1.8L turbocharged GDI engine fueled with gasoline and n-butanol, isobutanol, sec-butanol, and tert-butanol blends at ratios from 10% to 50% by volume showed that sec-butanol/gasoline yielded the lowest PN emissions, closely followed by n-butanol/gasoline, isobutanol/gasoline, and tert-butanol/gasoline. The mean particle diameter decreased with the addition of butanol due to the reduction in the fraction of larger particles. Sec-butanol blends shifted to smaller particle sizes, while tert-butanol blends to larger particle sizes. Storey et al. [68] found that the addition of 48% butanol (equivalent oxygen content as E30) in a GDI engine at 2600 rpm and 8 bar BMEP under rich conditions with advanced injection reduced smaller particles, while an important peak in accumulation mode particles could be observed compared to pure gasoline. The distributions showed a unimodal shape, and the butanol blend shifted the distribution to greater diameters, although butanol emitted fewer total particles than gasoline. Hergueta et al. [69] studied the effect of 33% butanol by volume in commercial gasoline containing 5% of ethanol and gasoline in a GDI engine. With the use of a transmission electron microscope (TEM), they showed that the butanol blend did not increase the primary particle diameters, due to the oxygen content of butanol that could limit the rate of soot formation and at the same time promote soot oxidation. Chan [70] tested a GDI vehicle fueled with iB16 (16% isobutanol) over the FTP and US06 driving cycles at standard and cold ambient temperatures. For the standard

FIGURE 5.3 Particle number emissions for the PFI-FFV and GDI-FFV over the FTP and LA92 cycles. The colors represent: E10 (gray), E51 (blue), E83 (red), and Bu55 (yellow). Adapted from Karavalakis et al. [57].

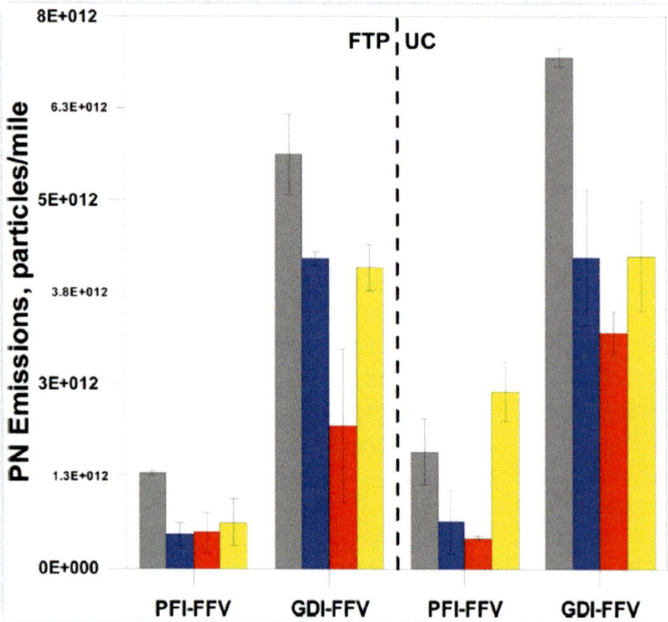

and cold ambient temperature FTP testing, the iB16 blend resulted in lower solid PN and black carbon emissions compared to gasoline. Lattimore et al. [71] carried out experiments with a single-cylinder GDI engine at an engine load of 8.5 bar and various compression ratios between 10.7 and 11.5 with B20 (20% by volume of n-butanol). They found that n-butanol addition to gasoline significantly reduced the accumulation mode PN emission, due to the earlier combustion phasing and thus increased post-combustion oxidation time. They also showed that Bu20 had more nucleation mode particles than gasoline, due to the lower soot accumulation mode particles that reduced nuclei adsorption. In a study by Karavalakis et al. [57], four alcohol blends were examined (E10, E51, E83, and 55% of isobutanol or Bu55) in GDI FFV over the FTP and LA92 driving cycles. While the isobutanol blend had lower total aromatics compared to E51 and E83 blends, it exhibited higher PN emissions than these fuels, as shown in Figure 5.3.

This finding was due to the lower oxygen content of the isobutanol blend and the higher content of heavier monoaromatic and naphtheno-aromatic hydrocarbons with high boiling points relative to E83. Additionally, the authors stated that branched butanols can produce intermediate higher molecular weight radicals, such as propene and butane, than ethanol, which react to produce higher molecular weight soot products. In a different study by Karavalakis et al. [72], when they tested low- and mid-level isobutanol blends in two GDI passenger cars over the FTP and LA92 driving cycles, they found PM mass, PN, and black carbon emissions reductions with increasing the alcohol level in the fuel.

This chapter summarized the most notable fuel factors influencing particulate emissions from GDI engines. In addition, this chapter focused on two of the most popular biofuels for SI application in the world, namely, ethanol and butanol, and their impact on particulate emissions from GDI engines. It was concluded that PM emission levels of GDI engines are strongly associated with the fuel physical and chemical properties, mixture composition, fuel spray strategy, ignition and combustion of operating cycle, and so on. It was highlighted that the use of ethanol and butanol fuels in GDI engines can reduce, in most cases, particulate emissions, including nanoparticles and black carbon. Given the historic trend of increasingly stringent emission standards in both the United States and EU for PM mass and PN emissions and the fact that black carbon is one of the most important anthropogenic global warming agents, the use of biofuels for particulate emissions reductions from GDI platforms is an attractive option.

References

1. Alkidas, A.C. "Combustion Advancements in Gasoline Engines," *Energy Conversion and Management* 48 (2007): 2751-2761.
2. Zhao, F., Lai, M.C., and Harrington, D., "Automotive Spark-Ignited Direct-Injection Gasoline Engines," *Prog. Energy Combust. Sci.* 25 (1999): 437–562.
3. Piock, W., Hoffmann, G., Berndorfer, A., Salemi, P. et al., "Strategies Towards Meeting Future Particulate Matter Emission Requirements in Homogeneous Gasoline Direct Injection Engines," *SAE Int. J. Engines* 4 (2011): 1455-1468, doi:10.4271/2011-01-1212.
4. Stevens, E. and Steeper, R., "Piston Wetting in an Optical DISI Engine: Fuel Films, Pool Fires, and Soot Generation," *SAE Int. J. Engines* 110 (2001): 1287-1294, doi:10.4271/2001-01-1203.
5. EPA, "Ethanol Waivers (E15 and E10)," Environmental Protection Agency, 2011.
6. Yanowitz, J. and McCormick, R.L., "Effect of E85 on Tailpipe Emissions from Light-Duty Vehicles," *Journal of the Air & Waste Management Association* 59 (2009): 172-182.
7. Hubbard, C.P., Anderson, J.E., and Wallington, T.J., "Ethanol and Air Quality: Influence of Fuel Ethanol Content on Emissions and Fuel Economy of Flexible Fuel Vehicles," *Environ. Sci. Technol.* 48 (2014): 861-867.
8. Yanowitz, J., Knoll, K., Kemper, J., Luecke, J. et al., "Impact of Adaptation on Flex-Fuel Vehicle Emissions When Fueled with E40," *Environ. Sci. Technol.* 47 (2013): 2990-2997.
9. Zhang, Z., Wang, T., Jia, M., Wei, Q. et al., "Combustion and Particle Number Emissions of a Direct Injection Spark Ignition Engine Operating on Ethanol/Gasoline and n-Butanol/Gasoline Blends with Exhaust Gas Recirculation," *Fuel* 130 (2014): 177-188.

10. Karavalakis, G., Short, D., Vu, D., Russell, R.L. et al., "The Impact of Ethanol and Iso-Butanol Blends on Gaseous and Particulate Emissions from Two Passenger Cars Equipped with Spray-Guided and Wall-Guided Direct Injection S.I. Engines," *Energy* 82 (2015): 168-179.

11. Irimescu, A., "Performance and Fuel Conversion Efficiency of a Spark Ignition Engine Fueled with Iso-Butanol," *Applied Energy* 96 (2012): 477-483.

12. Xue, C., Zhao, X.Q., Liu, C.G., Chen, L.J. et al., "Prospective and Development of Butanol as an Advanced Biofuel," *Biotechnology Advances* 31 (2013): 1575-1584.

13. Tao, L., Tan, E.C.D., McCormick, R.L., Zhang, M. et al., "Techno-Economic Analysis and Life-Cycle Assessment of Cellulosic Iso-Butanol and Comparison with Cellulosic Ethanol and n-Butanol," *Biofuels, Bioproducts, and Biorefining* 8 (2014): 30-48.

14. Velji, A., Yeom, K., Wagner, U., Spicher, U. et al., "Investigations of the Formation and Oxidation of Soot inside a Direct Injection Spark Ignition Engine Using Advanced Laser-Techniques," SAE Technical Paper 2010-01-0352, 2010, doi:10.4271/2010-01-0352.

15. Graskow, B., Kittelson, D., Abdul-Khalek, I., Ahmadi, M. et al., "Characterization of Exhaust Particulate Emissions from a Spark Ignition Engine," SAE Technical Paper 980528, 1998, doi:10.4271/980528.

16. Barone, T.L., Storey, J.M.E., Youngquist, A.D., and Szybist, J.P., "An Analysis of Direct-Injection Spark-Ignition (DISI) Soot Morphology," *Atmospheric Environment* 49 (2012): 268-274.

17. Sgro, L.A., Sementa, P., Vaglieco, B.M., Rusciano, G. et al., "Investigating the Origin of Nuclei Particles in GDI Engine Exhausts," *Combustion and Flame* 159 (2012): 1687-1692.

18. Richter, H. and Howard, J.B., "Formation of Polycyclic Aromatic Hydrocarbons and Their Growth to Soot - A Review of Chemical Reaction Pathways," *Prog. Energy Combust. Sci.* 26 (2000): 565-608.

19. Bockhorn, H., *Soot Formation in Combustion: Mechanisms and Models* (Springer Series in Chemical Physics), Vol. 59, (Berlin/New York: Springer-Verlag, 1994), ISBN:3-540-58398-X.

20. Frenklach, M. and Wang, H., "Detailed Modeling of Soot Particle Nucleation and Growth," *Twenty-Third Symposium (International) on Combustion*, The Combustion Institute, Pittsburgh, 1990, 1559-1566.

21. Heywood, J.B., *Internal Combustion Engine Fundamentals* (McGraw-Hill Series in Mechanical Engineering), (New York: McGraw-Hill, 1988), ISBN:0-07-100499-8.

22. Raj, A., Prada, I.D.C., Amer, A.A., and Chung, S.H., "A Reaction Mechanism for Gasoline Surrogate Fuels for Large Polycyclic Aromatic Hydrocarbons," *Combustion and Flame* 159 (2012): 500-515.

23. Choi, B.S., Choi, S.K., and Chung, S.H., "Soot Formation Characteristics of Gasoline Surrogate Fuels in Counterblow Diffusion Flames," *Proceedings of the Combustion Institute* 33 (2011): 609-616.

24. Kobayashi, Y. and Arai, M., "Characteristics of PM Exhausted from Pool Diffusion Flame with Gasoline and Surrogate Gasoline Fuels," SAE Technical Paper 2015-01-2024, 2015, doi:10.4271/2015-01-2024.

25. Burke, S.C., Ratcliff, M., and McCormick, R., "Distillation-based Droplet Modeling of Non-Ideal Oxygenated Gasoline Blends: Investigating the Role of Droplet Evaporation on PM Emissions," SAE Technical Paper 2017-01-0581, 2017, doi:10.4271/2017-01-0581.

26. Leach, F., Stone, R., Fennell, D., Hayden, D. et al., "Predicting the Particulate Matter Emissions from Spray-Guided Gasoline Direct Injection Spark Ignition Engines," *Proceedings of the Institution of Mechanical Engineers, Part D: Journal of Automobile Engineering* 231 (2017): 717-730.

27. Chapman, E., Winston-Galant, M., Geng, P., and Konzack, A., "Global Market Gasoline Range Fuel Review using Fuel Particulate Emission Correlation Indices," SAE Technical Paper 2016-01-2251, 2016, doi:10.4271/2016-01-2251.

28. Chapman, E., Winston-Galant, M., Geng, P., and Latigo, R., "Alternative Fuel Property Correlation to the Honda Particulate Matter Index (PMI)," SAE Technical Paper 2016-01-2250, 2016, doi:10.4271/2016-01-2250.

29. Aikawa, K., Sakurai, T., and Jetter, J.J., "Development of a Predictive Model for Gasoline Vehicle Particulate Matter Emissions," *SAE Int. J. Fuels Lubr.* 3, no. 2 (2010): 610-622, doi:10.4271/2010-01-2115.

30. Hellier, P., Ladommatos, N., Allan, R., Filip, S. et al., "The Importance of Double Bond Position and Cis-Trans Isomerisation in Diesel Combustion and Emissions," *Fuel* 105 (2013): 477-489.

31. Ogawa, T., Araga, T., Okada, M., and Fujimoto, Y., "Effects of Diesel Fuel Properties on Particulate Emissions from D.I. Diesel Engine – Part 2: Chemical Analysis and Characterization of Diesel Oil," SAE Technical Paper 9437494, 1994.

32. Karavalakis, G., Short, D., Vu, D., Russell, R. et al., "Evaluating the Effects of Aromatics Content in Gasoline on Gaseous and Particulate Matter Emissions from SI-PFI and SI-DI Vehicles," *Environ. Sci. Technol.* 49 (2015): 7021-7031.

33. Leach, F., Stone, R., and Richardson, D., "The Influence of Fuel Properties on Particulate Number Emissions from a Direct Injection Spark Ignition

Engine," SAE Technical Paper 2013-01-1558, 2013, doi:10.4271/2013-01-1558.

34. Dageförde, H., Koch, T., Beck, K.W., and Spicher, U., "Influence of Fuel Composition on Exhaust Emissions of a DISI Engine during Catalyst Heating Operation," SAE Technical Paper 2013-01-2571, 2013, doi:10.4271/2013-01-2571.

35. Khalek, I.A., Bougher, T., and Jetter, J.J., "Particle Emissions from a 2009 Gasoline Direct Injection Engine Using Different Commercially Available Fuels," *SAE Int. J. Fuels Lubr.* 3, no. 2 (2010): 623-637, doi:10.4271/2010-01-2117.

36. Kim, Y., Kim, Y., Kang, J., Rew, S. et al., "Fuel Effect on Particle Emissions of a Direct Injection Engine," SAE Technical Paper 2013-01-1559, 2013, doi:10.4271/2013-01-1559.

37. Zhu, R., Hu, J., Bao, X., He, L. et al., "Effects of Aromatics, Olefins and Distillation Temperatures (T50 and T90) on Particle Mass and Number Emission from Gasoline Direct Injection (GDI) Vehicles," *Energy Policy* 101 (2017): 185-193.

38. Chan, T.W., Lax, D., Gunter, G.C., Hendren, J. et al., "Assessment of the Fuel Composition Impact on Black Carbon Mass, Particle Number Size Distributions, Solid Particle Number, Organic Materials, and Regulated Gaseous Emissions from a Light-Duty Gasoline Direct Injection Truck and Passenger Car," *Energy and Fuels* 31 (2017): 10452-10466.

39. Jiao, Q. and Reitz, R.D., "The Effect of Operating Parameters on Soot Emissions in GDI Engines," SAE Technical Paper 2015-01-1071, 2015, doi:10.4271/2015-01-1071.

40. Fushimi, A., Kondo, Y., Kobayashi, S., Fujitani, Y. et al., "Chemical Composition and Source of Fine and Nanoparticles from Recent Direct Injection Gasoline Passenger Cars: Effects of Fuel and Ambient Temperature," *Atmospheric Environment* 124 (2016): 77-84.

41. U.S. Environmental Protection Agency, "Assessing the Effect of Five Gasoline Properties on Exhaust Emissions from Light-Duty Vehicles Certified to Tier 2 Standards: Analysis of Data from EPAct Phase 3 (EPAct/V2/E-89)," Final Report, EPA-420-R-13-002, April 2013, http://www.epa.gov/otaq/models/moves/epact.htm.

42. Butler, A.D., Sobotowski, R.A., and Hoffman, G.J., "Influence of Fuel PM Index and Ethanol Content on Particulate Emissions from Light-Duty Gasoline Vehicles," SAE Technical Paper 2015-01-1072, 2015, doi:10.4271/2015-01-1072.

43. Sobotowski, R.A., Butler, A.D., and Guerra, Z., "A Pilot Study of Fuel Impacts on PM Emissions from Light-Duty Gasoline Vehicles," SAE Technical Paper 2015-01-9071, 2015, doi:10.4271/2015-01-9071.

44. Karavalakis, G., Durbin, T.D., Yang, J., Ventura, L. et al., "Fuel Effects on PM Emissions from Different Vehicle/Engine Configurations: A Literature Review and Statistical Analysis," SAE Technical Paper 2018-01-0349, 2018, doi:10.4271/2018-01-0349.

45. Storch, M., Koegl, M., Altenhoff, M., Will, S. et al., "Investigation of Soot Formation of Spark-Ignited Ethanol-Blended Gasoline Sprays with Single- and Multi-Component Base Fuels," *Applied Energy* 181 (2016): 278-287.

46. Chen, L., Stone, R., and Richardson, D., "A Study of Mixture Preparation and PM Emissions Using a Direct Injection Engine Fueled with Stoichiometric Gasoline/Ethanol Blends," *Fuel* 96 (2012): 120-130.

47. Chen, L. and Stone, R., "Measurement of Enthalpies of Vaporization of Isooctane and Ethanol Blends and Their Effects on PM Emissions from a GDI Engine," *Energy & Fuels* 25 (2011): 1254-1259.

48. Di Iorio, S., Lazzaro, M., Sementa, P., Vaglieco, B.M. et al., "Particle Size Distributions from a DI High Performance SI Engine Fuelled with Gasoline-Ethanol Blended Fuels," SAE Technical Paper 2011-24-0211, 2011, doi:10.4271/2011-24-0211.

49. Catapano, F., Di Iorio, S., Lazzaro, M., Sementa, P. et al., "Characterization of Ethanol Blends Combustion Processes and Soot Formation in a GDI Optical Engine," SAE Technical Paper 2013-01-1316, 2013, doi:10.4271/2013-01-1316.

50. He, X., Ratcliff, M.A., and Zigler, B.T., "Effects of Gasoline Direct Injection Engine Operating Parameters on Particle Number Emissions," *Energy & Fuels* 26 (2012): 2014-2027.

51. Chan, T.W., Meloche, E., Kubsh, J., Brezny, R. et al., "Impact of Ambient Temperature on Gaseous and Particle Emissions from a Direct Injection Gasoline Vehicle and its Implications on Particle Filtration," SAE Technical Paper 2013-01-0527, 2013, doi:10.4271/2013-01-0527.

52. Esarte, C., Abián, M., Millera, Á., Bilbao, R. et al., "Gas and Soot Products Formed in the Pyrolysis of Acetylene Mixed with Methanol, Ethanol, Isopropanol or n-Butanol," *Energy* 43 (2012): 37-46.

53. Maricq, M.M., Szente, J.J., and Jahr, K., "The Impact of Ethanol Fuel Blends on PM Emissions from a Light-Duty GDI Vehicle," *Aerosol Science and Technology* 46 (2012): 576-583.

54. Storey, J.M., Barone, T., Norman, K., and Lewis, S., "Ethanol Blend Effects On Direct Injection Spark-Ignition Gasoline Vehicle Particulate Matter Emissions," *SAE Int. J. Fuels Lubr.* 3 (2010): 650-659, doi:10.4271/2010-01-2129.

55. Fatouraie, M., Wooldridge M.S., Petersen, B.R., and Wooldridge, S.T., "Effects of Ethanol on In-Cylinder and Exhaust Gas Particulate Emissions of a Gasoline Direct Injection Spark Ignition Engine," *Energy and Fuels* 29 (2015): 3399-3412.

56. Sakai, S. and Rothamer, D., "Effect of Ethanol Blending on Particulate Formation from Premixed Combustion in Spark-Ignition Engines," *Fuel* 196 (2017): 154-168.

57. Karavalakis, G., Short, D., Russell, R.L., Jung, H. et al., "Assessing the Impacts of Ethanol and Iso-Butanol on Gaseous and Particulate Emissions from Flexible Fuel Vehicles," *Environmental Science and Technology* 48 (2014): 14016-14024.

58. Mamakos, A., Martini, G., Marotta, A., and Manfretti, U., "Assessment of Different Technical Options in Reducing Particle Emissions from Gasoline Direct Injection Vehicles," *Journal of Aerosol Science* 63 (2013): 115-125.

59. Jin, D., Choi, K., Myun, C.L., Lim, Y. et al., "The Impact of Various Ethanol-Gasoline Blends on Particulates and Unregulated Gaseous Emissions Characteristics from a Spark Ignition Direct Injection (SIDI) Passenger Vehicle," *Fuel* 209 (2017): 702-712.

60. Oh, C. and Cha, G., "Influence of Oxygenate Content on Particulate Matter Emission in Gasoline Direct Injection Engine," *International Journal of Automotive Technology* 14 (2013): 829-836.

61. Wang, C., Xu, H., Herreros, J.M., Wang, J. et al., "Impact of Fuel and Injection System on Particle Emissions from a GDI Engine," *Applied Energy* 132 (2014): 178-191.

62. Vuk, C. and Vander Griend, S.J., "Fuel Property Effects on Particulates In Spark Ignition Engines," SAE Technical Paper 2013-01-1124, 2013, doi:10.4271/2013-01-1124.

63. Price, P., Twiney, B., Stone, R., Kar, K. et al., "Particulate and Hydrocarbon Emissions from a Spray Guided Direct Injection Spark Ignition Engine with Oxygenate Fuel Blends," SAE Technical Paper 2007-01-0472, 2007, doi:10.4271/2007-01-0472.

64. Wallner, T. and Frazee, R., "Study of Regulated and Non-Regulated Emissions from Combustion of Gasoline, Alcohol Fuels and their Blends in a DI-SI Engine," SAE Technical Paper 2010-01-1571, 2010, doi:10.4271/2010-01-1571.

65. Pechout, M., Dittrich, A., and Vojtisek-Lom, M., "Operation of an Ordinary PFI Engine on n-butanol and Iso-butanol and Their Blends with Gasoline," SAE Technical Paper 2014-01-2618, 2014, doi:10.4271/2014-01-2618.

66. Rankovic, N., Bourhis, G., Loos, M., and Dauphin, R., "Understanding Octane Number Evolution for Enabling Alternative Low RON Refinery Streams and Octane Boosters as Transportations Fuels," *Fuel* 150 (2015): 41-47.

67. Li, L., Sun, K., and Duan, J., "Effects of Butanol Isomers on the Combustion Characteristics and Particle Number Emissions of a GDI Engine," SAE Technical Paper 2017-01-2323, 2017, doi:10.4271/2017-01-2323.

68. Storey, J., Lewis, S., Szybist, J., Thomas, J. et al., "Novel Characterization of GDI Engine Exhaust for Gasoline and Mid-Level Gasoline-Alcohol Blends," SAE Technical Paper 2014-01-1606, 2014, doi:10.4271/2014-01-1606.

69. Hergueta, C., Bogarra, M., Tsolakis, A., Essa, K. et al., "Butanol-Gasoline Blend and Exhaust Gas Recirculation, Impact on GDI Engine Emissions," *Fuel*, 208 (2017): 662-672.

70. Chan, T.W., "The Impact of Isobutanol and Ethanol on Gasoline Fuel Properties and Black Carbon Emissions from Two Light-Duty Gasoline Vehicles," SAE Technical Paper 2015-01-1076, 2015, doi:10.4271/2015-01-1076.

71. Lattimore, T., Herreros, J.M., Xu, H., and Shuai, S., "Investigation of Compression Ratio and Fuel Effect on Combustion and PM Emissions in a DISI Engine," *Fuel* 169 (2016): 68-78.

72. Karavalakis, G., Short, D., Vu, D., Villela, M. et al., "Evaluating the Regulated Emissions, Air Toxics, Ultrafine Particles, and Black Carbon from SI-PFI and SI-DI Vehicles Operating on Different Ethanol and Iso-Butanol Blends," *Fuel* 128 (2014): 410-421.

About the Author

Georgios Karavalakis joined the University of California, Riverside, in July 2010 where he holds a research faculty position at the Center for Environmental Research and Technology and an adjunct professor position with the Department of Chemical and Environmental Engineering. He has a BEng (Hons) in engineering and an MSc in environmental engineering both from the University of Portsmouth and a Ph.D. from the National Technical University of Athens in chemical engineering.

He is a member of the Society of Automotive Engineers and a member of the American Association of Aerosol Research. He is currently leading and being involved in major research programs funded by California agencies, industry consortiums, and the federal government relating to alternative fuels, engine emissions, and secondary organic aerosol research from combustion systems. He has over 65 publications in high impact factor journals and over 100 publications in technical conferences.

Section 3

Particulate Emission Reduction Technologies

CHAPTER 6

Advanced Gasoline Combustion and Engine Controls

Mohammad Fatouraie, Kevin Peterson, Bopaiah Ittira Biddappa, Nikolay Livshiz, Jacob Larimore, and Michael Mosburger
Robert Bosch LLC

6.1 Gasoline Direct Injection

6.1.1 PN Formation Mechanisms

Particulates can form by a number of mechanisms, often with multiple mechanisms being present simultaneously at a given engine condition. Engine-out particulate emission is the result of particulate inception, growth, and oxidation mechanisms. The physics and chemistry that drive these mechanisms can be complex, nonlinear, interdependent, and engine-specific; thus, what is presented here is a general outline and not a prescriptive guide meant to apply to every engine, injector, and mode of operation. This also means that as a given engine parameter changes (e.g., engine speed), particulate emissions may go up or down, depending on a number of other factors.

In general, particulates are formed in regions with an over-abundance of fuel in the fuel/air mixture. This happens through either high-temperature thermochemical decomposition of the fuel by pyrolysis during the combustion or as a result of low-temperature chemistry pathways of radicals after the time of combustion. These rich zones are typically found either in the gas phase, where under-mixed regions of fuel lead to locally rich zones, or due to liquid films on the surfaces of the combustion chamber (e.g., piston, fuel injector tip, walls) [1, 2, 3].

The sources for particulates can also be independent of the engine fueling and instead related to engine oil, examples being oil introduction through the crankcase ventilation system (PCV), oil suction past loose piston rings, or oil leakage from the turbocharger. These oil-related mechanisms are generally indicative of engine malfunctions or poor engine design and won't be covered further here. The two general fuel-related mechanisms are described in more detail below.

6.1.1.1 Gas Phase

With regards to particulate formation, the ideal mixture is homogenous and contains no gas-phase rich zones capable of producing particulates. However, in direct-injection (DI) engines, liquid fuel is injected directly into the combustion chamber, requiring all the injected fuel to both vaporize and fully mix with the charge (air and EGR) present in the cylinder during the time available in the intake and compression strokes prior to combustion. If any locally rich zones persist until the time of combustion, they are likely to turn to soot. Sufficient mixing to prevent soot is driven by a combination of fuel injection characteristics (drop size, air entrainment in spray, penetration) and mixing characteristics (sufficient time and turbulence). High-speed imaging of the combustion chamber provides in situ spatial and temporal diagnostics of the combustion process. The incandescence of the soot particles are observed by imaging the combustion chamber through an endoscope. The still images of a DI combustion chamber with increasing gas-phase incandescence soot are presented in **Figure 6.1**. The images are taken at ~90° aTDC in an engine that is operating at room temperature to simulate an engine start. From left to right, the injection timing is retarded such that there is less time for mixing and evaporation. The result is that some liquid fuel is still present, and suspended in the charge, at the time of combustion. These rich pockets then burn as a result of pyrolysis and present themselves to varying degrees as incandescent light.

6.1.1.2 Liquid Films

Liquid films can form on any surface inside the combustion chamber and are a consequence of spray characteristics in DI engines. Frequent locations for liquid films are on the piston, combustion chamber walls, or intake valves. In these locations, a liquid film is formed due to improper spray targeting or injection timing. Liquid films can also form on the injector tip or within the nozzle holes and sac volume of the injector. These films are related to the injector needle movement characteristics and the internal geometry of the injector. These fuel films can lead to diffusion flames with rich combustion zones, which can significantly contribute to particulate emissions. Examples of diffusion flames initiated on the top of the piston (left panel) and tip of the injector (right panel) are

FIGURE 6.1 In situ high-speed imaging via endoscope can visualize the origin of particle formation in the combustion chamber. In these images, the piston is at ~90° aTDC and the injection timing is retarded from left to right. Additionally the engine is at room temperature to simulate an engine start. The result is varying levels of gas-phase soot incandescence as a result of insufficient time for mixing and evaporation.

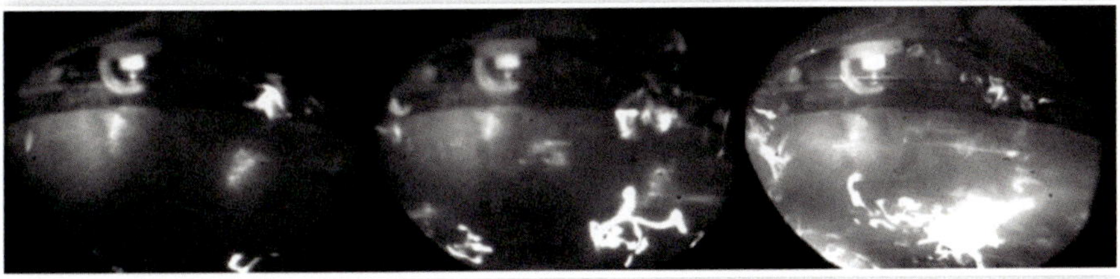

FIGURE 6.2 Left: Pool fire on the piston surface due to spray impingement and incomplete evaporation can be a significant source of particle emissions. Right: Soot incandescence from a diffusion flame due to liquid wall film on the injector tip.

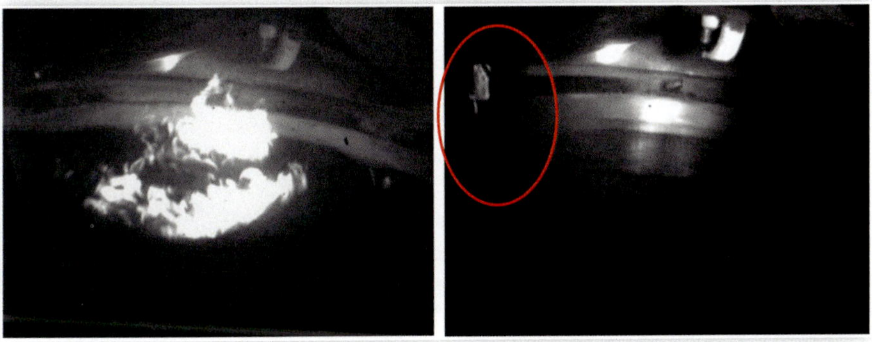

presented in Figure 6.2. In this case, an engine with side-mounted injector was used and the injector is located at the top left corner of the viewing area. The engine was operated at part load condition during the imaging and the start of injection (SOI) was varied from advanced (left panel) to retarded (right panel) timing. The still images used in the left and right panels are taken at 35 °aTDC and 85 °aTDC, respectively, during the expansion stroke, at which time the main combustion event has ended.

6.1.2 Operating Conditions

An engine must operate under a wide range of conditions, and the particular speed/load profile that an engine experiences can vary drastically based on the vehicle application, drive cycle, transmission selection, and other calibration parameters. For a given accelerator pedal position and external environment (wind, road condition, temperature, elevation), the engine will need to produce a given amount of power, which dictates a choice of speeds and loads for the engine operation. The physics happening inside the engine are intrinsically related to the operating conditions, and an overview of the relationship between some of the most important operating condition parameters and soot mechanisms is presented here. Changing one engine parameter often changes multiple physical parameters simultaneously. For example, increasing RPM changes the component temperatures, air flow, and volumetric efficiency. Throughout the engine calibration process, the best control response for a given operating point is optimized, given the constraints in emissions, fuel economy, and engine responsiveness.

6.1.2.1 Engine Speed

Engine speed affects fundamental properties that influence the charge induction, fuel spray development, combustion, and engine-out emissions. Higher engine speeds reduce the time available for mixing and evaporation, which can have a negative effect on charge homogeneity. However, turbulence and mean air velocity increase as the engine speed increases, which improves mixing and enhances the charge homogeneity, thus mitigating the locally rich zones in the gas phase and reducing the particulate production. The resulting effect on mixing is engine and condition specific. Regarding fuel impingement on the piston, because piston speed increases at higher RPM, the piston moves away from the spray more quickly and reduces the piston impingement for a given SOI.

Increased engine speed generally leads to higher component temperatures due to more friction and a shorter time for heat transfer. Higher temperatures accelerate the evaporation of the liquid fuel films on the components but can also promote low-temperature soot formation reactions. Depending on the temperature range and the location of the fuel film, higher temperatures can exhibit non-monotonic emission behavior. For example, while non-monotonic particulate behavior can be observed for injector tip temperature [4], pistons generally benefit from being hotter in regards to particulate emissions, although the relationship is complex [5, 6].

The cumulative effect of all these changes indicates that PN (particle number) could go up or down as a function of engine speed, which is shown in **Figure 6.3**.

FIGURE 6.3 The effect of engine speed on PN emission is complex and depends on various other parameters such as engine load.

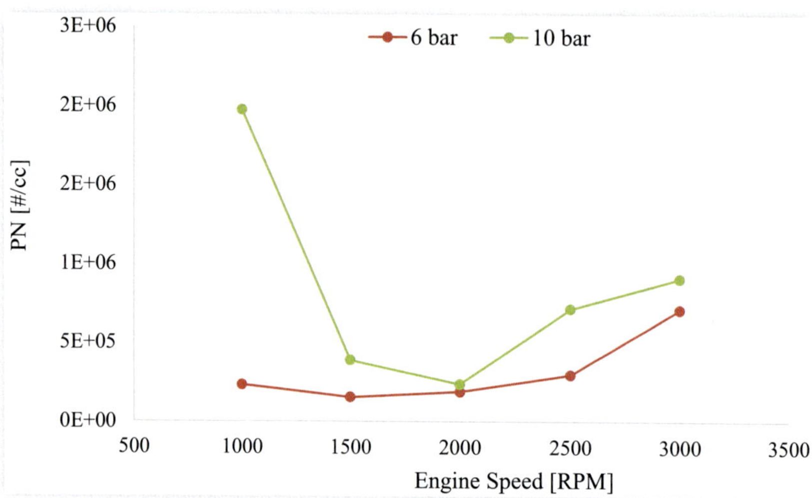

Sweeps of engine speed at two loads are presented for a turbocharged engine with a side-mounted injector. The tests were performed under warm engine conditions with a single injection per cycle at SOI = 280 °bTDC. While the engine showed insensitivity to engine speed in terms of PN emission for the load point of 6 bar BMEP between 1000 and 2500 RPM, at higher load (10 bar BMEP) significant sensitivity to engine speed is observed. The effects of engine load will be discussed next.

6.1.2.2 Engine Loads

As the engine load increases, the amount of injected fuel mass increases proportionally and PN typically increases. Keeping all other factors constant, as injected fuel mass increases, the amount of fuel which requires mixing increases, exacerbating under-mixing mechanisms. Impingement mechanisms also get worse since more fuel mass impacts surfaces like pistons or valves. At higher engine loads, the liquid film on the injector tip can also increase.

In general, the engine component temperatures increase at elevated engine loads. As more fuel is burned per cycle, a greater heat flux through the combustion chamber leads to increased internal surface temperatures, such as in the piston, walls, and valves. For surface impingement—related PN mechanisms, the increase in temperature is generally helpful in promoting evaporation. A possible exception to this is the Leidenfrost effect. The Leidenfrost effect refers to the cases when the temperature of a surface is significantly higher than the boiling point of the fuel, which prevents the evaporation of the liquid film due to the instant creation of a thin vapor film below the liquid surface. This phenomenon alters the rebound and splash characteristics of the impinged fuel as well as the time for evaporation [7].

Engine-out PN emission as a function of the engine load is presented in Figure 6.4. The data represent a load sweep at 2000 RPM during warm engine operation. Optimized injection timing in terms of lowest PN emission was selected for each load point. The non-monotonic trends have been observed at different engine speeds and other engine tests.

FIGURE 6.4 The effect of engine load on PN can be non-monotonic but generally increases at high engine loads.

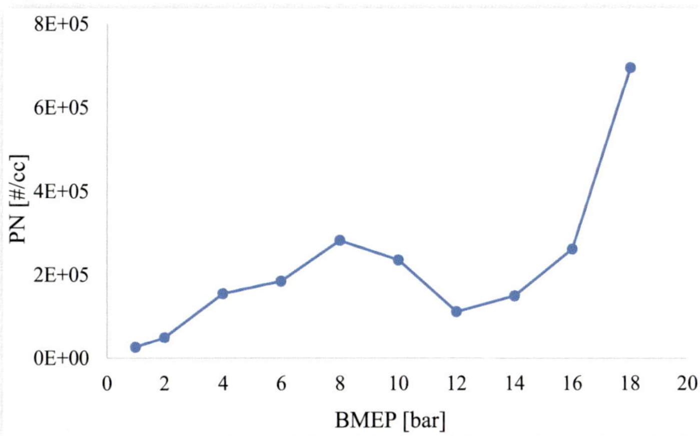

6.1.2.3 **Oil and Coolant Temperature**

The temperature of the oil and coolant of the engine also affects the fuel and component temperatures. Fuel temperature is initially at ambient (in the fuel tank), but while traveling through the fuel rail, it is heated to near coolant temperature. Because gasoline is a multi-component mixture of varying composition, the fuel temperature has a significant influence over the portion of the volatile components of the fuel. Colder fuel temperatures inhibit vaporization of the fuel and increase spray penetration and typically lead to an increase in PN. Colder component and temperatures also inhibit the evaporation of fuel films on internal engine surfaces. Emissions during cold start are a major issue, with often 50% or more of emissions being produced in the first few minutes of a drive cycle while the engine is still cold. The PN emissions during a steady-state engine warm-up process can be seen in **Figure 6.5**. The engine was operated at a constant load of 10 bar BMEP and 2000 RPM, starting from the ambient temperature condition until reaching the warm operation temperature.

6.1.2.4 **Engine Speed and Load Transients**

During real world driving, the engine is frequently changing both speed and load due to vehicle acceleration and deceleration. In particular, increases in engine load are associated with increased PN. The increase in engine load is accompanied by an increase in injected fuel mass per pulse, and the added fuel mass penetrates further, possibly impacting the piston. Because the piston surface is colder from running at a lower load, evaporation is inhibited and soot production due to the surface diffusion flames is increased. This often results in a spike in particulate production during transient increases in engine load, which typically levels out to a steady-state level after tens of seconds. These tip-in events can be simulated at

FIGURE 6.5 PN emissions typically reduce with increasing engine temperature.

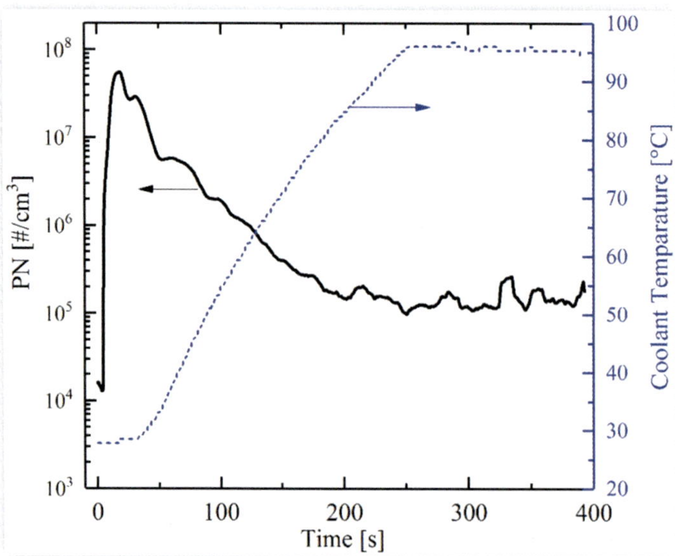

FIGURE 6.6 Engine load transients are typically accompanied by a temporary spike in PN emission. The transient calibration refinement during the transients have potential of order of magnitude reduction.

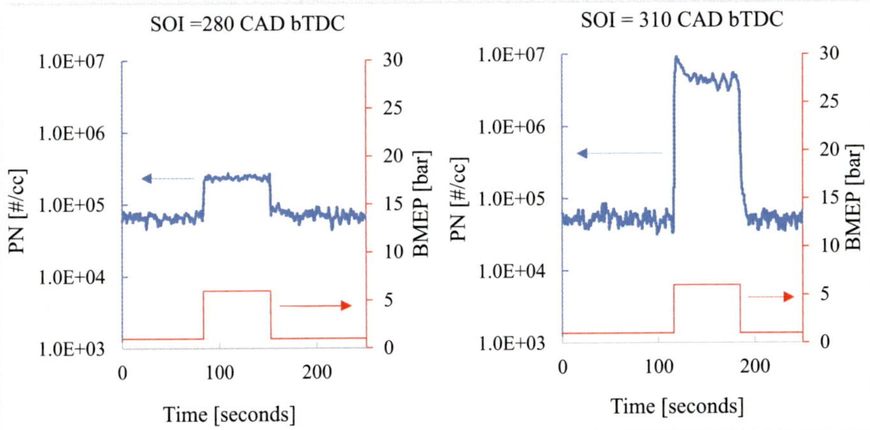

the engine dyno tests by load jumps. The PN emission results of a load jump from idle to mid-load (1 bar to 6 bar BMEP) at 1500 RPM during warm engine operation is shown in **Figure 6.6**. The two panels presented in the figure are indicating the same load jump at two different injection timings. The retardation of the injection timing resulted in a significant decrease of steady-state PN emission during the high load portion of the load jump.

6.1.3 Engine Calibration

While the operating conditions of the engine are mostly dictated by outside influences, such as the driver and road conditions, the calibration parameters are optimized to pick the best engine operation for a given demand. While many parameters are considered for the calibration parameters, such as gaseous emissions, fuel economy, and engine responsiveness, the calibration parameters influencing the PN emissions are discussed in greater detail below.

6.1.3.1 Injection Timing

Injection timing is a key parameter for influencing PN mechanisms. For homogeneous-charge engine operation (which is the norm for most of the speed/load operating points of most production DI engines), injection happens during the intake stroke. Injection timing is often a compromise between fuel consumption and PN. Fuel consumption is generally improved as injection timing is moved earlier (presumably due to better mixing), but the likelihood of spraying fuel on the piston and causing PN due to piston wetting also increases with earlier injection timing, so injection timing should be chosen to be sufficiently late to avoid the piston. The injection timing also determines how much time is available for the charge to mix and for any liquid films to dry. Injection timing is often chosen to be as early as possible while still avoiding piston impingement, with an additional safety factor. This trade-off timing is significantly different depending on the engine geometry (i.e., side- vs. centrally mounted injectors) and the injector spray targeting.

6.1.3.2 Fuel Pressure

Increased fuel pressure is a valuable tool for PN mitigation. Increasing fuel pressure shortens the time needed to inject a given amount of fuel, increasing the time available for mixing and drying. Higher fuel pressure also increases the ratio of the inertial to viscous forces, thus enhancing the spray turbulence and improving the break-up and atomization of the spray, speeding vaporization and mixing, and increasing in-cylinder turbulence. Spray penetration often stays roughly constant with increased fuel pressure. Despite the higher spray momentum at higher injection pressure, the fuel droplets are smaller and thus have a smaller momentum-to-drag ratio and entrain more air into the spray plume, decreasing penetration. Therefore, for in-cylinder combustion characteristics, increased fuel pressure is beneficial. The primary drawbacks to increasing fuel pressure are the need for a more robust valve train to drive the fuel pump, the associated increase in the parasitic losses, and the need for advanced control strategies to control fuel mass during short injection events.

The effects of injection timing and the fuel injection pressure are shown in **Figure 6.7** for a mid-load steady-state operating point at 2000 RPM. The sharp increase in the PN emissions by advancing the SOI beyond 310 °bTDC indicates

FIGURE 6.7 Increase in fuel pressure typically reduces particle formation. Very early SOI timings lead to spray impingement on the piston and lead to a sharp increase in PN, whereas very late injection timing reduces the available time for mixture formation and also affect PN negatively.

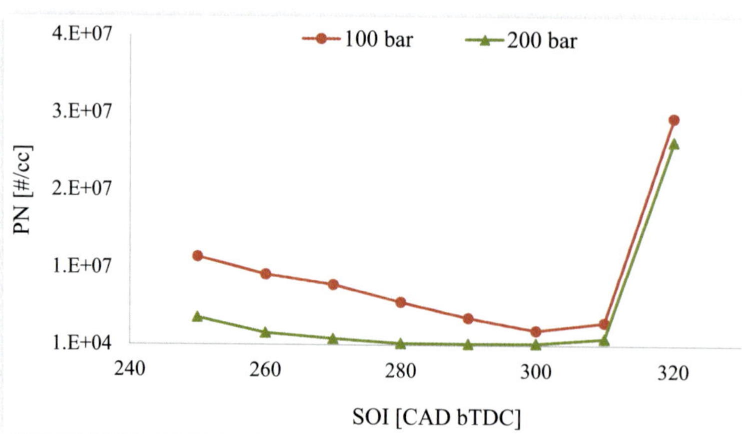

FIGURE 6.8 Multiple injection strategies have the potential to reduce PN emission due to piston impingement when used in the right context.

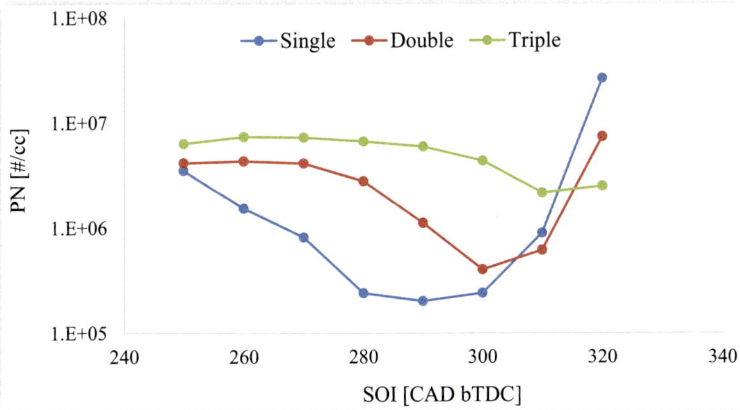

piston impingement. Higher fuel pressure decreased the PN emissions regardless of the injection timing, but the magnitude of improvement was dependent on the injection timing.

6.1.3.3 Number of Injection Events

Employing multiple injection events during a single combustion cycle is an effective method for mitigating some PN mechanisms. Fuel spray penetration increases with increasing fuel mass, so breaking the injection into multiple events decreases the penetration, minimizing wetting of the piston or other surfaces. One of the primary advantages of multi-injection is to begin the spray event at an earlier timing without wetting the piston than would be possible for single injection. The tip wetting of injector can increase or decrease depending on injector design and engine operating conditions. The effect of multiple injection at mid-load operation is shown in Figure 6.8. The fuel rail pressure was held constant at 200 bar. Finding the optimum injection timing, number of injections, and split of fuel masses is often a matter of experimental trial and error.

6.1.4 Summary Formation Mechanism

Particulates can form due to multiple mechanisms. An understanding of which mechanisms are present/dominant at a given condition is the first step toward mitigating those mechanisms. The dominant mechanism is a function of engine operating condition, and the operating condition is often dictated by driver input and external conditions such as road grade and ambient temperature. For a given operating condition, appropriate calibration settings must be chosen to mitigate PN, where the dominant control parameters are injection timing, fuel pressure, and multi-injection. In general, a combination of injection timing that avoids piston wetting, high fuel pressure, and the opportunity for multi-injection at certain engine conditions constitutes a good baseline for PN mitigation.

References

1. Tree, D.R. and Svensson, K.I., "Soot Processes in Compression Ignition Engines," *Progress in Energy and Combustion Science* 33, no. 3 (2007): 272-309, doi:10.1016/j.pecs.2006.03.002.

2. Drake, M.C., Fansler, T.D., Solomon, A.S., and Szekely, G.A., "Piston Fuel Films as a Source of Smoke and Hydrocarbon Emissions from a Wall-Controlled Spark-Ignited Direct-Injection Engine," SAE Technical Paper 2003-01-0547, 2003, doi:10.4271/2003-01-0547.

3. Stevens, E. and Steeper, R., "Piston Wetting in an Optical DISI Engine: Fuel Films, Pool Fires, and Soot Generation," SAE Technical Paper 2001-01-1203, 2001, doi:10.4271/2001-01-1203.

4. Xu, H., Wang, C., Ma, X., Sarangi, A.K. et al., "Fuel Injector Deposits in Direct-Injection Spark-Ignition Engines," *Progress in Energy and Combustion Science* 50:63-80, 2015, doi:10.1016/j.pecs.2015.02.002.

5. Köpple, F., Seboldt, D., Jochmann, P., Hettinger, A. et al., "Experimental Investigation of Fuel Impingement and Spray-Cooling on the Piston of a GDI Engine via Instantaneous Surface Temperature Measurements," *SAE Int. J. Engines* 7, no. 3 (2014): 1178-1194, doi:10.4271/2014-01-1447.

6. Köpple, F., Jochmann, P., Kufferath, A., and Bargende, M., "Investigation of the Parameters Influencing the Spray-Wall Interaction in a GDI Engine—Prerequisite for the Prediction of Particulate Emissions by Numerical Simulation," *SAE Int. J. Engines* 6, no. 2, (2013): 911-2925, doi:10.4271/2013-01-1089.

7. Han, Z., Xu, Z., and Trigui, N., "Spray/Wall Interaction Models for Multidimensional Engine Simulation," *International Journal of Engine Research* 1, no. 1 (2000): 127-146, doi:10.1243/1468087001545308.

About the Authors

Dr. Mohammad Fatouraie joined Robert Bosch's Advanced System Engineering division in January 2016. He is currently a principal engineer in combustion and emissions research. Prior to joining Bosch, Mohammad worked as an assistant research scientist in the Mechanical Engineering department of the University of Michigan. In 2010 he received his M.Eng. in automotive and manufacturing engineering and in 2013 he received his Ph.D. in mechanical engineering, both from the University of Michigan, Ann Arbor.

Dr. Kevin Peterson has worked in the automotive industry for more than 5 years and is a senior engineer at Robert Bosch LLC, focusing on engine testing and optical diagnostics for emissions mitigation. Peterson holds a Ph.D. and M.S. in mechanical engineering from the University of Michigan, as well as an M.S. in electrical engineering. He also has a B.A. in physics from Carleton College.

Bopaiah Ittira Biddappa is a very passionate automotive engineer in the systems engineering group at Robert Bosch LLC, Michigan. He graduated with a master's degree in automotive engineering from Clemson University, South Carolina, and has been with Bosch for 3 years. He has a bachelor's degree in mechanical engineering from Visvesvaraya Technological University. Bopaiah's background is in IC engines, combustion and Emissions. He has worked extensively in particulate emission reduction, conducting tests on the engine and chassis dynamometers. He has previously published papers in the Journal of Applied Polymer Science. When he is not saving the polar bears, he loves to go mountain climbing, play soccer with his dog, and play cricket.

Dr. Jacob Larimore is currently a Customer Chief Engineer in the Powertrain Solutions division at Robert Bosch LLC where he focuses on the advancement of powertrain technology, the reduction of emissions, and the improvement of fuel economy. Larimore holds a Ph.D. and an M.S. in mechanical engineering from the University of Michigan, Ann Arbor, as well as a B.S. from Missouri University of Science and Technology, also in mechanical engineering.

Nikolay Livshiz is a graduate of U of M - Dearborn with a B.S. in mechanical engineering and a B.S. in engineering mathematics. Nikolay has been with Bosch since 2012 working in the systems and advanced engineering team on various topics related to ICEs and vehicle powertrain solutions.

Dr. Michael Mosburger is an engineering manager at Robert Bosch LLC. His group focuses on system and advanced engineering aspects in the field of internal combustion engines and powertrain systems development. Dr. Mosburger received a Ph.D. degree in mechanical engineering from the University of Michigan in 2012, as well as an M.S. in physics in 2011. Prior to coming to the United States, Michael studied industrial engineering at Karlsruhe Institute of Technology in Germany and received an M.S. degree in 2007.

CHAPTER 7

Gasoline Particulate Filter Design, Fundamentals, and Function

Thorsten Boger
Corning GmbH

7.1 Introduction

As discussed in the previous chapters, we observe an increased global awareness of particulate emissions from gasoline engines as well as their impact on human health. This is reflected in tightened particulate emissions regulations for gasoline vehicles in many regions of the world. Improved combustion technologies as well as gasoline particulate filters (GPF), sometimes also referenced in Germany as Otto Partikelfilter (OPF), will be applied to modern powertrain architectures to meet these new regulations. While the first low-volume platform with GPF was commercialized by Mercedes-Benz in 2014 on the S500 limousine, the first high-volume GPF applications began in 2017 at several OEMs to meet EU6d. Examples of announcements or published applications can be found in references [1, 2, 3, 4, 5]. By 2019, we expect GPFs to be an essential element for aftertreatment systems across all European gasoline direct injection (GDI) vehicles and in the early stages of a broad introduction to direct injection as well as port fuel injection gasoline vehicles in China (e.g., [6, 7]).

While diesel particulate filters (DPF) have been in use in passenger vehicle applications since the early 2000s and are by now a standard component in the aftertreatment toolbox, the development of GPFs did not really start until the 2008–2010 timeframe. Early "prototypes" of GPFs were basically derived from DPF technologies, evaluated on gasoline vehicles. While these early prototypes helped to reduce the tailpipe particulate emissions, their overall performance with respect to other critical parameters was far from optimal, pressure drop being one example among others. Since then, significant progress has been made, and a robust application understanding has been developed. In this chapter, the basic requirements as well as fundamental design principles for GPFs will be summarized. Examples of advanced, commercialized filter technologies will be described, and their application in gasoline engine systems will be discussed.

7.2 Aftertreatment Architectures Using GPF

A wide range of different GPF architectures exist, as GPFs are a new aftertreatment component introduced into existing systems. The early systems brought to market, starting in 2017 in Europe, comprise almost every configuration that one could imagine. Two primary differentiating characteristics for these architectures are, with respect to the location of the GPF in the exhaust system, close-coupled (CC) or underfloor (UF), and if a catalyst function is integrated into the filter or not. This is summarized in **Figure 7.1**.

With respect to the location of the GPF in the system, we use the term "close-coupled" for architectures in which the GPF is upstream of the flex-pipe and "underfloor" for installations downstream of the flex-pipe. In the CC position, one can further differentiate between systems in which the GPF is located in first position

FIGURE 7.1 Aftertreatment architectures using a gasoline particulate filter.

GPF Location	Comment	Uncoated GPF	Coated cGPF
Close-coupled 1st Component CC1	Directly downstream Turbo-charger/ Manifold	N/A	
Close-coupled 2nd Component CC2	Downstream close-coupled TWC but before flex pipe	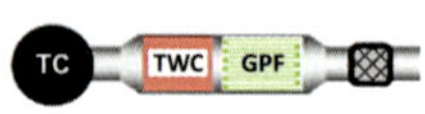	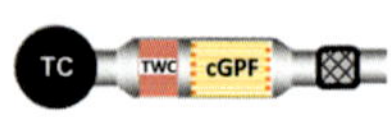
Underfloor UF	Downstream of flex pipe		

TC Turbo-charger · Flex pipe · GPF UncoatedGPF · cGPF CoatedGPF · TWC Three-way Catalyst (TWC)

(CC1) or in second position (CC2) downstream of the three-way catalyst (TWC). While the decision to install the GPF in a CC or UF position is in most cases primarily driven by the available packaging space, the placement does have significant impact on the exhaust gas temperature at the GPF and, related to this, the management of collected soot. This will be discussed in a later section.

The decision between a coated or an uncoated GPF is mostly driven by considerations related to how far an existing aftertreatment system can or should be changed, that is, an uncoated GPF in UF can typically be combined with an existing TWC system. The TWC addresses the gaseous emissions and the GPF provides the filtration function only. Differences in pressure drop and filtration behavior influence this decision as well, as will be discussed below. For coated GPF applications, the level of catalyst integration depends on the system and catalyst design. Typically, the level of integrated catalyst mass decreases from CC1 > CC2 ≥ UF positions. The actual washcoat amount, however, depends also on the TWC technology used, as will be discussed in more detail in the subsequent Chapters 9 and 10. With respect to the filter substrate design, the primary coating parameter to be considered is not necessarily the precious metal content but rather the total washcoat loading, which represents the volume that has to be integrated into the porous filter structure, either on or within the filter walls. Typical values are 50–150 g washcoat per liter of filter.

With respect to sensors, these can be the same sensors as in a conventional gasoline aftertreatment system, namely, temperature and lambda or oxygen sensors. In addition, absolute pressure or differential pressure sensors are likely applied to monitor the pressure change across the filter, as well as to identify complete failure or removal of the filter component. Additional temperature sensors are possible as well, especially in UF applications. Typical sensor structures and architectures will be discussed in the chapters on application examples and on-board diagnostics (OBD) (Chapters 11, 12, and 14).

7.3 General Filter Design and Application Requirements

The basic design of all GPFs in use today is the same as for diesel particulate filters, shown schematically in **Figure 7.2**. The filters are based on a refractory, porous ceramic honeycomb structure. The channels are plugged in a checkerboard pattern at alternate ends. This results in a structure that forces the exhaust gas to flow from the inlet channels, through the porous wall, and back out into the outlet channels. Within the porous wall structure, particles are removed by filtration resulting in a significant reduction in particulate matter emissions. This type of filter is commonly referenced as a wall-flow filter.

The main design parameters for these filter technologies, aside from material selection, are those describing the macroscopic structure of the honeycomb and those that describe the microstructure of the porous filter wall. The honeycomb structure is usually characterized by its cell density, that is, the number of cells or channels per

FIGURE 7.2 Wall-flow particulate filter design.

unit area, and the thickness of the walls or webs. The unit commonly used to describe the cell density is cells per square inch, cpsi, or in some cases as cells per cm^2. The web thickness, t_w, is usually provided in 1/1000 of an inch (a customary U.S. unit called "mil"), or in metric units in mm or µm. Two key characteristics of the honeycomb structure, which can be derived from the aforementioned values, are the open frontal area (OFA) and the hydraulic diameter, d_h. For square channels, these values can be calculated by

$$d_h = (\mathrm{CPSI})^{-1/2} - t_w$$

$$\mathrm{OFA} = d_h^{\,2} / (d_h + t_w)^2 .$$

The microporous wall material is typically described by its porosity and its median pore size (D50) as determined by mercury porosimetry. Common values used for GPF designs and materials are summarized in Table 7.1 (values from [8, 9, 10, 11]).

Examples of commercialized GPFs are those within the Corning® DuraTrap® GC and GC HP product families, with 200 cpsi/8 mil/55% and 300 cpsi/8 mil/65%, respectively, both in a number of different pore sizes [8, 9].

TABLE 7.1 Range of typical gasoline particulate filter properties

Property	Uncoated GPF	Coated GPF
Porosity	45%-55%	55%-65%
Median pore size	10 µm-15 µm	15 µm-25 µm
Cell density	200 cpsi-360 cpsi	300 cpsi
Web thickness*	6 mil-12 mil	8 mil-12 mil

* mil = 1/1000 inch or 25.4 µm.

The selection of the optimal filter design and filter material is determined by a number of application requirements:

- Filtration performance: The absolute value of filtration efficiency required is determined by the engine-out raw emissions and the regulatory standards to be met.
- Pressure drop: The pressure drop, or resistance to flow, has impact on the fuel efficiency as well as the peak power of an engine and typically has to be as low as possible.
- Catalyst activity and utilization: Applicable only for coated GPFs, with the main parameter from the filter substrate being the washcoat capacity and catalyst efficiency.
- Mechanical and thermomechanical robustness: The filter has to withstand the mechanical stress applied during canning as well as the thermomechanical stress from temperature gradients in operation.
- Passive soot oxidation and soot mass limit: The filter has to be robust enough to withstand potentially severe soot regenerations, but also enable a maximum of passive soot regeneration.

Unfortunately, these requirements generally do not lead to one unique, optimal product design. Instead, a trade-off is required based on the relative importance of each attribute for a given application. The details and fundamentals that drive these trade-off decisions will be discussed in the following two sections.

7.4 Filter Design Fundamentals and Considerations

7.4.1 Material Selection

Initially, cordierite, aluminum titanate (AT), silicon carbide (SiC), and mullite were proposed as suitable ceramic materials for GPFs, analogous to the materials that are used for DPF [8, 9, 10, 11, 12, 13]. The requirements for GPF performance are challenging, particularly as it relates to (a) thermal shock resistance, which are for GPFs similar to those for flow-through TWC substrates, (b) the need for monolithic structures to maximize the accessible filter volume and frontal area due to a sometimes very compact packaging space, as well as (c) the desire to have a low thermal mass and low thermal conductivity material that quickly responds to temperature changes to take advantage of passive soot regeneration potential. Together, these challenges and their respective operating conditions resulted in the selection of cordierite as the dominant material in use [4, 6, 8, 9, 10, 11]. This is analogous to the flow-through substrates, used for nearly all TWC for the last 40 years. The key characteristics of the advanced cordierite ceramics are their low thermal expansion ($\sim 10^{-6}$ m/mK), their high temperature capability (>1300°C), their low intrinsic material density (2500 kg/m^3), and their high intrinsic strength. Especially with respect to the latter, significant developments have been made enabling today's filter technologies with very high porosity and high OFA (little ceramic per unit volume) at appropriate strengths, which were unthinkable just a few years ago.

7.4.2 Filtration

The particles found in gasoline exhaust are typically in the range of roughly 10–400 nm (described by their electrical mobility diameter), several orders of magnitude smaller than the typical pore sizes (10–25 μm) used in GPFs (see Table 7.1). Hence, filtration is not the result of a sieving effect but is rather primarily based on deposition due to Brownian motion and interception. Inertia effects can also play a role, but usually to a lesser extent. These mechanisms are summarized schematically in **Figure 7.3**.

The filtration based on Brownian motion is related to the random movement or diffusion of small particles relative to the fluid stream lines. Once a particle gets close enough to a solid surface, it can deposit and is removed from the fluid stream. This mechanism is dominant for small particles, for example, those <100 nm, and its contribution decreases as the particles become larger, since the Brownian motion decreases with particle size. Interception occurs when a particle's trajectory passes within one particle radius of the collecting body. In such a case, the particle traveling along that stream line makes contact with the body and may be collected there. As one would expect, this collection mechanism decreases in relevance as particles become smaller. It is therefore most effective for particles of intermediate sizes. Deposition as a result of inertia is relevant for large, heavy particles only, which do not follow the fluid steam lines. In the following paragraphs, only a brief description of the physics describing the two most relevant mechanisms for GPF applications will be provided to enable a basic understanding of the relevant parameters. More exhaustive information can be found, for example, in [14, 15, 16, 17] and the references therein.

Most models represent the porous filter wall as an assemblage of unit collectors, onto which the soot particles are deposited. The size and number of the unit collectors can be derived from the mean pore size and porosity, for example, by assuming spherical shape:

$$d_c = \frac{3}{2} \cdot \frac{1-\varepsilon}{\varepsilon} \cdot \hat{d}_{Pore}$$

FIGURE 7.3 Relevant filtration mechanisms.

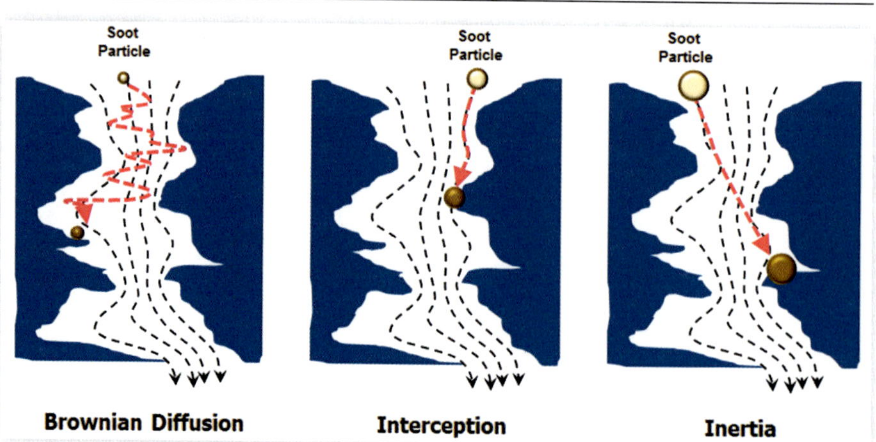

For these unit collectors, the efficiency can be determined using established correlations, as given, for example, in [14]. The collection efficiency of particles by interception can be described by [14]

$$\eta_{Int} = 1.5 \cdot N_r^2 \frac{\left[g(\varepsilon)\right]^3}{\left(1+N_r\right)^s}$$

Where $g(\varepsilon)$ and the power s are functions of the porosity. The dependence on soot particle and collector size is considered by $N_r = d_s/d_c$. From the correlation, it can be observed that η_{Int} is independent of the flow rate and temperature. For Brownian motion, the unit collector efficiency η_{BM} can be described by [14]

$$\eta_{BM} = 4 \cdot \frac{A_s^{1/3}}{Pe_i^{2/3}} \cdot \left(1-\varepsilon\right)^{2/3}$$

with A_s being a parameter, primarily dependent on the porosity ε, and Pe_i being the Peclet number. The Peclet number is proportional to the fluid velocity inside the pore space u_w/ε and the ratio between collector diameter d_c and diffusion coefficient for Brownian motion D_{BM}:

$$Pe_i = \frac{u_w}{\varepsilon} \cdot \frac{d_c}{D_{BM,i}}$$

The particle size d_s and temperature T dependence of this collection mechanism are introduced via the Brownian diffusion coefficient, $D_{BM} \sim (T/d_s^2)$.

To better understand the effect of the different design and operating parameters, examples are shown in **Figure 7.4** for the filtration efficiency, as function of the particle size, calculated based on the above correlations from [14]. First of all, one can observe that the filtration efficiency is generally very high for small particles and for large particles. In the intermediate range, a minimum can be observed. This "V" shape behavior is due to the overlapping effect, with deposition due to Brownian motion

FIGURE 7.4 Examples for the influence of different parameters on the filtration efficiency: (a) pore size, (b) porosity, (c) web thickness, (d) wall flow velocity, and (e) temperature. Note that particle size is described as electrical mobility diameter.

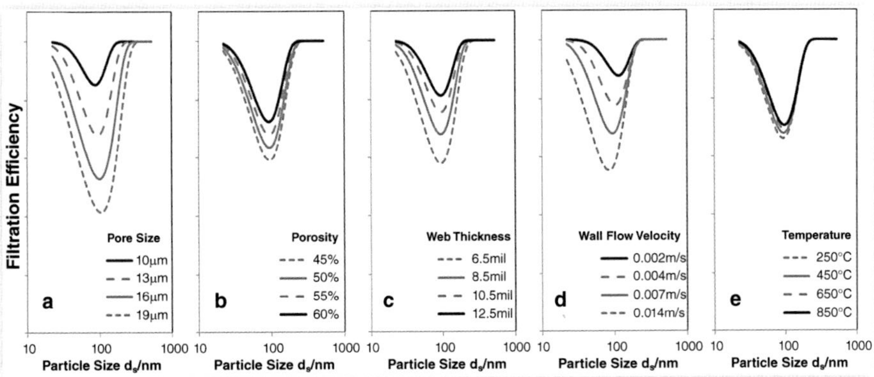

dominating in the small particle range and decreasing as particle size gets larger, and interception (and inertia) dominating for large particles but decreasing as particles become smaller. The minimum is due to both mechanisms having lower efficiency. From Figure 7.4(a) and (c), one can see the expected trend with small pores or thick walls enabling higher filtration efficiency over the entire size range. The porosity dependence is at first counterintuitive, with higher efficiencies for higher porosities. The reason for this is that as porosity is increased, the residence time within the pore space increases, providing more time for particles to be collected. This trend breaks down as we approach porosities of $\varepsilon \rightarrow 1$. For the flow velocity and temperature effect, one can observe that only the Brownian motion collection is impacted, as described above.

Figure 7.4 shows that for GPFs, where the particles in the exhaust are expected to be primarily in the range of 10–100 nm, the collection based on Brownian motion is dominant. A simplified model based on Brownian motion only, which is useful as a practical engineering tool, had been introduced by Zhong et al. [18]. It is based on a number of mathematical simplifications, which enable establishment of a simple correlation for the filtration efficiency based on a filtration characteristic parameter, A_{Filt}:

$$A_{Filt} \sim \frac{\varepsilon^{0.43} t_w (\mathrm{CPSI})^{1/3}}{d_{50,pore}^{5/3} \mathrm{SV}^{2/3}}$$

This parameter covers all basic filter design variables. The correlation was validated and calibrated using data from a large number of laboratory experiments with filters of very different properties and test conditions. In **Figure 7.5**, the correlation is

FIGURE 7.5 Correlation of the filtration efficiency measured on laboratory equipment of clean research filter samples with the filtration characteristic parameter A_{Filt}. Symbols: experimental data, Line: simple correlation FE = $f(A_{Filt})$.

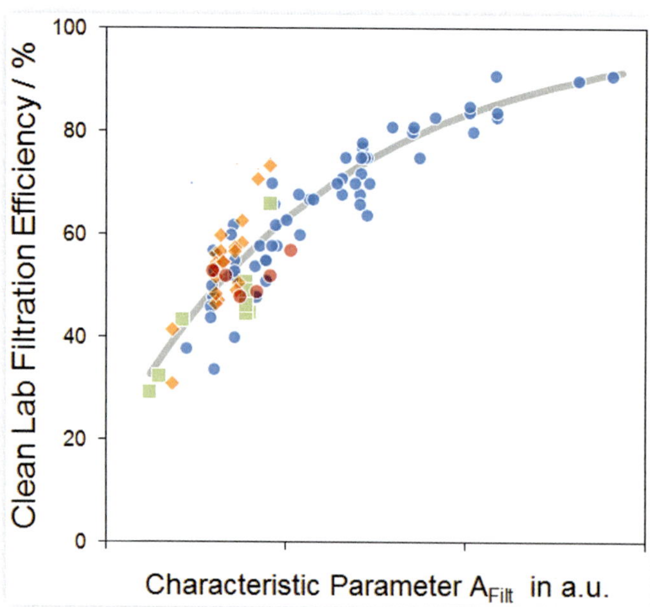

shown using data from [18] as well as a large number of additional data. The experimental data include filters covering porosities between 45% and 67%, median pore sizes between 9.8 μm and 28 μm, cell densities between 180 cpsi and 350 cpsi, and web thicknesses between 7 mil and 19 mil. In addition, a wide range of part sizes and space velocities were used. With the help of the definition of A_{Filt}, one can understand the sensitivity of filtration to changes in any of the design and process variables. Just considering the different powers in the correlation, one can observe the following order: pore size > web thickness > space velocity > porosity > cell density.

During operation, the filter collects particulates, which then modify the effective microstructure of the filter, and can act as very efficient new collectors for additional particulates [19]. An example of this behavior measured on a laboratory filtration bench is shown in **Figure 7.6**. Plotted is the filtration efficiency as function of the soot load in the filter. The experiment always starts with a clean filter (e.g., soot load equal to 0 g/L) followed by the accumulation of soot collected by filtration. From the results in Figure 7.6, one can see that as soot is accumulated, the filtration efficiency rapidly increases, approaching values close to 100%. The different data sets represent the same filter tested after different exposure to on-road vehicle testing, starting with the fresh reference and then tests after 45,000 km, 105,000 km, 160,000 km, and 240,000 km. During the mileage accumulation, the filter collects not only soot but also inorganic particles referenced as ash, which cannot be removed by oxidation (note that prior to the laboratory test, the filter always has been heat treated in an oven to remove any accumulated soot). Comparison of the different measurements illustrates the effect of the accumulated ash. With progressing mileage or ash quantity, the clean filtration improves and the 100% saturation is reached earlier. More data showing the same effect measured in particulate emission measurements on vehicles will be provided in the next chapter.

FIGURE 7.6 Evolution of filtration efficiency as a function of soot load obtained in laboratory measurements. Data for a filter with different mileage (0 km to 240,000 km) or ash load.

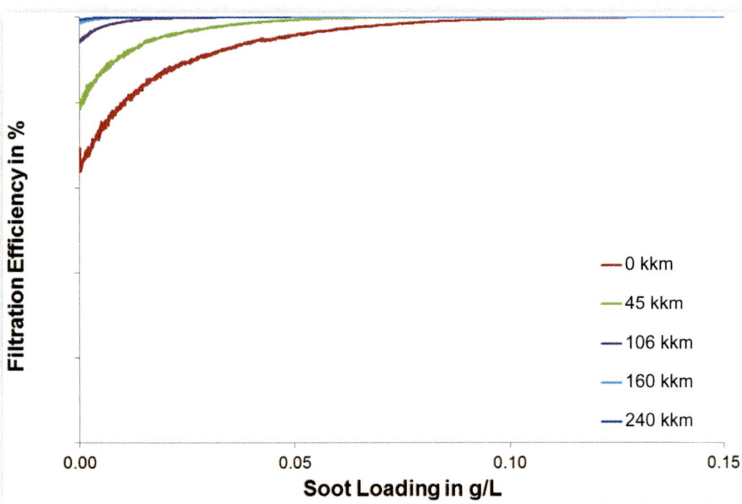

Catalyst coating on the particulate filter also changes its filtration characteristics. Catalyst particles, which are coated into the porosity of the filter wall, usually called in-wall coating, consume some of the pore space and change the effective porosity, as well as the effective pore size. The reduction in effective pore volume always reduces the amount of soot required to reach the saturation in filtration efficiency. The change in effective pore size can result in enhanced filtration, as well as decreased filtration efficiency, depending on the coating technology as well as the substrate microstructure properties. In some cases, coatings may also be applied onto the filter walls, usually referenced as on-wall coating. In this case, the coating acts like a membrane with effective properties that can be derived from its particle size and typical packing densities for random particles.

7.4.3 Pressure Drop

The pressure drop of a filter is an application critical parameter of significant practical relevance. It impacts engine power as well as fuel consumption. In this section, a brief description of the most relevant factors will be provided. There is rich literature describing the modeling of the pressure drop of particulate filters in more detail. Examples are noted in references [20, 21, 22] and references therein.

The pressure drop of particulate filters is composed of five primary contributions, shown schematically in **Figure 7.7**. The inlet and outlet effects, shown as (1) and (5) in Figure 7.7, are due to the contraction and acceleration as the gas enters the inlet channels and the expansion and deceleration of the gas as it exits the channels, respectively. The turbulent entrance effects that occur as a result of flow development inside the channels are typically lumped into these contributions. The inlet and outlet contributions are described by terms proportional to the kinetic energy, with the proportionality constant ζ_j:

$$\Delta p_j^{i/o} = \zeta_j \cdot \rho_j \cdot \left(u_j^{i/o}\right)^2$$

The index j represents filter inlet or outlet, and i/o indicates the condition at the inlet or outlet of the filter. Correlations for ζ are empirical and typically include the OFA as variable. Examples are provided in reference [23, 24].

For the frictional losses along the inlet and outlet channels (Figure 7.7, index 2 and 4), existence of a laminar flow profile is generally assumed within the channels.

FIGURE 7.7 Schematic of individual pressure drop contributions of wall flow filters.

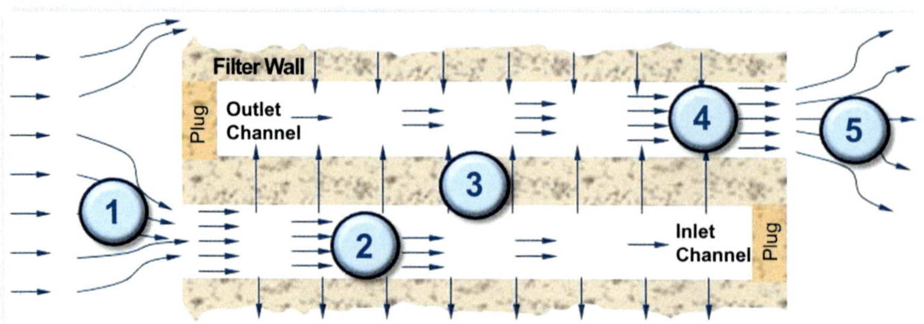

The pressure drop along the channels can therefore be described analogous to flow-through substrates:

$$\Delta p_j^{ch} = 2 \cdot f_0 \cdot \eta \cdot \frac{L_{ch}}{d_h^2} \cdot u_j^{ch,eff}$$

The index j again represents the inlet and outlet channel. η, L_{ch}, and d_h represent the viscosity, effective channel length, and hydraulic diameter, respectively. The friction factor f_0 is equal to 14.2 for a square channel. With respect to the velocity to be used, it has to be considered that at the inlet side (region of inlet plugs), all the flow is in the inlet channels, whereas the opposite is true on the outlet side (region of outlet plugs), with all the gas flow exiting via the outlet channels. Hence, the local flow rate in the inlet and outlet channels varies along the length, which has to be considered in the choice for $u_j^{ch,eff}$.

The last contribution to pressure drop comes from the resistance to flow across the filter wall (index 3 in Figure 7.7), from the inlet to the outlet channels. It is worth noting that this is the only contribution determined by the porous microstructure of the filter. In general, the pressure drop across the wall can be described by Darcy's law:

$$\Delta p^w = \eta \cdot \left(\frac{\Delta L_w}{\kappa} \right)_{eff} \cdot u^w$$

with ΔL_w as effective thickness, κ as permeability, and u^w as wall-flow velocity. The permeability of a clean wall is in general a function of the porosity, the pore size distribution, and the general morphology of the porous wall, considering, for example, the connectivity of the pore network. The simplest approaches to estimate the permeability are based on correlations utilizing the porosity and mean pore size only, as described, for example, in [16, 25]:

$$\kappa_w = \frac{3}{200} \cdot \varepsilon \cdot \hat{d}_{pore}^2$$

Here κ_w is the wall permeability, ε the porosity, and d_{pore} the median pore diameter. These correlations can provide reasonable estimates for the permeability of clean, uncoated filter walls. A comparison with some experimental values is shown in **Figure 7.8**. Typical permeabilities for uncoated filter technologies are in the range of $\kappa_w \approx 0.8 - 4 \times 10^{-12}$ m^2.

Coating changes and decreases the permeability of the filter wall. For in-wall coatings, the catalyst particles consume some of the pore volume, reducing the effective porosity as well as pore size. In general, higher washcoat loading results in lower permeability. The impact in absolute terms depends significantly on the coating technology. The use of pore size and porosity data from mercury porosimetry measurements and the correlation provided above usually does not provide for sufficient results. The most effective way to describe the impact of in-wall coatings on the permeability is to use semi-empirical correlations, calibrated to experimental data from coated filters. The impact of on-wall coatings can be described reasonably well if a resistance in series approach of the coating layer and the porous wall is applied. The permeability of the coating layer itself is described by its particle size, porosity, and thickness.

CHAPTER 7

FIGURE 7.8 Permeability of filters. Experimental data and simple correlation.

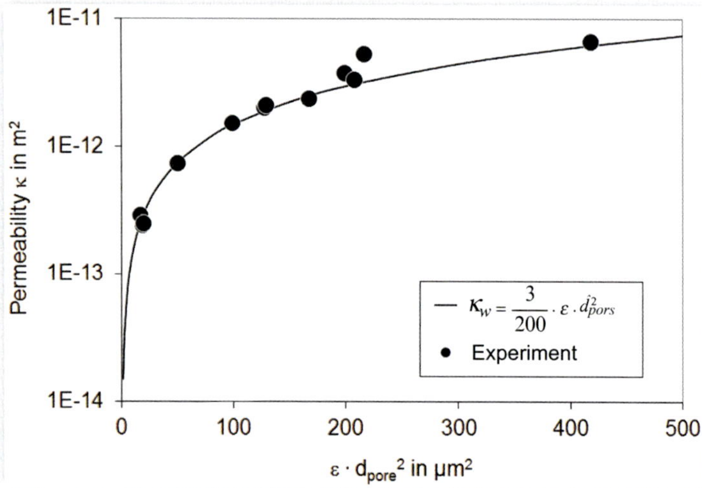

The contribution of each of the individual pressure losses described above depends on the flow conditions, as well as the filter technology. In Figure 7.9, examples are shown for an uncoated as well as a coated GPF under different operating conditions. For the uncoated filter example, it can be observed that while the share of the resistance of the flow across the wall remains moderate and roughly constant, the contribution

FIGURE 7.9 Share of the different pressure drop contributions for uncoated and coated GPFs at different operating conditions. Simulations for filter size of 4.66″ × 6″, no soot and no ash.

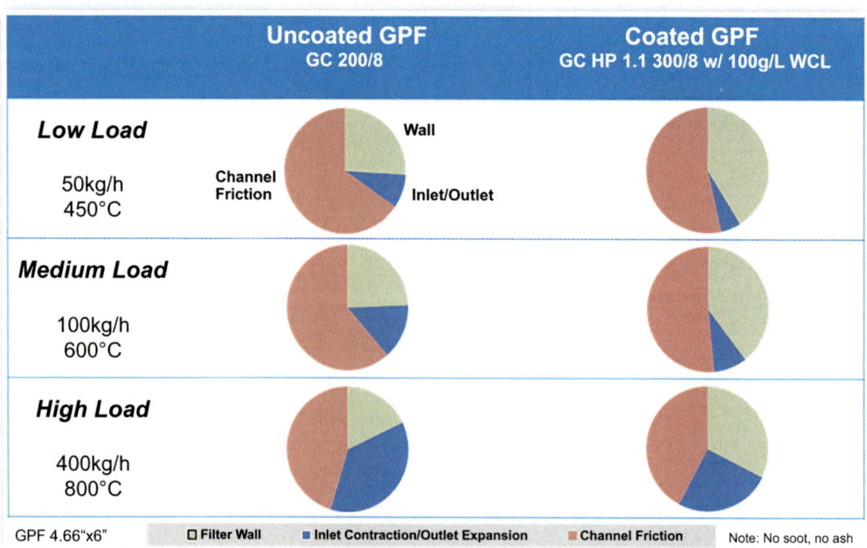

FIGURE 7.10 Pressure drop as function of soot load for different GPFs with different coating technologies. All filters with same technology and 300 cpsi/8 mil design.

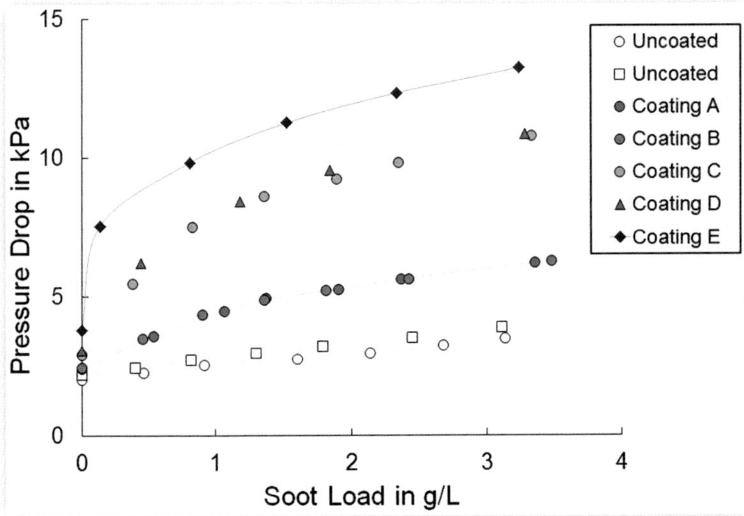

of the inlet and outlet effects increases significantly as the flow rate increases. For coated filters, the contribution of the wall is under all conditions higher compared to the uncoated filter technology due to the lower wall permeability. With respect to the distribution between the two other areas of pressure loss, the same holds as for uncoated filters.

As for filtration, the accumulation of soot and ash also changes the permeability of the filter wall, resulting in an increase in pressure drop. **Figure 7.10** shows laboratory measurements of pressure drop as function of the soot load, measured for a number of different GPF examples. The different data sets represent samples of a high-porosity GPF technology, all in the same size, two without coating and five with different coating technologies.

For all cases, one can observe a steep initial increase at low soot loadings, then leveling off to a more moderate slope of pressure drop vs. soot load. The moderate slope is similar for all examples. It is driven by the increase in soot cake thickness, which is primarily related to the cell design of the filter (e.g., filtration surface area), which was 300/8, and the same for all samples tested. The initial steep increase is the result of some soot accumulating in the pore structure, decreasing the local permeability. Since the local velocities inside the pores can be high due to the local porosity, this effect is very pronounced. As the examples show, this effect is a strong function of the coating applied to the filter. The different coatings shown vary in washcoat level as well as coating technology. The washcoat load of coatings A, B, and C is the same and at a medium level, but different technologies were applied. The washcoat loading for coating D is higher compared to A–C and very high for coating E. The effect of the washcoat loading level is to some extent driven by the fact that more coating results in greater reduction of open pore volume. The absolute effect, however, depends also on the selected coating technology, as can be seen by comparing coatings A–C.

FIGURE 7.11 Evolution of clean pressure drop without soot as a function of ash load. Symbols: Data measured on a cold flow bench on different gasoline particulate filters operated on durability vehicles up to 240,000 km and on engine bench. Lines: Simulations using different assumptions.

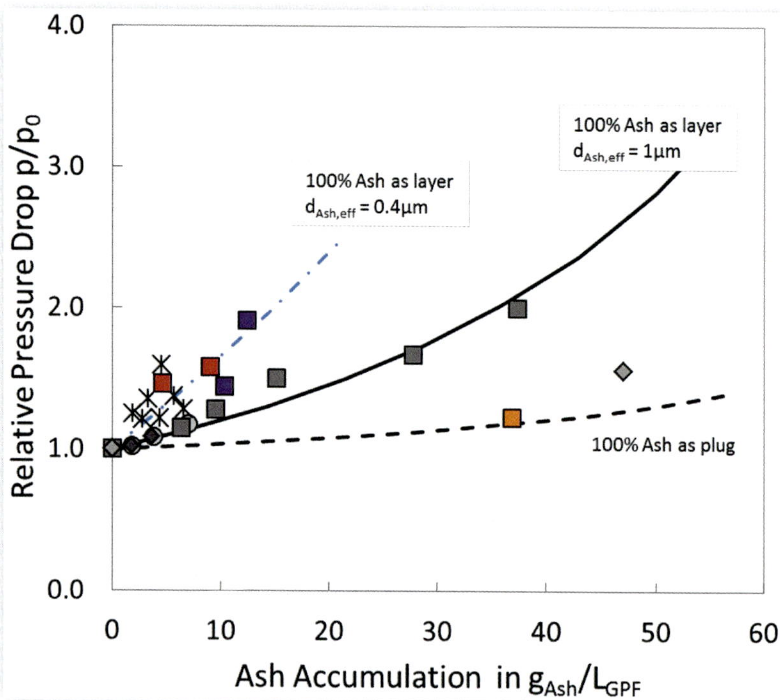

Examples showing the effect of ash on pressure drop are shown in Figure 7.11. Plotted is the change in pressure drop without any soot, relative to the fresh value. The symbols represent experimental data from a large number of filters, on which ash had been accumulated, either during operation on vehicles (majority of data) or on engine bench. The pressure drop values were measured on a cold flow bench after the soot had been burned off. For all examples, one can observe that the pressure drop increases as ash is accumulated. The level of pressure drop increase depends on the individual example and its exposure. Also illustrated in Figure 7.11 are three lines obtained by simple 0-D simulations using different extreme assumptions. All simulations were based on an ash density of 300 g/l. The dashed line represents the limiting case, when all ash would be accumulated as plug in the rear of the filter. This is usually a reasonable assumption for diesel filter applications. For GPF, however, it appears to be a too optimistic assumption. The other two lines represent conditions where it is assumed that all ash is deposited as an axially uniform layer on the channel walls. The difference between both lines is the assumed effective ash particle size, which is 1 μm in one case and 0.4 μm in the other (note that a similar effect could be observed for a change in the assumed density). The assumption of the ash being deposited on the wall seems to reflect the experimental data better, which is also in line with the microscopic observations made during detailed post-test analysis of the filters, as will be discussed in a dedicated section below.

7.4.4 Catalyst Functionality

Since the impact of the particulate filter on the catalytic performance will be discussed in detail in subsequent chapters, only a brief overview will be provided in this section. In general, the filter substrate impacts the catalytic function with respect to (a) the temperature during transient changes in the exhaust conditions, driven primarily by the thermal inertia, and (b) how the catalyst is presented to the exhaust gas, determining the catalyst efficiency and utilization. The effects during transient cold start conditions were discussed in detail by Opitz et al. [26]. They observed a somewhat inferior cold-start performance of a GPF with 300/8 cell geometry and 65% porosity compared to a standard 600 cpsi flow-through catalyst substrate. The study showed that the thermal inertia (or mass) of the filter has a crucial effect, as expected. The plugs play a separate role as they are unique to filters and influence especially the thermal conditions in the first few millimeters. In modern GPFs designed for TWC coatings, the substrate mass is already addressed. The majority of applications are using filters with 300 cpsi, as the example in [26], but the web thickness is reduced from 12 mil in [26] to 8 mil (e.g., [8, 9]). This reduces the matrix density by more than 30% and significantly improves the heat-up and light-off behavior. The need for low thermal mass is also one of the reasons why cordierite has been selected for GPFs, since its material density is roughly 25% lower compared to aluminum titanate or SiC.

Other significant differences between the filter and a flow-through substrate are related to the heat transfer as well as the internal and external mass transfer. The latter is one reason why it is advantageous to use GPFs with 300 cpsi for coated applications vs. the typical 200 cpsi for uncoated applications, where this is less relevant (and the design can be done for lowest clean pressure drop). While [26] showed that during the cold start, the performance of a given catalyst applied to a filter is inferior compared to a standard TWC, other operating conditions exist under which a wall-flow structure can perform at essentially at the same level [27]. At high reaction rates, where significant mass transfer limitations exist for flow-through catalysts, a filter might even perform better.

7.4.5 Mechanical Robustness and Packaging

For the application, particulate filters are packaged into a metal can that can be welded into exhaust systems, which are then installed onto the vehicle (see, e.g., [28]). A ceramic mat is used in between the filter and the metal can to provide protection against vibrations, compensate for dimensional tolerances, and thermally insulate the external shell of the can, analogous to TWCs. To allow for sufficient retention forces, high compressive pressures are applied, typically in the range of 5–11 bar, depending on the application. As a first requirement, the filter has to have sufficient strength to survive this packaging process. The metric usually used to describe the part strength with respect to canning is the isostatic strength, measured by applying a uniform (isostatic) pressure to the external filter surface until first damage is observed. Key parameters driving the isostatic strength of a honeycomb structure are the strength of the porous material, the cell density, and the web thickness. The skin thickness and quality can have some effect as well. The strength of the porous material is determined by the intrinsic material strength, which can be influenced by the material technology, and the porosity. Higher porosities mean less solid material to carry the load; the dependence of strength upon porosity is highly nonlinear. Examples of correlations can be found, for example, in [29]. The cell density and web thickness impact the OFA, as described above. While high values of OFA are desirable for low pressure drop, the opposite is true with respect to strength since 1-OFA describes the area and volume fraction of the solid (porous) walls, available to carry mechanical loads. If the walls close to the skin are viewed as load-bearing beams, a higher cell density means a larger number of beams available, and a thicker wall means more area and stiffer properties. For the isostatic strength, the web thickness plays a key role with a highly nonlinear relationship, often limiting practical designs.

In the application, the filter is exposed to conditions very similar to normal TWCs today—high temperatures, rapid temperature changes and gradients, and significant vibrations. In addition, GPFs can be exposed to extreme thermal conditions during severe soot burning events, with the potential for extreme local heat release as a result of the exothermic reaction, similar to what we know from DPFs. All of this requires a high resistance to thermal stress from thermal gradients inside the part. The thermal stress is generally proportional to the product of local temperature gradient and thermal expansion coefficient, leading to thermal strain. The strain results in stress. The level of stress is dependent on the elastic modulus, which is a function of both the intrinsic material properties and the porosity [29]. Generally, higher porosities lead to a lower elastic modulus and a favorable, more elastic behavior. Hence, some of the disadvantages of porosity in strength are compensated in thermomechanical stress.

While the basic fundamentals are fairly straightforward, the application to real-world driving conditions is challenging. The highly transient and spatially non-uniform temperature field inside a particulate filter in operation, combined with the complex material properties of low thermal expansion cordierite compositions and the fracture behavior of a brittle ceramic, make it still extremely challenging to quantitatively predict the theoretical thermal robustness. The assessment for thermomechanical robustness is therefore still primarily done in extensive, empirical test programs. Examples will be discussed in Chapter 8.

7.5 Application Fundamentals

In this section, a few selected examples from practical applications will be presented to illustrate the meaning and relevance of the fundamentals described in the previous section. More examples and topics related to the durability and long-term use of GPFs will be covered in the next chapter and some of the application examples in Chapters 11 and 12.

7.5.1 Filtration over Different Emission Cycles

Compared to diesel particulate filter applications, which today boast significant experience and understanding, filtration in gasoline applications is performed to a large extent by the filter structure only, with little help from soot. The accumulated ash enhances the filtration but takes some time or mileage to accumulate in an appreciable quantity. As a result, for new filters, we seldom observe filtration efficiencies measured on gasoline vehicles over different cycles to be in the >90%-99% range, which is common for diesel

FIGURE 7.12 Examples of filtration efficiencies measured on different test vehicles over different cycles. All filters are fresh with a maximum of roughly 100 km prior to the test. Note: in the system layouts, red represents TWCs, green represents uncoated GPFs, and orange represents coated GPFs.

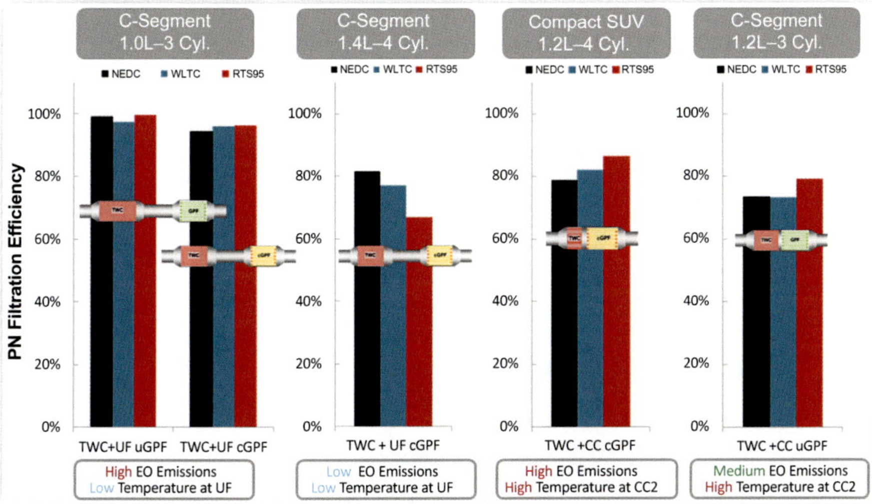

applications. For new GPFs, filtration efficiency is more commonly in the range of >50%-80%. The efficiencies observed depend on the filter sizing and filter properties, as expected, but also on the conditions in the exhaust system. Examples for filtration efficiencies measured over different cycles on four different vehicles, using different aftertreatment architectures and filter technologies, are shown in **Figure 7.12**.

Applications with lower temperatures usually enable the storage of small amounts of soot, which is favorable for filtration. As a result, for a given vehicle and filter, UF installation typically enables higher filtration efficiency. Higher engine raw emissions can enable a faster build-up of some soot in the filter, which is favorable for filtration, but at the same time require higher initial filtration performance. With respect to filtration, higher values are achieved when larger filter sizes are used, as expected based on the fundamentals described above. In practical applications, however, the filter size is often determined by packaging limitations as well as pressure drop requirements. For the first GPF platforms coming to market in 2017 and 2018, the average filter size is roughly the same volume as the engine displacement. However, individual applications can be higher or lower.

For future regulations, it is expected that particles smaller than the current lower limit of 23 nm will be included in the particle count. This increases the count of the total number of particulates emitted by the engine, as can be seen in **Figure 7.13** (left chart) by comparing the number from the CPC or the DMS500 accumulation mode with the total number measured by the sum of DMS500 nucleation and accumulation mode. Since it is hard to read from a logarithmic diagram, the added pie charts show the relative contributions of nucleation and accumulation mode particles, indicating for the example shown an increase of roughly 25% when smaller particles are considered as well. With the GPF, the small particles are also effectively removed

FIGURE 7.13 Examples of PN emissions measured on a T-GDI (EU6b) test vehicle over NEDC and WLTC test cycles using the standard particle counter (CPC) and a DMS 500 for particle size distribution and a count of particles smaller than 23 nm. Left: Cumulative particle number count per km (TWC only and w/ coated GPF in CC2); Right: Particle size distribution over entire cycle (WLTC) for TWC only and with coated GPF in CC2.

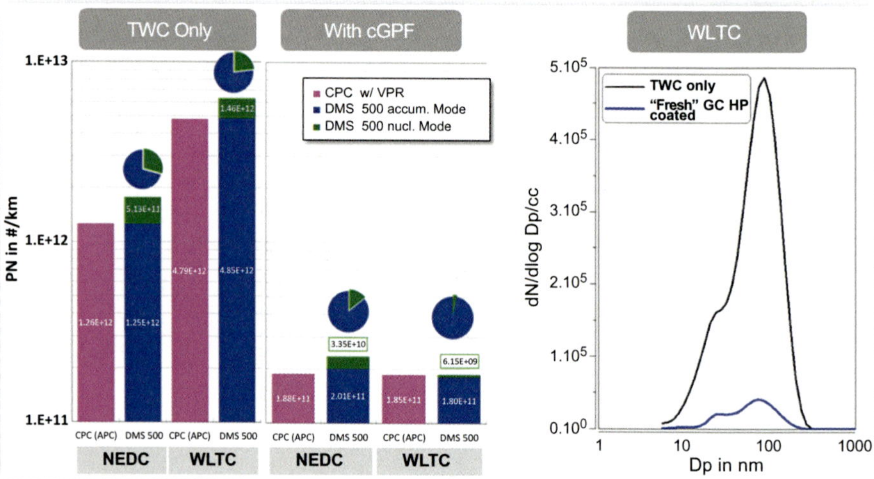

because they are in the range where collection based on Brownian motion is very efficient (see also Figure 7.4). This helps to keep the total tailpipe emissions low, as can be seen from the bar chart in the middle, as well as the particle size distribution with and without GPF on the right of Figure 7.13.

7.5.2 Ash Accumulation

As mentioned above, particulate filters collect not only soot but also inorganic ash particles. These increase the pressure drop as was shown in Figure 7.11. The effect on pressure drop depends on how the ash is collected and where it is deposited, as was already mentioned above. Key questions, which can be answered by evaluating filters after testing or operation, are: (a) Where does ash accumulate (on the wall or in the rear as ash plug)? And (b) does ash penetrate the porosity and change the wall permeability?

Figure 7.14 shows a CT scan of a filter operated over a high cumulative mileage of 200,000 km on public roads in China. The total mass of accumulated ash was about 80 g. CT scans are useful in determining which (volume) fraction of the accumulated ash is deposited as ash plug in the rear of the filter. Ash deposited as a thin layer onto the channel walls cannot be seen due to limitations in resolution. From the image in Figure 7.14(a), one can observe only short ash plugs, too short to represent all the ash accumulated (and much shorter compared to what we know from DPFs). This is in agreement with the assumption that the ash particles follow the gas stream lines inside the filter, deposit on the filter walls, and remain there due to the lack of high soot concentrations, which could lead to a migration of ash towards the rear of the filter during oxidation events [31]. The latter is the dominant effect in diesel particulate filters,

FIGURE 7.14 Examples of (a) a CT scan of a filter operated over 200,000 km on public roads in China (left) and (b) optical microscopy images of filter sections along the length of a filter (right). Note: color of the optical image was modified to enhance the contrast between wall and ash layer [30].

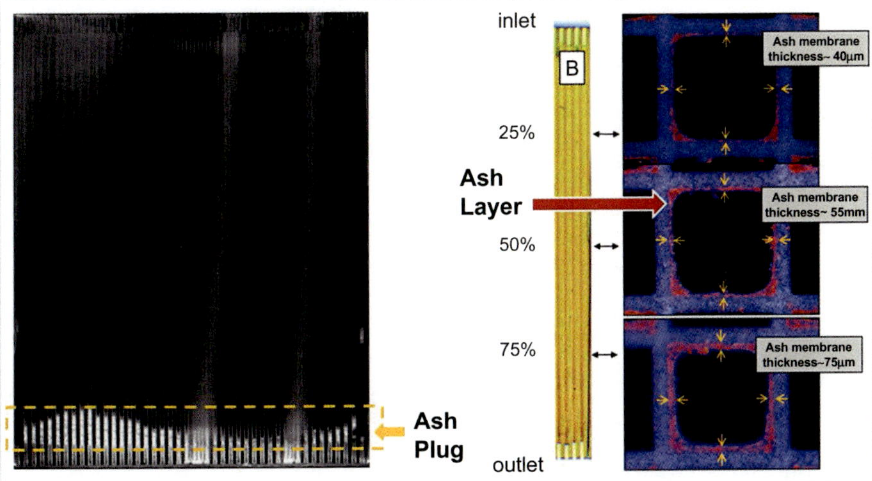

where the exhaust has significantly higher soot-to-ash ratios, but apparently not for gasoline filter applications. For gasoline filter applications, a significant fraction of ash is collected as a thin layer on the channel walls. Characterization of such ash collected on the filter walls can be done using destructive methods combined with optical (OM) and scanning electron microscopy (SEM). Examples of images showing the ash layer at three different locations along the length of the filter are shown in Figure 7.14(b). To better illustrate the ash layer, the color of the ash and filter wall has been changed by post-processing of the images.

A closer look at the ash layer, as presented in **Figure 7.15**, demonstrates that the ash is not penetrating into the porous filter structure but rather stays as a fine, highly porous layer on top of the filter wall. Essentially, almost no ash particles can be found inside the pores and the concentration in fine particles decreases rapidly from the pore entry into the microstructure, as shown in detail in Figure 7.15, right-hand side. From a modeling perspective, this suggests that a treatment as layers in series of the "clean" wall and an ash layer is appropriate. The ash acts as a highly efficient filtration medium, as has already been shown earlier, for example, in Figure 7.6 and related text. As a result, the penetration of soot into the microstructure of the filter is suppressed, changing the pressure drop vs. soot load behavior. Examples of this are shown in **Figure 7.16**, left and middle. As can be seen, the clean, soot-free pressure drop increases as ash is loaded. However, the initial steep increase in pressure drop with soot load is reduced. In practical vehicle applications, where we usually always have at least a small amount of soot (based on our experience, typically about 0.2 g to 1 g), this can lead to the surprising observation that the pressure drop is initially decreasing. This is counterintuitive but explained by the effect of the ash. Once a sufficient ash layer is formed, a roughly linear pressure drop response vs. soot load is observed.

FIGURE 7.15 SEM images of a gasoline particulate filter wall with collected ash. Left: polished cross section with ash on top of an uncoated GPF wall; right: high-resolution fractured surface SEM showing the ash layer/filter wall interface and the decrease in small ash particles collected on the pore walls along the initial depth of the pores.

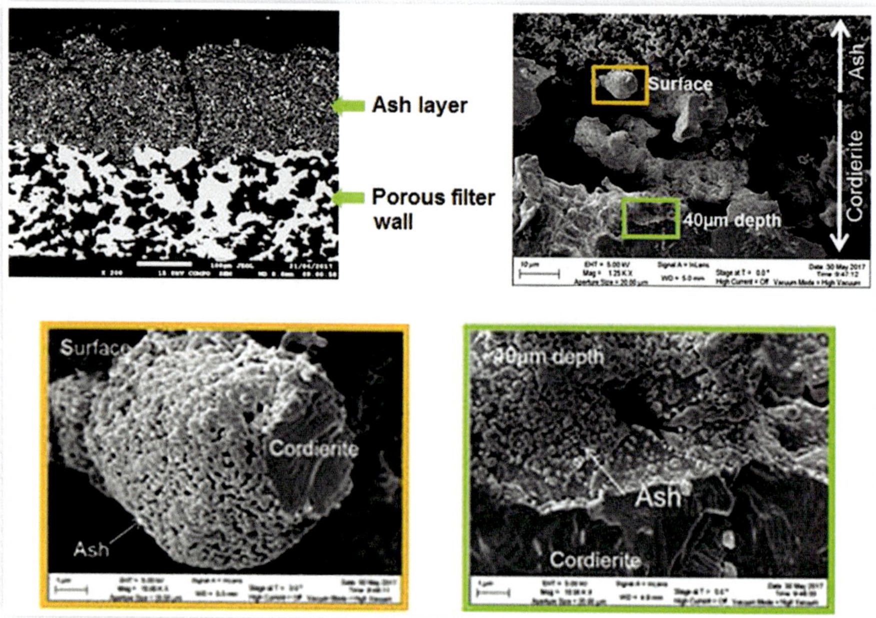

FIGURE 7.16 Pressure drop as a function of soot for filters with different mileage or ash load (left coated GPF, middle uncoated GPF) and SEM showing soot deposited on top of the ash layer. Note: the size of the filters shown is different and the vehicles also generated different ash.

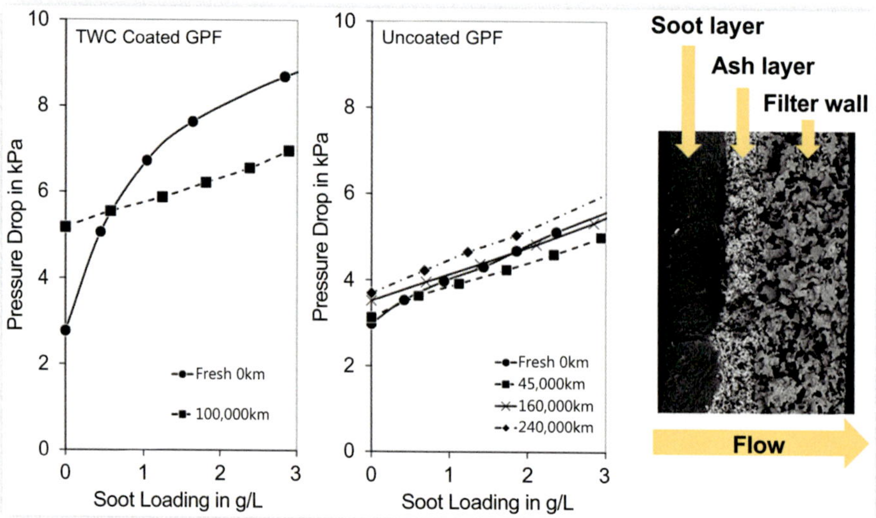

7.5.3 Filter Design Based on Pressure Drop

Minimizing filter pressure drop is a key objective for GPF product design and optimization. Besides the filter material properties, porosity and pore size, which are optimized to minimize the intrinsic resistance to flow, the cell design parameters, cell density and web thickness, are the key variables. As has been shown above in Section 7.4.3, for advanced filter technologies, about 60%-80% of the pressure drop is controlled by cell design contributions. In addition, the part volume and frontal area play a key role [8].

Examples showing the impact of different filter designs, as well as the filter length on pressure drop, are shown in **Figure 7.17**. For designs A–C, the properties have been taken from an early publication [10, 11, 32, 33, 34]. The properties for designs D–F are those of DuraTrap® GC filter technology with 55% porosity. From Figure 7.17(a), it can be seen that shorter filters don't always exhibit a lower pressure drop, in difference to flow-through catalysts, where short length means low pressure drop. The reason is that filtration surface area is needed to minimize the resistance to flow of the walls. Filters that are too long show an increase in pressure drop, in this case due to the increase in the frictional losses. Therefore, typically an optimal filter length with respect to pressure drop exists. Filter lengths observed in commercialized GPF applications are mostly in the range of 4″-6″, with a few exceptions. The comparison of the data for the different designs explains the evolution over the last few years. Original designs in 300/12, derived from DPFs, were initially replaced by designs with reduced web thickness and increased cell density, for example, 360 cpsi [33, 34]. Due to the high frictional losses of these designs and the fact that very low soot quantities are found in GPFs, the designs quickly changed to lower cell densities in the range of 200 cpsi, reducing the frictional losses at no practical penalty for other performance. In addition, improvements were made to the microstructures to minimize the sensitivity to soot. A further reduction in cell density, for example, to 100 cpsi (Design F), has some minor potential for a further pressure drop reduction in the fresh state if the filter length is large. However, as Figure 7.17(b) shows, this benefit is consumed and quickly turned to a disadvantage after only small ash quantities or mileage is accumulated. Therefore, such designs have not been pursued so far. Higher cell densities theoretically could have some benefits at very high ash loads due to the higher filtration area (see Figure 7.17(b)). However, the benefit is realized only at extreme ash loads. These ash loads are equivalent to mileages well above 250,000-300,000 km, based on the typical ash finding rates we observed in a large number of field trials so far (see Chapter 8). Hence, a vehicle would have to operate under highly disadvantaged conditions with respect to power and fuel economy for the majority of its useful life before a breakeven and potentially a slight benefit could be observed. Further thinning of the web thickness is, in general, possible. However, due to the significant impact of web thickness on isostatic strength, this requires changes to the material porosity and microstructure, which lead to a lower permeability and higher sensitivity to soot. In practical applications, this often consumes the potential benefit. Reducing the web thickness also has a significant effect on filtration, which is why a compromise is usually made [7].

For coated filters, the permeability of the wall is usually lower and higher cell densities, for example, 300 cpsi, can be of benefit. As described above, besides the pressure drop aspect, the higher cell densities are also favorable for the catalyst function. Most coated GPFs used for the first generation of commercialized systems have a cell density of 300 cpsi and 8 mil web thickness.

FIGURE 7.17 Pressure drop as a function of (a) filter length and (b) ash load for different uncoated GPF designs. Designs A–C with 300/12, 360/8, and 360/5 in 48% porosity from [10, 11, 32, 33, 34], Designs D–F in 300/8, 200/8, and 100/12 with porosity of 55% for DuraTrap® GC filter technology. Conditions used for simulations: 5.2″ diameter, 370 kg/h, 800°C, no soot, ash as uniform layer.

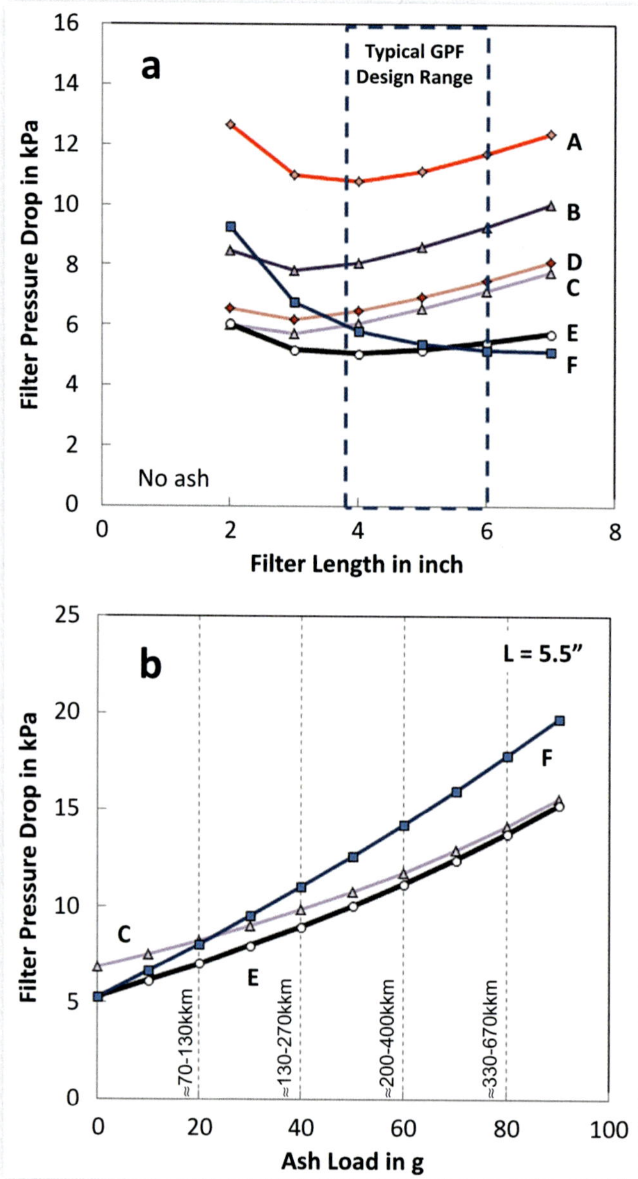

7.5.4 Soot Oxidation and Regeneration

In vehicle applications, the soot collected in the filter has to be managed to prevent excessive exhaust pressure drop, requiring oxidation of the soot or regeneration of the filter. In general, different oxidants are available in an exhaust gas, with O_2, NO, and NO_2 being the most effective ones under normal operating temperatures. In normal stoichiometric $\lambda = 1$ applications, essentially no oxygen is available. In case of a TWC upstream, the levels of NO_x are likely also too low to enable sufficient rates of soot oxidation. Hence, it can be assumed that during normal operation, the rate of soot burning will be minimal. One condition under which soot can be oxidized is fuel cuts, during which the fueling is stopped and essentially intake air containing oxygen is passed through the engine into the exhaust system. These fuel cuts occur frequently, every time the driver lets off the gas pedal with the gear engaged. Fuel cuts usually last only a few seconds but they are frequent [35]. Hence, they offer the potential for a passive regeneration method, without any modification in the operation of the engine. Active measures to force the oxidation of soot would require temporary operation in a lean mode or the addition of a secondary air pump. In Chapter 8, as well as in some of the later chapters, more data will be provided with respect to the soot regeneration under vehicle application conditions. In this section, a brief summary of the soot oxidation kinetics with oxygen will be provided.

The oxidation of carbon and soot has been studied extensively, and a wide spectrum of publications exists; examples are [36, 37, 38, 39]. In the context of combustion engines, we typically describe as soot all oxidizable carbonaceous material that is collected. Gasoline soot consists of elemental carbon and some organic material adsorbed on the carbon. The fraction of elemental carbon depends strongly on the operating conditions and is typically in the range of 50%-90% [40]. Since the organic material is generally volatile, it can be evaporated under higher load conditions, with elevated exhaust temperatures in the range of 500°C-900°C. The elemental carbon can only be oxidized. In general, the oxidation of carbon with oxygen follows one of the two following reaction paths:

$$C + \tfrac{1}{2} O_2 \rightarrow CO$$

$$C + O_2 \rightarrow CO_2$$

The carbon is either totally oxidized to CO_2 or only partially to CO. The distribution between both reactions is determined by the temperature and the availability of oxygen. Typically, a simple global reaction kinetic as provided by the following equation can be used to describe the soot oxidation [41, 42]:

$$-\frac{dm_c}{dt} = A \cdot \exp\left(-\frac{E_a}{RT}\right) \cdot m_c^n \cdot p_{O_2}^r$$

Here E_a is the activation energy for the reaction and n and r are the reaction order in soot (carbon) and oxygen, respectively. The actual reactivity depends on the properties of the soot and can vary. In **Figure 7.18**, examples of thermogravimetric analysis (TGA) soot oxidation measurements in air of soot samples collected on

FIGURE 7.18 TGA soot oxidation measurements of different GDI soots collected on engine or vehicle and Printex U as reference soot.

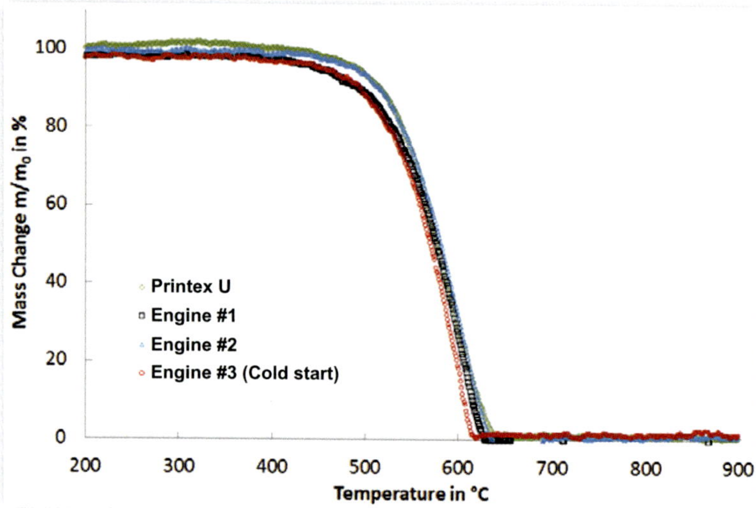

very different GDI engines are shown. In addition, Printex U is shown as reference soot, providing fair agreement with the three engine soot examples tested. Under the conditions used, soot starts to oxidize around 400°C, with high rates above roughly 550°C. Temperatures above 550°C are the conditions typically targeted in diesel applications, where engine-out raw emissions are very high and the soot in the filter has to be regenerated within a few minutes. In gasoline applications, under most operating conditions aimed at passive regeneration during fuel cuts, the engine-out raw emissions are significantly lower. Therefore, lower burn rates and lower temperatures can be acceptable as well.

Soot oxidation rates at lower temperatures down to 300°C have been evaluated, for example, in [35]. The example of Printex U is shown as a representative reference in **Figure 7.19**. At high temperatures, an activation energy of 152 kJ/mol was found, which is in line with the typical values used for diesel studies at high temperatures. At low temperatures, the results suggest a lower activation energy, which was also observed by Koltsakis et al. [41]. The measured kinetics allowed for a good description of soot oxidation experiments performed on vehicles operated at low load and exhaust temperature [35, 43].

The effectiveness of short fuel cuts of a few seconds is shown in **Figure 7.20** for different samples tested in a laboratory bench. The fuel cut was simulated by changing the feed gas from nitrogen to synthetic air for a defined period of time. The details of the procedure are described in [35]. Under the conditions used, roughly 50% of the soot has been oxidized after 20 s and essentially all soot is regenerated after 40–50 s. This has been confirmed in vehicle experiments reported in [8]. The effectiveness, even at low temperatures, has been demonstrated in extended vehicle experiments (e.g., [6, 35, 43]).

FIGURE 7.19 Soot oxidation rates from isothermal TGA soot oxidation measurements of Printex U model soot (data from [35]).

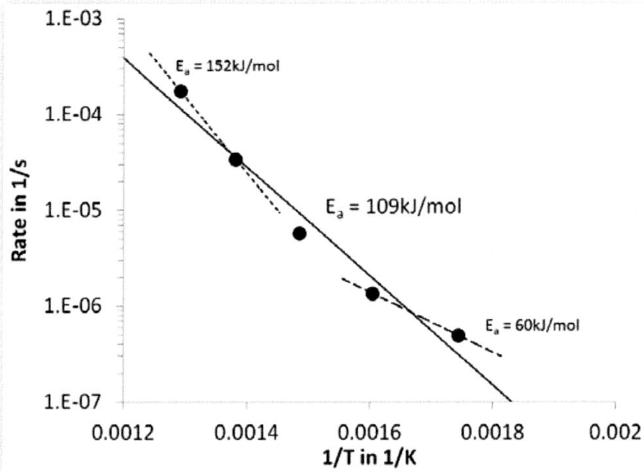

FIGURE 7.20 Laboratory reactor experiments simulating fuel cuts on soot-loaded GPF samples. Experimental setup as described in [35], different symbols represent samples of different filters with 2″ diameter, temperature 700°C.

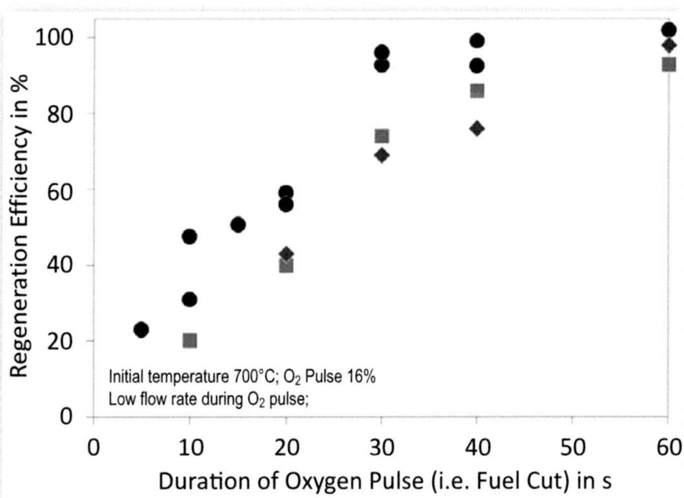

7.6 Summary

In this chapter, the basics of GPFs have been discussed. GPFs are introduced to meet EU6d and China 6 regulations starting in 2017. The selected system architectures show a wide variety, depending on application and OEM, which is not uncommon for new technologies. The design rules and fundamentals build on the knowledge gained from DPFs, with some practical differences. The latter are mainly due to the differences in soot raw emissions and operating conditions.

In the coming chapters, practical examples will be discussed, demonstrating the robustness and practical aspects of the new GPF technology, as well as examples for the integration into modern vehicle applications, to be introduced onto the road to meet new emissions standards.

References

1. Daimler Communications, www.media.daimler.com, Stuttgart, May 27, 2016.
2. Lanzerath, P., Wunsch, R., and Schön, C., "Der erste Serienpartikelfilter für Benzinmotoren von Mercedes-Benz," *17. Int. Stuttgarter Symposium*, 2017.
3. Volkswagen Konzernkommunikation Nr. 264/2016, https://www.volkswagen-media-services.com/, Wolfsburg, 2016.
4. Hentschel, L., Demmelbauer-Ebner, W., Theobald, J., Wendt, W. et al., "The New 1.0 l TSI with 85 kW and Petrol Particulate Filter – Clean, Efficient Performance for the Up! GTI," *Proceedings of 26th Aachen Colloquium Automobile and Engine Technology*, 2017, 971.
5. Gouzonnat, F., Dessarthe, S., Goursot, N., Souhaite, P. et al., "PSA Group's Proposals to Improve the Engine of the Year 2015 & 2016," *SIA Powertrain Conference*, Versaille, 2017.
6. Zhang, Q., Li, J., Ren, X., Li, Q. et al., "GEELY 18TD GPF System Design and Application," *SAE International Powertrains, Fuels & Lubricants Meeting*, Beijing, China, October 16-19, 2017
7. Colignon, C., Zhang, K., and Coulet, B., "Bare Gasoline Particulate Filter System Design and Application," *SAE International Powertrains, Fuels & Lubricants Meeting*, Beijing, China, October 16-19, 2017
8. Boger, T., Gunasekaran, N., Bhargava, R., and Bischof, C., "Particulate Filter Technologies for DI Gasoline Engines," *MTZ* 06 (2013): 452-458.
9. Rose, D., Boger, T., and Nicolin, P., "Optimization of Gasoline Exhaust Systems for EU6c and Beyond - Considering the Impact of a Particulate Filter Integration," *Proceedings of 23rd Aachen Colloquium Automobile and Engine Technology*, 2014, 937.

10. Ito, Y., Shimoda, T., Aoki, T., Yuuki, K. et al., "Next Generation of Ceramic Wall Flow Gasoline Particulate Filter with Integrated Three Way Catalyst," SAE Technical Paper 2015-01-1073, 2015, doi:10.4271/2015-01-1073.

11. Ito, Y., Shimoda, T., Aoki, T., Shibagaki, Y. et al., "Advanced Ceramic Wall Flow Filter for Reduction of Particulate Number Emission of Direct Injection Gasoline Engines," SAE Technical Paper 2013-01-0836, 2013, doi:10.4271/2013-01-0836.

12. Ogyu, K., "Requirement of Exhaust Emission Control on Direct Injection Gasoline Engines – A New Approach on Gasoline Particulate Filters," *IPQC Conference Advanced Emission Control Concepts for Gasoline Engines*, Stuttgart, 2012.

13. Mikulic, I., Koelman, H., Majkowski, S., and Vosejpka, P., "A Study about Particle Filter Application on a State-of-the-Art Homogeneous Turbocharged 2L DI Gasoline Engine," *Proceedings of 19. Aachener Kolloquium Fahrzeug- und Motorentechnik*, 2010, 449.

14. Tandon, P., Heibel, A., Whitmore, J., Kekre, N. et al., "Measurement and Prediction of Filtration Efficiency Evolution of Soot Loaded Diesel Particulate Filters," *Chemical Engineering Science* 65 (2010): 4751-4760.

15. Konstandopolous, A.G., Kostoglou, M., Vlachos, N., and Kladopoulou, E., "Progress in Diesel Particulate Filter Simulation," SAE Technical Paper 2005-01-0946, 2005, doi:10.4271/2005-01-0946.

16. Tien, C. and Ramarao, B.V., *Granular Filtration of Aerosols and Hydrosols*, (Elsevier, 2007).

17. Bollerhoff, T., Markomanolakis, I., and Koltsakis, G., "Filtration and Regeneration Modeling for Particulate Filters with Inhomogeneous Wall Structure," *Catalysis Today* 188 (2012): 24-31.

18. Zhong, D., He, S., Tandon, P., Moreno, M. et al., "Measurement and Prediction of Filtration Efficiency Evolution of Soot Loaded Diesel Particulate Filters," SAE Technical Paper 2012-01-0363, 2012, doi:10.4271/2012-01-0363.

19. Sanui, R. and Hanamura, K., "Scanning Electron Microscopic Visualization of Bridge Formation inside the Porous Channels of Diesel Particulate Filters," SAE Technical Paper 2016-01-9079, 2016, doi:10.4271/2016-01-9079.

20. Konstandopoulos, A.G. and Johnson, J.H., "Wall-Flow Diesel Particulate Filters – Their Pressure Drop and Collection Efficiency," SAE Technical Paper 890405, 1989, doi:10.4271/890405.

21. Konstandopoulos, A.G., "Flow Resistance Descriptors for Diesel Particulate Filters: Definitions, Measurements and Testing," SAE Technical Paper 2003-01-0846, 2003, doi:10.4271/2003-01-0846.

22. Mansoudi, M., Heibel, A., and Then, P.M., "Predicting Pressure Drop of Wall-Flow Diesel Particulate Filters – Theory and Experiment," SAE Technical Paper 2000-01-0184, 2000, doi:10.4271/2000-01-0184.

23. Konstandopoulos, A.G., Skaperdas, E., and Masoudi, M., "Inertial Contributions to the Pressure Drop of Diesel Particulate Filters," SAE Technical Paper 2001-01-0909, 2001, doi:10.4271/2001-01-0909.

24. Gaiser, G. and Mucha, P., "Prediction of Pressure Drop in Diesel Particulate Filters Considering Ash Deposit and Partial Regenerations," SAE Technical Paper 2004-01-0158, 2004, doi:10.4271/2004-01-0158.

25. Schermerhorn, A.P., Khodosevich, K., Joshi, A., and Boger, T., "Detailed Simulation of Exhaust Flow and Deposition of Soot Particles in Porous Particulate Filter Walls," *Filtration* 12, no. 4 (2012): 246-256.

26. Opitz, B., Drochner, A., Vogel, H., and Votsmeier, M., "An Experimental and Simulation Study on the Cold Start Behaviour of Particulate Filters with Wall Integrated Three Way Catalyst," *Applied Catalysis B: Environmental* 144 (2014): 203-215.

27. Opitz, B. and Votsmeier, M., "A Simulation Study on the Conversion Efficiency of Catalytically Active Particulate Filters," *Chemical Engineering Science* 149 (2016): 117-128.

28. Jean, E., "GPF, Just a DPF for Gasoline Engines?," *Proceedings of 26th Aachen Colloquium Automobile and Engine Technology*, Aachen, 2017, 987.

29. Rice, R.W., *Porosity of Ceramics*, (New York: Marcel Dekker Inc., 1998).

30. Shimizu, M., Park, J., He, S., Lv, J. et al., "Performance Evolution of Gasoline Particulate Filters over Extended Vehicle and Engine Bench Operation," *17th Hyundai-Kia International Powertrain Conference*, Gyeonggi-do, Korea, October 24-25, 2017.

31. Sappok, A., Govani, I., Kamp, C., Wang, Y. et al., "In-Situ Optical Analysis of Ash Formation and Transport in Diesel Particulate Filters during Active and Passive DPF Regeneration Processes," *SAE Int. J. Fuels Lubr.* 6, no. 2 (2013): 336-349, doi:10.4271/2013-01-0519.

32. Vogt, C.-D., "Ceramic Wall Flow Filter Material to Meet EU6 Emissions Standards and Low CO_2 Targets," *CAPOC 9*, Brussels, 2012.

33. Vogt, C.-D., Toyoshima, T., Takahashi, A., Sakamoto, H. et al., "Low Pressure Drop Filter Concept for Direct Injection Gasoline Engines to Reduce Particulate Numbers and to Meet CO_2 Targets," *11th Hyundai-Kia International Powertrain Conference*, Namyang, Korea, October 19-20, 2011.

34. Saito, C., Nakatani, T., Miyairi, Y., Yuuki, K. et al., "New Particulate Filter Concept to Reduce Particle Number Emissions," SAE Technical Paper 2011-01-0814, 2011, doi:10.4271/2011-01-0814.

35. Boger, T., Rose, D., Nicolin, P., Gunasekaran, N. et al., "Oxidation of Soot (Printex® U) in Particulate Filters Operated on Gasoline Engines," *Emiss. Control Sci. Technol.* 1 (2015): 49–63.

36. Stanmore, B.R., Brilhac, J.F. and Gilot, P., "The Oxidation of Soot: A Review of Experiments, Mechanisms and Models," *Carbon* 39 (2001): 2247-2268.

37. Messerer, A., Niessner, R. and Pöschl, U., "Comprehensive Kinetic Characterization of the Oxidation and Gasification of Model and Real Diesel Soot by Nitrogen Oxides and Oxygen under Engine Exhaust Conditions: Measurement, Langmuir-Hinshelwood, and Arrhenius Parameters," *Carbon* 44 (2006): 307-324.

38. Dernaika, B. and Uner, D., "A Simplified Approach to Determine the Activation Energies of Uncatalyzed and Catalyzed Combustion of Soot," *Applied Catalysis B: Environmental* 40 (2003): 219–229.

39. Yezerets, A., Currier, N.W., Eadler, H.A., Suresh, A. et al., "Investigation of the Oxidation Behavior of Diesel Particulate Matter," *Catalysis Today* 88 (2003): 17–25.

40. Wang, C., Xu, H., Martin Herreros, J., Lattimore, T. et al., "Fuel Effect on Particulate Matter (PM) Composition and Soot Oxidation in a GDI Engine," *2014 Cambridge Particle Meeting.*

41. Kalogirou, M., Schejbal, M., and Koltsakis, G., "Experimental Procedures for Efficient After-Treatment Model Calibration," *2010 DOE Crosscut Workshop on Lean Emissions Reduction Simulation (CLEERS)*, Dearborn, MI, 2010.

42. Konstandopolous, A.G., Kostoglou, M., Housiada, P., Vlachos, N. et al., "Multichannel Simulation of Soot Oxidation in Diesel Particulate Filters," SAE Technical Paper 2003-01-0839, 2003, doi:10.4271/2003-01-0839.

43. Nicolin, P., Rose, D., Kunath, F., and Boger, T., "Modeling of the Soot Oxidation in Gasoline Particulate Filters," SAE Technical Paper 2015-01-1048, 2015, doi:10.4271/2015-01-1048.

About the Author

Dr. Thorsten Boger currently serves as commercial technology director for light-duty products at Corning Environmental Technologies (CET). He is responsible for leading the interface between customer and technology for global Corning customers of gasoline and diesel technologies as applied to passenger cars and chassis-dynamometer-certified light commercial vehicles. In addition, he serves as the regional technology director for Europe, CET. Boger joined Corning in 1997, working in different roles focused on new technologies for mobile emission control as well as industrial applications in the chemical and refining industry. He graduated with a degree in process engineering (Dipl.-Ing.) and received his doctorate (Dr.-Ing.) from the University of Stuttgart.

CHAPTER 8

Gasoline Particulate Filter Application and Durability

Dominik Rose and Thorsten Boger
Corning GmbH

8.1 Introduction

Gasoline particulate filters (GPFs) are a new technology and were introduced to the market for high-volume applications in 2017. A generic challenge for these new GPF applications are the changing performance characteristics, as a function of mileage due to collected ash and/or soot deposits. Gasoline-powered vehicle applications equipped with traditional three-way catalysts (TWCs) only, for example, without a GPF, have to consider (only) the change of catalytic performance due to thermal and chemical aging [1]. The industry developed a broad understanding of design to apply robust technical solutions that comply with different global emissions requirements over the full useful life of a vehicle. The application of GPFs to the exhaust after treatment system adds new variables to consider. The performance characteristics of a GPF will change over the vehicle lifespan due to an accumulation of soot and ash, with implications on pressure losses over the entire exhaust system, filtration efficiency, and potentially conversion efficiency for gaseous criteria pollutants. In addition, new requirements with respect to the management and protection of the new component arise, which also add new requirements with respect to the component and system durability.

In the previous chapter, some of the fundamentals related to filters and their design have been introduced. In this chapter, we will focus on the practical application and operation of GPFs. In the first section, the impact of application conditions on filtration as well as the evolution of the filtration behavior over accumulated mileage will be discussed. In the second and third sections, the impact of ash and soot accumulation and the related management of the filter will be discussed. Finally, we will evaluate conditions and requirements with respect to the thermal stress the filter is exposed to, as well as procedures to assess the robustness.

8.2 Filtration Efficiency in Practical Applications

The primary function of the GPF is to reduce the tailpipe emissions with respect to particle number (PN) and particle mass (PM) below the relevant regulated limit. In this section, we will briefly summarize the measurement procedures for particulate emissions (a detailed description of the equipment and procedures is provided in Chapter 13) and then discuss the impact of the application conditions and the test cycle on the achieved filtration efficiency. This is followed by examples demonstrating how the filtration performance evolves over accumulation of extended mileage.

8.2.1 Procedures for Particulate Emission Measurement

For vehicles certified as passenger cars or light-duty vehicles, the emission measurements are performed on representative vehicles. This is in contrast to heavy-duty applications, where the majority of testing is done on an engine dynamometer, using the engine only. Starting with EU6d in Europe (September 2017), two main types of procedures are applied for vehicle tests [2, 3, 5]:

- Testing on chassis dynamometer following a prescribed driving profile and using laboratory measurement equipment
- Testing the vehicle while driving on public roads for real driving emissions (RDE) using portable emission measurement equipment (PEMS)

In Europe and China, the two markets where GPF will be first introduced, the test cycle and procedure to be used on the chassis dynamometer are the World Harmonized Light Duty Test Cycle and Procedure (WLTC) and (WLTP), respectively [2, 3, 4]. The test is performed at 23°C, with the vehicle being soaked at this temperature range between 6 h and 36 h prior to the test (coolant temperature at 23°C), and emissions are measured and recorded starting from the beginning, to include the cold start. The cycle contains four phases covering different driving patterns, from urban driving to fast highway driving. The total cumulative emissions measured at the tailpipe during the test cycle are normalized for the total mileage covered, 23.3 km, and represented as average emissions per kilometer.

In Europe starting in September 2017 and in China with the start of CN6, vehicles also have to be tested under real-world driving conditions (measuring RDE), on public roads and over wider climatic and altitude conditions [4, 5]. The emission measurement equipment, usually referenced as PEMS, is installed onto the test vehicle. A number of boundary conditions are applied to assure tests are properly representative. Some of the conditions most relevant to particulate emissions, due to the potential for more severe transients in engine operation, include faster accelerations, more dynamic conditions, and wider temperature ranges including tests at very low temperatures down to 0°C, or even −7°C for the extended conditions [5]. As described in previous chapters, the particulate output during such conditions is significantly higher compared to "normal" operations, especially if paired with the full range of possible fuel compositions.

While the RDE certification tests have to be performed on a real road testing as opposed to a test facility, for development purposes, it is generally helpful to establish surrogate RDE test cycles that can be performed in a repeatable manner on a chassis dynamometer. One example of such a test that is known quite broadly across the industry is the RTS-95, representing severe and dynamic driving behavior. In **Figure 8.1**, this cycle and the WLTC are shown, together with examples for the exhaust temperature and coolant temperature for a European test vehicle with a 1.2L TGDI (turbocharged gasoline direct injection) engine. Both tests were performed with a cold start of a vehicle soaked for several hours at room temperature.

FIGURE 8.1 WLTC (left) and the RDE surrogate test RTS-95 (right). Shown are exhaust and coolant temperature (top), modal PN emissions (middle), and vehicle velocity (bottom) for a 1.2L TGDI vehicle.

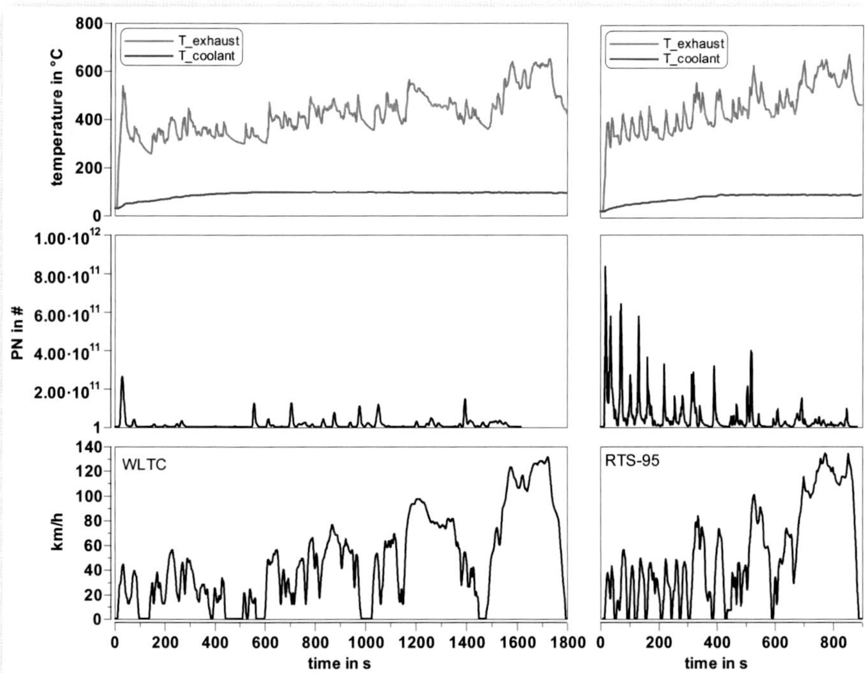

8.2.2 Effect of Application Conditions on Filtration Efficiency

The filtration efficiency of a filter is a function of the filter properties, but also of the operating conditions, as described in the previous chapter. Because the operation during an emissions test cycle is highly transient, the operating conditions change constantly. The most relevant variables, with respect to filtration, are the flow rate (space velocity) and the temperature. The particulate concentration also influences the tailpipe emissions, but not the filtration efficiency, except as soot accumulates inside the filter. In **Figure 8.2**, data are recorded for a 1.2L TGDI vehicle operated over the WLTC and the RTS-95. Plotted are the cumulative PN emissions with the TWC only (dashed lines) and with a GPF added in close-coupled (CC) position, located directly downstream of the same TWC (solid lines). The reduction in tailpipe PN emissions as a result of the GPF clearly occurs for both test cycles.

Due to the different application conditions and engine-out emissions, the tailpipe PN emissions (PN) obtained with a specific filter depend strongly on the test cycle. The cumulative filtration efficiency, however, is typically less cycle dependent, as seen in **Figure 8.3**, where the filtration efficiency on the RTS-95 cycle is plotted versus the one obtained for the WLTC. The data shown was obtained on three different vehicles, tested with different filter technologies and after treatment architectures. For all three vehicles, one can observe that the filtration efficiency is primarily determined by the filter technology and vehicle and less so by the test cycle.

While the examples discussed above show that the effect of the test cycle is moderate only, it does not mean that a certain filter technology delivers a fixed value of filtration efficiency. The filter sizing, relative to the engine, and exhaust conditions have a significant influence. This is shown in **Figure 8.4**, where the same coated filter technology, in identical component size, was tested over WLTC in two very different vehicle applications. The two vehicles and engines used had about the same cumulative engine-out raw PN emissions. However, the transient flow conditions and PN

FIGURE 8.2 Particle number emission test data for 1.2L TGDI vehicle measured over WLTC and RTS-95. Data were obtained with TWC only and with TWC + GPF in close-coupled position.

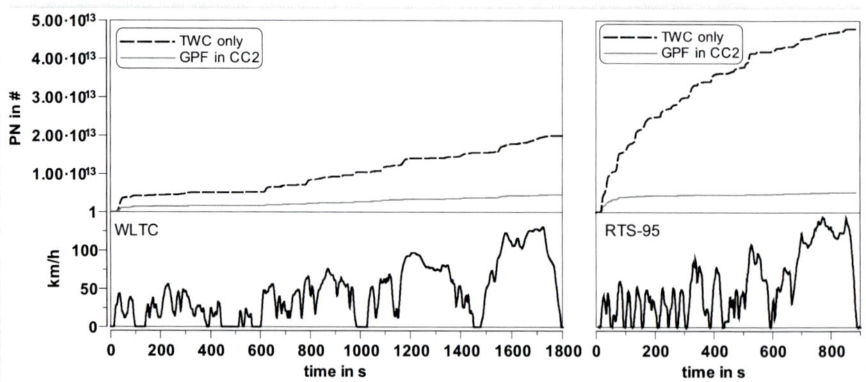

FIGURE 8.3 Filtration efficiency obtained over RTS-95 vs. filtration efficiency over WLTC for different filter-vehicle combinations.

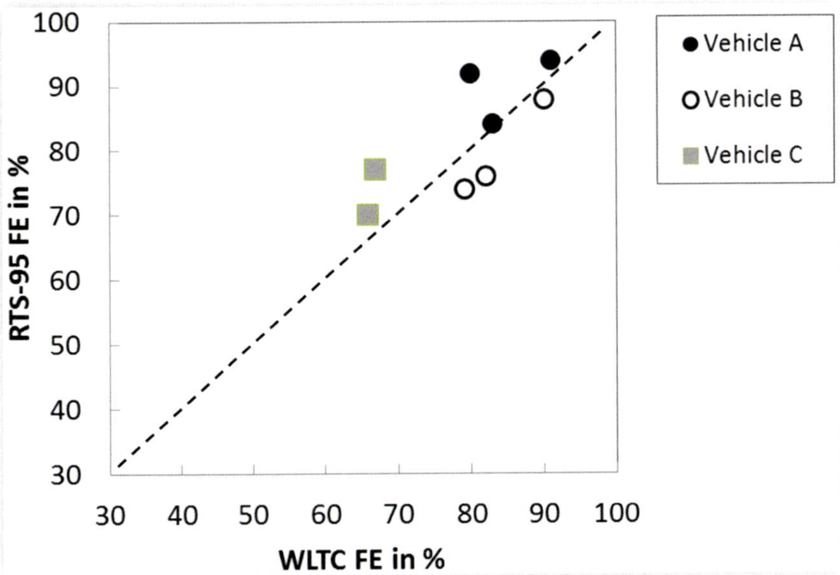

FIGURE 8.4 Influence of the application conditions on the filtration performance. Data obtained with the same filter technology and component size tested on two different vehicles and engines. Shown from top to bottom are space velocity (SV), cumulative particle number emissions (PN), and vehicle speed. Dashed lines are simulations using identical filter properties.

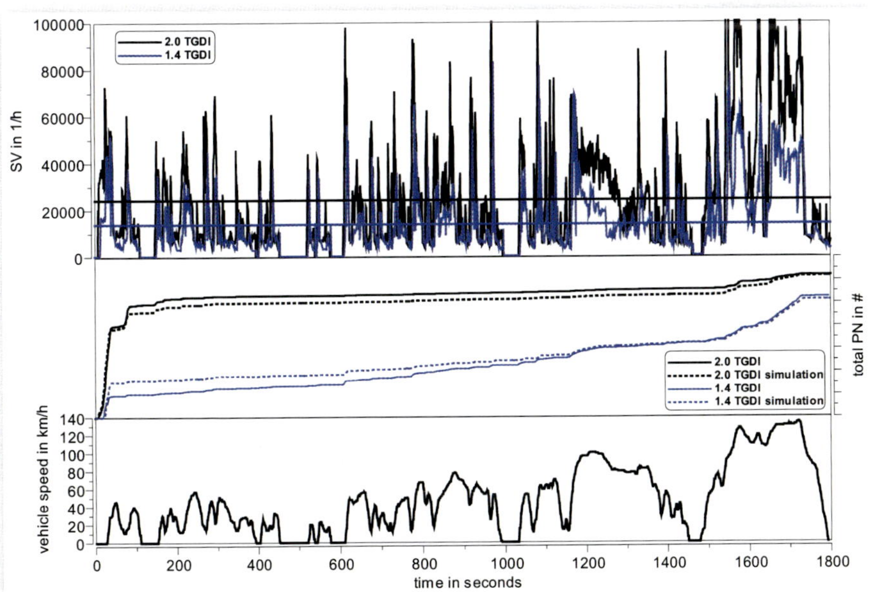

CHAPTER 8

emissions differed significantly. The filtration efficiency was about 7% higher for the smaller engine with overall lower flow rates and lower emissions during the initial phase, resulting in the lower cumulative PN emissions.

Also shown in Figure 8.4 is data obtained with a simple filtration model according to the one described in Section 7.4.2, omitting any potential accumulation of soot. The same filter parameters were used in both cases; only the boundary conditions were set according to the difference in vehicles. The agreement with the measurements is reasonable. Using this model allows us to perform a simple case study, illustrating the effect of the filter sizing for the different application examples. This is shown for three vehicle examples in **Figure 8.5**. The boundary conditions for the WLTC were obtained on vehicles with different inertia and different engine size and calibration. The transients of cumulative engine-out PN, exhaust temperature at the GPF inlet, and mass flow are shown in the diagrams on the right for reference. The filtration simulations used identical filter properties for all three application examples. The results suggest that, as mentioned above, for a given filter technology, the application and the filter sizing have a significant impact on the filtration efficiency that can be achieved.

Another factor having a significant impact on the achievable filtration efficiency of a given filter technology is the potential for soot accumulation during the cycle. The fact that even small amounts of soot can already change filtration significantly has already been discussed in Chapter 7. For a given filter, the amount of soot that will accumulate

FIGURE 8.5 Case study on GPF size impact on filtration efficiency for different applications. Simulations using experimental data from three vehicles as boundary conditions (right-hand charts).

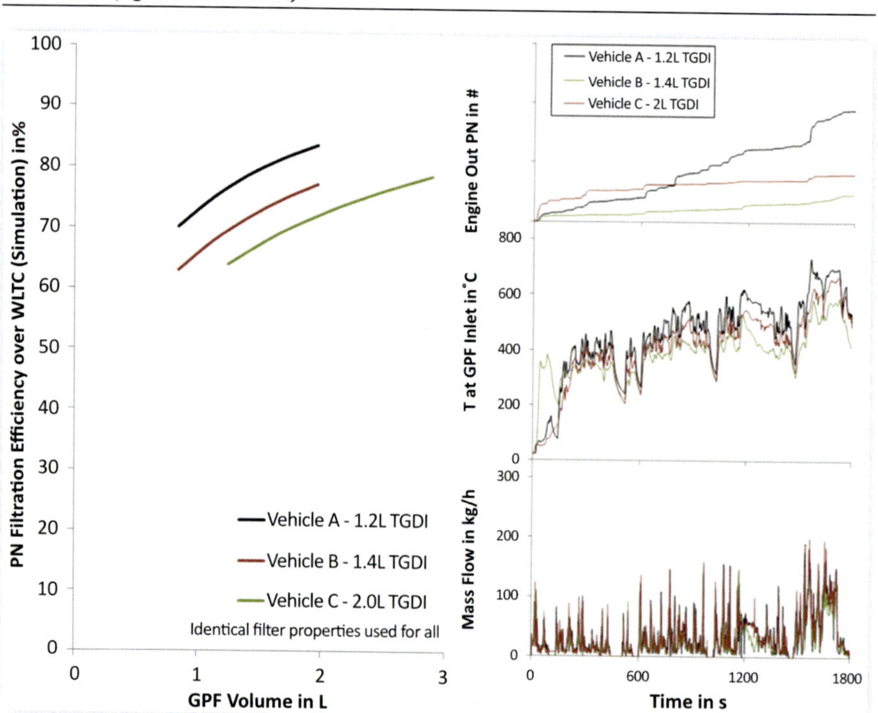

FIGURE 8.6 Impact of filter location (temperature) on filtration. Data is obtained with the same filter tested on the same vehicle in two locations.

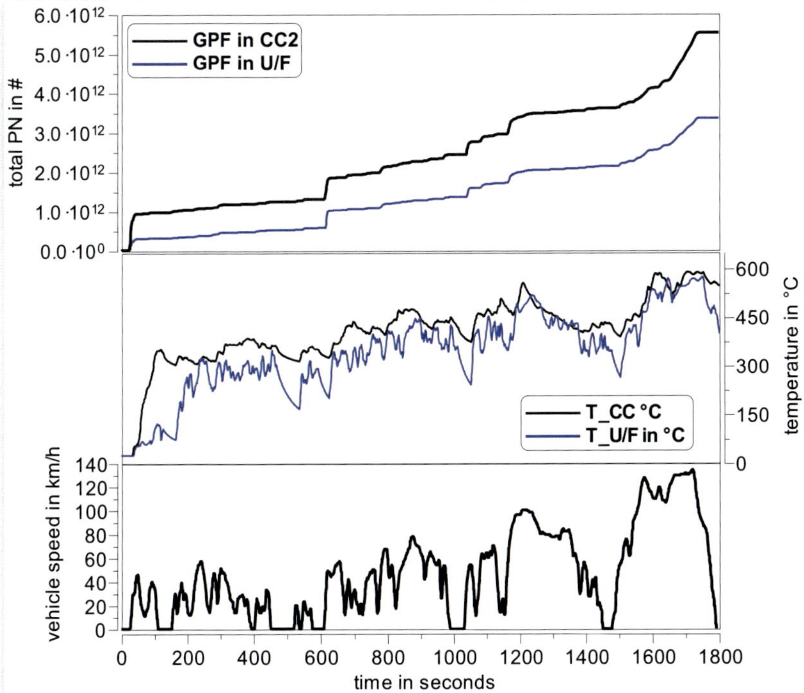

depends primarily on (a) the engine-out soot emissions (i.e., higher soot emissions increase the potential) and (b) the temperature level and oxygen availability (i.e., the rate of soot oxidation). In general, this often means that for a given filter, the achievable filtration efficiency is lower in a CC position compared to a colder underfloor (UF) installation. Another factor that is influencing the filtration in CC versus UF locations is the higher temperature and higher volume flow in the CC application. This leads to shorter residence time in the filter and a lower probability for particles to be collected by Brownian motion. This effect is usually not compensated for by a higher particle diffusion coefficient. Examples of the differences between the same filter sample tested on the same vehicle in UF and CC2 position are shown in Figure 8.6.

The vehicle velocity, the cumulative particle emissions (PN) at the tailpipe and the exhaust gas temperature at the filter inlet are all shown. The difference in temperature between the two locations is especially pronounced during the initial phase of the WLTC. This is where the main difference in particle slip is observed in these examples.

8.2.3 Evolution of Filtration Efficiency over Initial Mileage

Like soot, small amounts of ash also affect the filtration efficiency of a given filter. This is especially relevant during the initial break-in period, where in some applications,

FIGURE 8.7 Initial evolution and increase in PN filtration efficiency over WLTC.

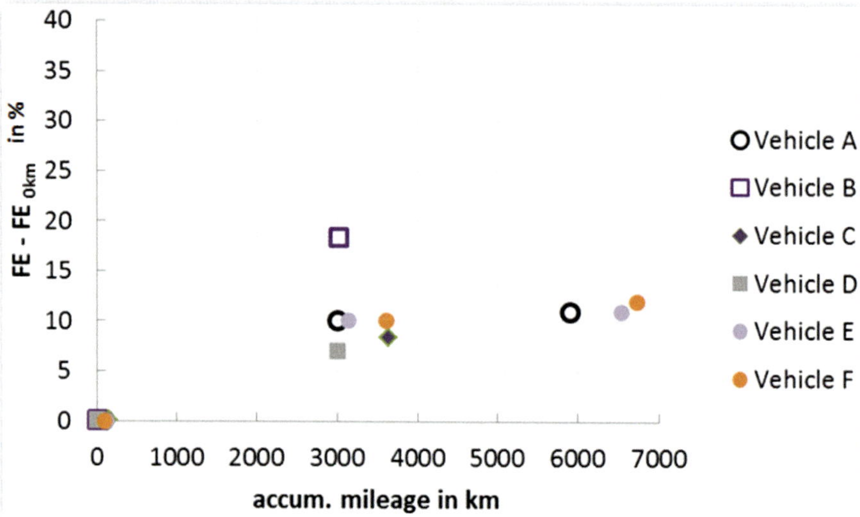

a significant evolution and increase in filtration efficiency can be observed. Examples of this evolution and behavior are observed in a number of our test vehicles and shown in Figure 8.7 (note that these are examples and the actual evolution will be application specific).

In Figure 8.7, the improvement in PN filtration efficiency is plotted and is calculated as the difference between the filtration efficiency measured after a given mileage and the initially measured efficiency of the totally fresh filter (0 km). The data shown were obtained on a number of different European vehicles, retrofitted with different filter technologies and architectures. For the examples shown the filtration efficiency increased by 7-10% after 3000 km. In one case, the filtration efficiency increased even more. The change in pressure drop over this initial period was typically small (not shown).

8.2.4 Evolution of Filtration Efficiency over Extended Mileage

Tailpipe emissions have to be maintained below the regulated limits for the entire useful life of the vehicle, with the required mileage defined in the individual regulations, 160 kkm for EU6 and CN6a and 200 kkm for CN6b. Since GPFs are a new technology and durability testing is time consuming and expensive, limited data are available in the public domain today. Examples of filtration efficiency data for extended mileage accumulation programs from some of our recent programs as well as data that have been published in references [6-11] are summarized in Figure 8.8. Common to all these studies is that filtration efficiency and resulting tailpipe emissions have improved with increasing mileage. The rate of enhancement is very rapid for some and more gradual for others. The data shown also suggest that the selected filter technology can have an effect on the actual rate of improvement, while the location of GPF installation or the presence or lack of a catalytic coating seems to have a smaller effect.

FIGURE 8.8 Evolution of filtration efficiency over mileage on the GPF. Data for examples C–K adapted from [6, 9, 10, 11].

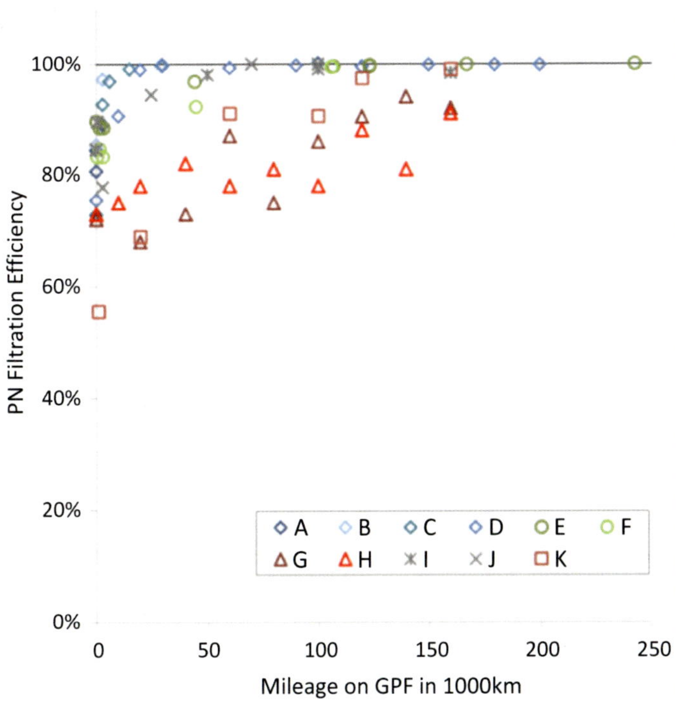

	Ref.	Vehicle Type	Test Cycle	Gasoline Particulate Filter				
				Technology	CPSI/Web	Porosity	Coated	Location
A	new	C-SUV	WLTC	GC HP	300/8	65%	Coated	CC2
B	new	C-SUV	WLTC	GC HP	300/8	65%	Coated	CC2
C	new	D-SUV	WLTC	GC HP	300/8	65%	Coated	CC2
D	[6]	E-SUV	WLTC	GC HP	300/8	65%	Coated	UF
E	[6]	C	WLTC	GC	200/8	55%	Uncoated	CC2
F	[6]	C	WLTC	GC	200/8	55%	Uncoated	CC2
G	[9]	D	WLTC	Comp.	360/5	45-50%	Uncoated	CC2
H	[9]	D	WLTC	Comp.	360/5	45-50%	Uncoated	UF
I	[10]	n.a.	NEDC	n.a.	n.a.	n.a.	Coated	UF
J	[10]	n.a.	NEDC	n.a.	n.a.	n.a.	Coated	CC2
K	[11]	n.a.	WLTC	Comp.	300/10	65%	Coated	UF

The improvement in filtration with mileage is expected and related to the formation of an ash layer with high filtration efficiency (see Chapter 7) and holds as long as the filter does not get damaged. Hence, in contrast to the catalyst performance, which tends to deteriorate over extended operation, due to thermal aging and poisoning of the catalyst, the filtration function does "not age." This is reflected in the regulations, which usually assume a deterioration factor of 1 (e.g., no deterioration) for particulate emissions. With respect to the application work and testing, this means that it is key to focus on the initial filter performance and ensure proper management of the filter to prevent its damage over life.

8.3 Ash Accumulation in Vehicle Applications

As already mentioned in previous sections and chapter, the particulate filter removes and collects not only soot particles but also inorganic particles, usually referenced as ash. Unlike soot, these inorganic ash components cannot be removed by oxidation or other means of gasification. They remain in the filter and accumulate over the life of the filter. In the previous section, it has been shown that this has the desirable effect of improving the filtration performance of the filter. This benefit, however, does not come for free, but has its price in the form of an increased component pressure drop [6, 7, 8, 9, 10, 11, 12, 13, 14]. The fundamentals of ash accumulation in the filter and impact of pressure drop have been discussed in Chapter 7. This section will focus on observations around ash accumulation rates and the impact on fuel consumption and power.

Ash represents the accumulation of any inorganic material that is trapped and collected in the filter, regardless of its origin. This is summarized schematically in **Figure 8.9**.

In general, contributions can come from (a) inorganic sources in the fuel and the oil (engine and turbocharger), (b) particles resulting from corrosion of the exhaust line, or (c) losses in washcoat or mat material of upstream catalyst components [12, 15]. The contribution from metal corrosion is dependent on the material selection and potential contributions during the breaking-in of the engine. This has been observed

FIGURE 8.9 Potential sources for "ash" and inorganic particles.

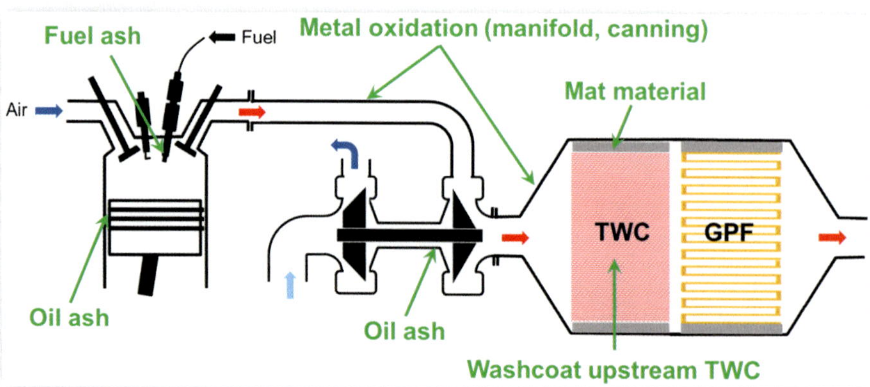

primarily when prototype exhaust systems had been used and operated at high loads (e.g., high temperatures). We have generally observed only minor amounts of materials originating from upstream TWC washcoats or mat, in most of our studies.

However, in reference [12] these contributions accounted for 30% of the inorganic material collected. Because both of these types of particles have a relatively high density, they usually consume less channel volume per mass accumulated and also influence the average "ash density." Furthermore, especially particles and fibers detached from the upstream TWC are relatively large with high inertia and usually can be found in the back of the filter, where their impact on pressure drop is smaller.

Under most conditions, the dominant source for ash components originates from inorganic additives in the lubricant oil. Lubricant oils contain a certain amount of additives to enhance the stability, adjust the viscosity, and reduce friction and wear [16]. Examples known to dominate ash deposits are Zn-containing anti-wear additives, for example, ZDDP, and Ca-containing detergents [15]. Their content in the oil is usually described as sulfated ash content (SAPS) in weight percent. For example, 1 kg of oil with 1% SAPS would theoretically result in 10 g of sulfated ash. Higher SAPS content generally means that more ash can be generated. The other parameter directly influencing the oil ash generation is the oil consumption of the engine. Often, for a given vehicle platform, there is a distribution of oil consumption for the vehicles sold. For filter design purposes, it is of significant importance to know the amount of ash that is accumulated over the operation of the vehicle, for example, in grams of ash per kilometer of operation. The most common approach is to ignore contributions from corrosion and debris from the TWC and calculate the theoretical amount of sulfated ash based on the oil consumption. Then a finding rate with value less than one is applied to this theoretical value. For diesel applications, the ash accumulation rate has been in the range of 0.4-0.8 mg/km, with a typical finding rate of 60–70% [17]. The difference to unity is assumed to be due to components that remain in the oil sump (increased concentration) or that are deposited along the exhaust system or upstream catalyst components [15]. For gasoline applications, a range of finding and ash accumulation rates have been observed and reported, summarized in **Table 8.1**.

In **Figure 8.10**, these values, together with some more details for the lumped values from Table 8.1, are plotted as a function of the engine displacement used in the various vehicle studies. No clear trend with engine size is apparent. The average ash accumulation rate is 0.18 ± 0.1 mg/km. This value is lower than the value commonly found for diesel vehicles. The finding rates shown in Table 8.1 range from 10% to 60% and are in many cases lower compared to the diesel experience (60-70%). Although it is not entirely clear why this difference exists, potential explanations could be due to larger quantities of ash material already trapped in the upstream TWC. In addition, the ash collected might not be in the sulfated form [15]. For filter design and sizing purposes, the data reported suggests that the typical ash mass per 100,000 km will be in the range of 8 g to 28 g. For the example with the highest accumulation rate from Table 8.1, the value would be 40 g.

Looking at the composition of the accumulated ash, we can find relatively consistent observations in the reported studies [11, 12, 14, 15] as well as in our own work. In several studies, a significant fraction of Fe-containing species have been observed (often combined with some Cr, Mo), likely originating from metal wear. The next dominating species are Ca, P, and S based, originating from the lubricant oil. Often only small

TABLE 8.1 Ash finding and accumulation rates in GPF vehicle test programs

Vehicle/engine	Oil SAPS content	Finding rate in % of SAPS	Ash accumulation rate in mg/km	Reference
C-SUV, 1.2L TGDI	0.77%	50-60	0.15-0.16	Internal
D-SUV, 2.0L TGDI	0.90%	60	0.14-0.22	Internal
D-SUV, 2.0L TGDI	0.90%	34-60	0.07-0.13	Internal
E-SUV, 2.0L TGDI	n.a.	n.a.	0.40	Internal
Different vehicle (N = 7), C-E segment, 1.2-3L TGDI	n.a.	n.a.	0.25 ± 0.05	Internal
C-segment, 1.2L TGDI	0.56%	19	0.05	[7]
C-segment, 1.2L TGDI	1.26%	10	0.05	[7]
Different vehicles (N = 12), C-E Segment, 1.4-3L TGDI	n.a.	n.a.	0.02-0.3	[6]
Ford Taurus, 3.5L TGDI	n.a. Ca 1760mg/kg, Zn 835mg/kg	50	0.29[a]	[12]
Ford Flex, 3.5L TGDI	n.a. Ca 2320mg/kg, Zn 939mg/kg	50	0.24[a]	[12]
Opel Zafira, 1.6L TGDI	0.80%	n.a.	0.08-0.13	[9]
Vehicle, 1.8L TGDI	n.a.	n.a.	0.14	[11]
Engine bench, 1.6L TGDI	1.17%	48	n.a.	[15]
Engine bench, 1.5L TGDI	0.6% and 1.25%	27-30	n.a.	[14]

[a] Including about 30% of TWC material, 20% corrosion material.

FIGURE 8.10 Ash accumulation rates per kilometer from internal vehicle test data (gray) and published data (blue, references as in Table 8.1.).

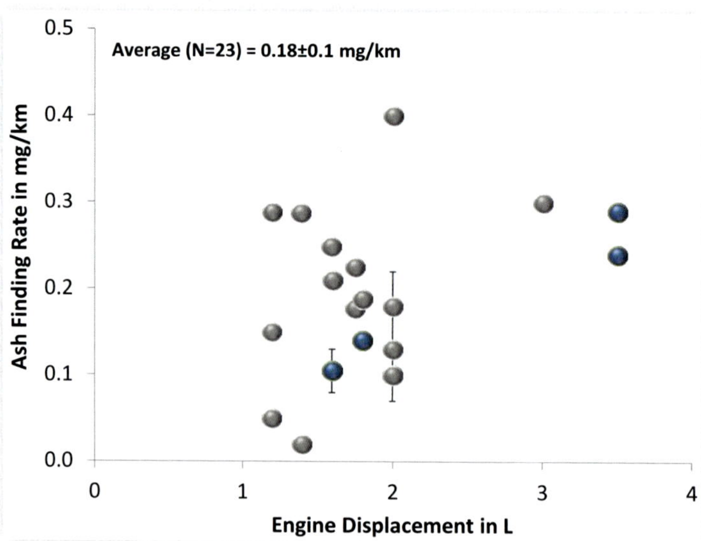

quantities of the Zn from the lubricant are found in the GPF. In [14, 15] it is also shown that the amount of ash scales with the level of Ca-based additives in the lubricant oil.

With respect to the location of the ash deposits inside the GPF, we can usually separate between the ash deposited as a layer onto and along the channel walls and ash that is collected as a plug in the rear of the filter. The latter is the dominating deposition location in diesel particulate filters (DPFs) [17, 18]. For GPFs, most studies have found the majority of ash collected as a layer along the channel walls. Only a minority of ash collects as a short plug in the rear. In [12], 60-67% of the ash was located on the channel walls after 130,000 and 150,000 miles, about 60 g of ash in both cases. The average ash layer thickness was 12.4 μm (10-14 μm range), and the ash plugs were on average 6-7 mm long. In [9], only very short ash plugs were observed after 160,000-200,000 km, representing 17-21 g of ash, estimated to represent about 40% of the total ash (e.g., 60% would be on the channel wall). In our own studies, we also observed primarily ash along the channel walls. In **Figure 8.11**, computed tomography scans are shown for a large number of filters which had been operated on real road for 12,000-100,000 km [6]. The filters shown had been retrofitted to vehicles operated in a non-accelerated, "normal" user profile over the period of 18-31 months with normal cold start frequency and driving patterns. Most of the filters were installed in the UF position.

For all filters, no or very short ash plugs have been observed. The majority of ash was deposited onto the channel walls. Similar observations were made on other filters we operated in durability programs up to 160,000-240,000 km [6, 7, 8].

FIGURE 8.11 Tomography scans of GPFs which had been retrofitted to vehicles and operated in a "normal" user profile for 12,000-100,000 km. Ash accumulation ≈0.02-0.3 mg/km. Note: Images are not to the same scale [6].

8.4 Soot Management in Vehicle Applications

The introduction of particulate filters to gasoline powertrains is introducing new application challenges, similar to those implemented in most modern diesel applications today. Understanding and managing the soot loading state of the GPF is essential for several reasons. As has been shown in the previous chapter, the resistance to flow of a GPF is sensitive to its soot loading, impacting the exhaust back pressure. The exhaust system pressure drop increases with the amount of collected soot and will reduce engine efficiency due to higher engine pumping losses. For turbocharged engines, it also has influence on the turbine pressure ratio P_3/P_4, that is, pressure ratio pre- to post-turbo. Considering events during which the soot is oxidized (regenerated), the heat that will be released is proportional to the amount of soot oxidized. The protection of the components of the exhaust system is therefore another key objective. This leads to the overall need to understand the soot loading state of the GPF at any time and to trigger potential active regeneration measures, if needed.

8.4.1 Soot Load Monitoring

The change in the soot loading state is the result of the differences between the soot emitted by the engine and removed by filtration on the one hand and the amount of soot oxidized on the other hand. This is captured in the following simple soot mass balance:

$$\underbrace{\frac{dm_c}{dt}}_{\substack{\text{Change in}\\ \text{soot load}}} = \underbrace{\eta_f \cdot w_{C,in} \cdot \dot{m}_G}_{\substack{\text{Soot removed}\\ \text{from exhaust gas}}} - \underbrace{M_C \cdot R_C}_{\substack{\text{Soot}\\ \text{oxidized}}}$$

Here

m_c is the mass of soot

$\dot{m}_G$ is the exhaust mass flow

w_C is the soot mass fraction

η_f is the filtration efficiency

M_C is the molecular mass of carbon

R_C is the soot oxidation rate

One approach to monitor the soot loading state is to continuously perform a soot mass balance.

To obtain information about the soot emitted by the engine, different options exist, differing in their detail and level of sophistication. One could use combustion models, taking into account mixture formation, temperature, and pressure in the cylinder. However, such an approach would require a high computational effort with the potential need for additional sensors. Alternatively, simplified phenomenological engine models, primarily map-based, are considered [19]. The simplified models usually take into account the engine load, which is proportional to

FIGURE 8.12 Engine-out particulate emissions map based on engine load and speed [20].

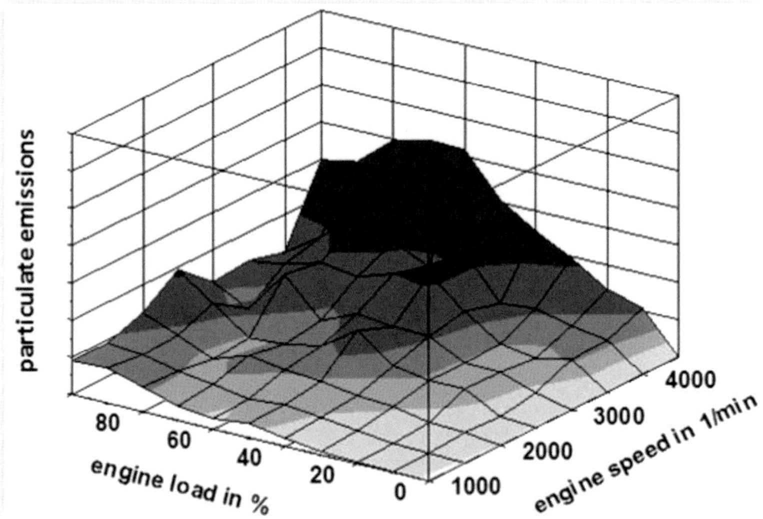

fuel quantity, and the engine temperature, for example, for cold-start emissions. Another example for a simplified approach was provided in reference [20], where the engine-out particle concentration has been described by an engine load—engine speed (rpm) map. Cold start and low engine temperatures were considered by corrections related to the coolant temperature. The resulting particle emissions map is shown in **Figure 8.12**.

To obtain the fraction of the PM that is actually collected within the filter, information about the filtration efficiency is needed. For this purpose, either simplified models, as those presented in Chapter 7, or simplified assumptions can be applied, for example, a fixed average value, increasing with mileage to reflect the enhancement in filtration with ash. The rate of soot oxidation, as a second part of the mass balance, can be assessed by using kinetic rate expressions as those described in Section 7.5.4. The input to these expressions are the oxygen concentration, obtained with the lambda sensor present in every gasoline application, and information about the filter or exhaust gas temperature. The latter value can be obtained either by using a temperature sensor or be based on a model value. Finally, the rate of reaction is proportional to the soot mass, which is calculated. The determination of the oxidation rates can be done either by direct calculation or map based. An example of the combination and implementation of a soot mass balance model, as it has been described in [20], is shown in **Figure 8.13**.

In addition to the open-loop-based soot mass models discussed above, sensors can be used to supplement the soot detection in GPFs [19]. The accumulated soot is changing the flow restriction, which is defined as pressure drop divided by the volumetric flow. This value could be correlated to the specific soot load in the filter and be used to provide closed-loop control.

FIGURE 8.13 Simple GPF mass balance model (map based) [20].

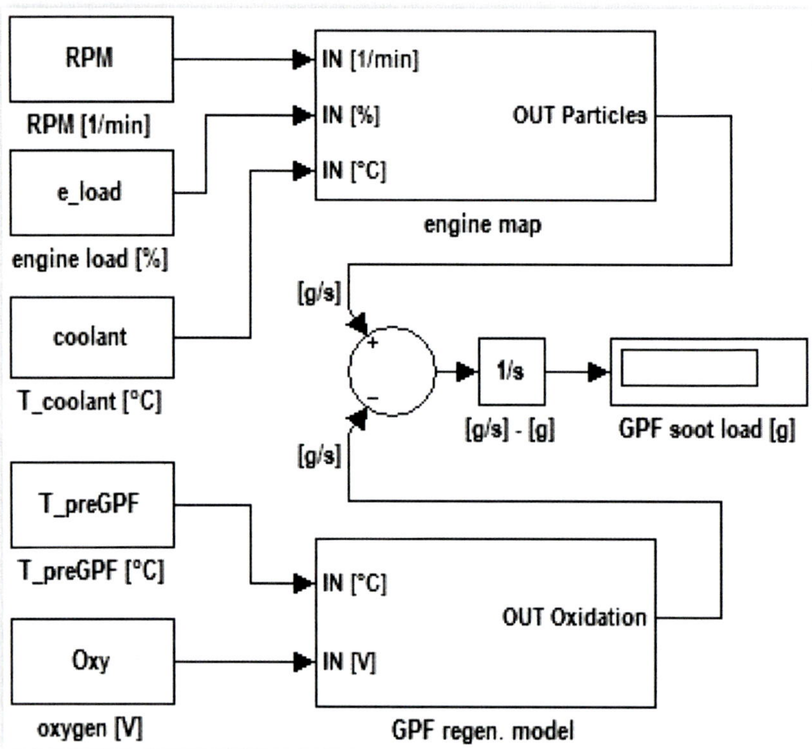

8.4.2 Examples of GPFs Operated in Purely Passive Applications

To illustrate the evolution of soot loading in GPFs, a number of examples will be discussed in this section. The examples are based on vehicle tests performed with EU5 and EU6 vehicles, which were retrofitted with a GPF. No functionality for active filter regeneration has been added. The operation was purely passive, meaning soot oxidation occurred only as a result of the "natural" conditions present during the drive profile. As can be expected, actual customer behavior will be one very important parameter influencing the soot accumulation and oxidation rates. This can be reflected by a variety of drive cycles with different maximum and average speeds, as well as different dynamics related to load changes. The speed variation can cover different regional (e.g., pure urban usage) and different commuter patterns. Possible scenarios are of course infinite, but the approach described in this chapter can help to understand the system limitations. Other important aspects would be the environmental and geographical boundary conditions.

The potential of passive regeneration has been evaluated in a number of vehicle experiments. Examples from these experiments are shown in **Figure 8.14**, with data from three different vehicles and tests. The vehicles differed in weight-to-power ratio, as well as their engine-out particulate raw emissions, increasing from vehicle I (left)

FIGURE 8.14 Vehicle testing in an urban drive pattern with GPF loaded with soot prior to the test. Filter location was in underfloor position. Examples represent three different vehicle applications with different weight-to-power ratios and different engine-out raw emissions.

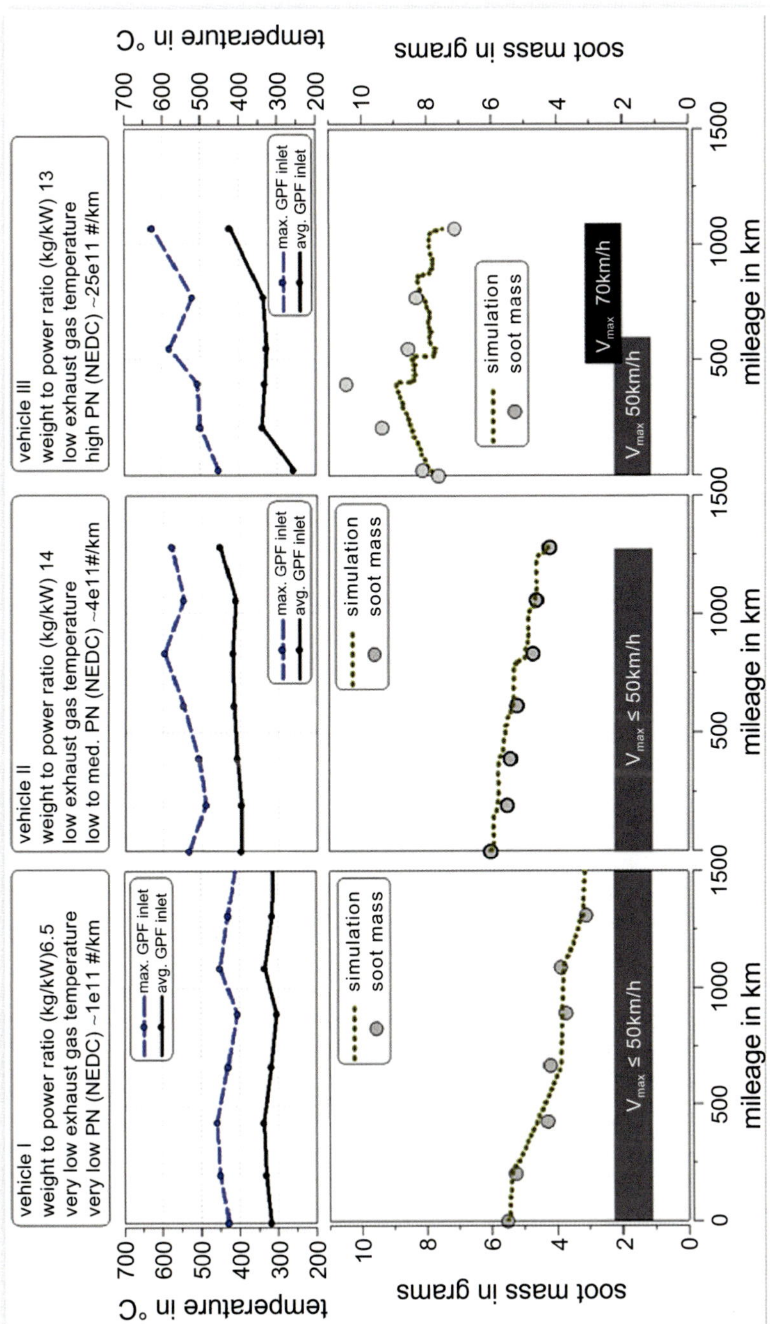

to vehicle III (right). The vehicles were operated on public roads in a purely urban drive pattern with a maximum velocity of 50 km/h. The mileage per day was typically around 250 km, performed in a two-shift operation. The data shown in Figure 8.14 therefore represent about 70-90 h of driving for each vehicle. The vehicles had been instrumented with data acquisition systems, recording values available from the ECU via the OBD interface as well as from additional sensors that have been added (see reference [20] for details). All relevant data was recorded in 1 Hz frequency.

The top diagrams in Figure 8.14 show the maximum and average temperature of the exhaust gas just upstream of the GPF (note that shown are the daily maximum and average values, not the actual recorded second-by-second data). Due to the low load drive profile and the UF location, the average temperature is very low for all three vehicles and in the range of 300-400°C. The maximum temperature, which could represent just short excursions during a few seconds, is higher. However, the maximum temperature for vehicle I ranges between 400°C and 460°C and for vehicles II and III mostly between 500°C and 550°C. In all cases, the temperature is at a level where one might assume that no soot oxidation will occur, since the general rule of thumb from DPFs has been that a minimum of 550°C-600°C is needed to regenerate a particulate filter.

The lower diagrams in Figure 8.14 show the soot load in the GPF throughout the test sequence. The accumulation of an appreciable mass of soot in the GPF has been found very challenging on our test vehicles and under the conditions present (traffic, climate, fuel, etc.). Therefore, the GPFs had been preloaded ex situ with 5-8 g of high elemental carbon content soot (Printex U) prior to the test and were then installed onto the test vehicles. In the examples shown, the soot mass starts with a relatively high load at 0 km. The soot mass change of each GPF was evaluated on a daily basis, with the weight measured one or two times per day (for the weighing, the filters were removed from the vehicles). These measured soot loads are shown as gray circles in Figure 8.14. It is interesting to note that for vehicles I and II, the soot mass continuously declines, despite the low exhaust temperatures. For vehicle III, we observed an initial increase in soot mass until the drive profile was changed to also allow for a higher load and speed up to 70 km/h. After this change, the soot mass decreases, or at least appears to reach a stable value. Also shown in Figure 8.14 is soot mass data obtained by simulation (dashed lines) based on the approach discussed above, for example, in Figure 8.13. The engine-out raw emissions were simulated by a map-based approach as described above in Figure 8.12, with the recorded transient values for engine speed and load as input. For the simulation of the soot oxidation, an advanced 1D filter model was used, with soot oxidation kinetics as described in [20, 21]. Input to the filter and soot oxidation simulations were the transient exhaust gas temperature, the mass flow, and the oxygen concentration at the inlet to the GPF recorded on the vehicles. In general, the simulations are capable of following and describing the experimental soot mass. The decrease for vehicles I and II is predicted, as well as the general behavior for vehicle III. For the latter, the initial steep increase in soot mass is underpredicted, likely due to the simplicity and insufficient accuracy of the engine-out model. The fact that the simulations can follow the unexpected experimental trends, with soot being passively regenerated at very low temperatures, confirms that just looking at the temperature is insufficient. Rather, one has to consider the entire mass balance and the timescale. Vehicle III shows an initial increase in soot mass since its raw emissions are much higher than for the other vehicles, about 25 times that of vehicle I. The soot that

FIGURE 8.15 An example of vehicle operation in an extended city mode with maximum velocity of <70 km/h. Data for oxygen, vehicle velocity, and GPF inlet temperature were obtained on a test vehicle, the soot mass was obtained in 1D simulations using vehicle data as boundary conditions.

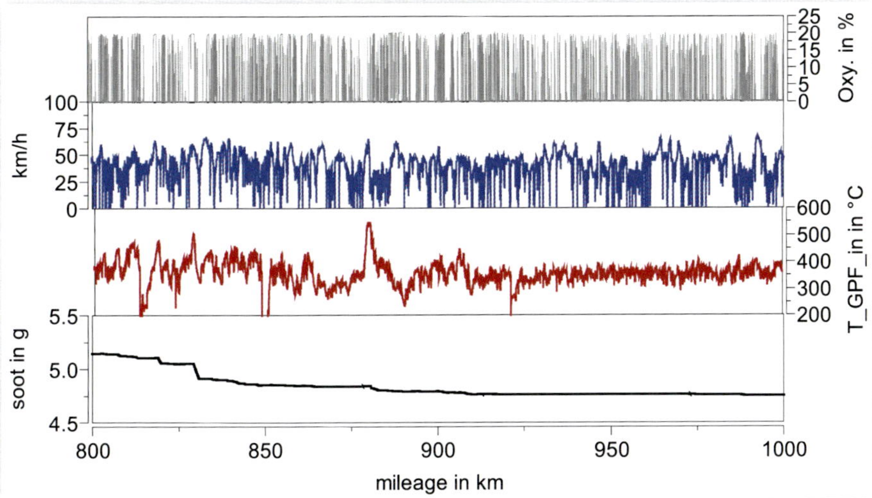

can be regenerated at the conditions present is insufficient to balance the high level of soot emitted by the engine. For vehicles I and II, the rate at which soot is oxidized exceeds the amount of soot added from the engine. To rationalize that passive regeneration works at the very low temperatures, one has to keep in mind that the timescale of the experiment is days (70-90 h), whereas the rule of thumb for DPF regenerations are based on regenerations performed within 10-20 min and at 25-50 times higher engine-out soot mass emissions.

An example to illustrate the passive soot regeneration in gasoline applications in some more detail is shown in Figure 8.15.

The data for oxygen concentration, vehicle velocity, and exhaust gas temperature were obtained on a test vehicle during city driving with an extended maximum velocity range up to ≤70 km/h. The extremely transient nature of city driving, with frequent velocity changes and resulting fuel cuts, accompanied with oxygen spikes, is apparent. The soot load trace shown on the bottom diagram was obtained using a detailed 1D simulation with the vehicle data as boundary conditions. As can be seen, the soot mass steadily decreases over the period plotted. Looking at the soot mass trace in more detail shows that there are phases with no change or very minor changes, and then there are phases during which a step change can be observed. These steps can be linked to short excursions in temperature in combination with oxygen pulses from the fuel cuts. Hence, the observed passive soot regeneration is the accumulation of a large number of short events during which some soot is oxidized.

The examples discussed above have all been obtained with GPFs located in an UF location. The overall effectiveness of passive soot regeneration can be enhanced if the exhaust temperature is increased, which can be achieved by moving the filter into a CC position. In Figure 8.16, the results from a simulation study are shown, in

FIGURE 8.16 Modeling study on the impact of the filter position on passive soot oxidation for three different initial soot loads. Exhaust boundary conditions from a vehicle operated in city mode at very low maximum velocity of <30 km/h [20].

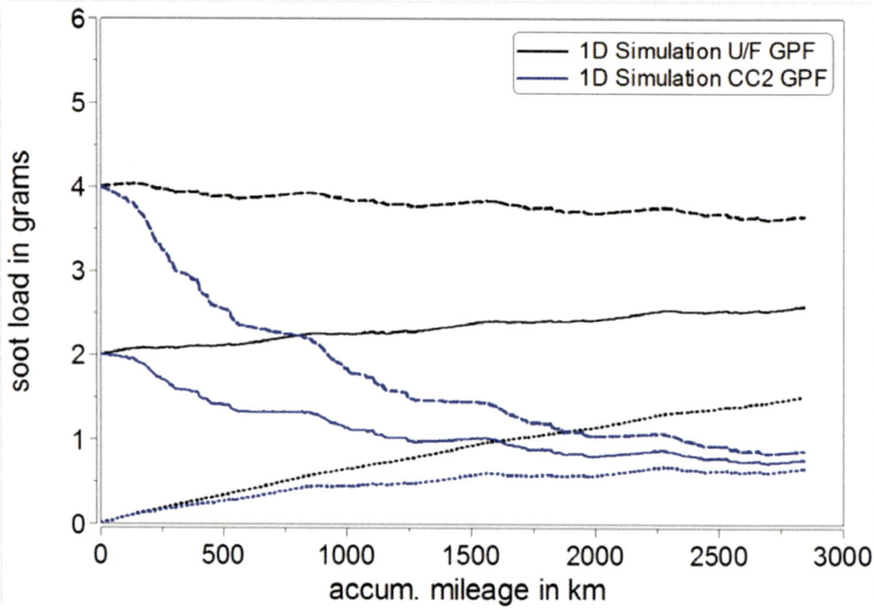

which the impact of the GPF location was evaluated (details are described in [20]). The boundary conditions were taken from data recorded on a vehicle operated in a very slow city driving mode with maximum velocity of <30 km/h and average velocity around 20 km/h. The simulations cover about 2800 km or roughly 140 h of driving in this profile. Since the soot oxidation rate is also a function of the soot mass, the simulations were performed for three initial soot loads of 0 g, 2 g, and 4 g. Blue lines show the change of soot mass for the GPF in CC2 position, and black lines represent the GPF being in the UF position.

For each of the examples with different initial soot loadings, the benefit in terms of passive oxidation rate of the CC location can be seen. Starting with the data for the initially clean filter, for example, with 0 g soot load at 0 km, both filters show an increase in soot load indicating that the initial rate of soot emitted by the engine exceeds the rate at which the soot is oxidized. This is due to the reaction rate being proportional to the quantity (or better surface area) of soot available for reaction, while the rate of soot coming from the engine is "constant." The slope for the filter in the UF position is steeper, due to the lower exhaust temperature and the temperature dependence of the reaction rate. For the very high initial soot load of 4 g, both locations show a steady decrease in soot load over the distance considered. This indicates that under the conditions applied, such a very high soot load would not be sustainable. The rate at which the soot is reduced, however, is again dependent on the location. In this case, the dependence of temperature dominates, and a large difference is observed for the two conditions. Finally, the condition observed in our experiments with a load of 2 g represents an intermediate condition. In this case, the soot load decreases for the CC conditions, meaning the temperature effect dominates over the effect of the soot mass on reaction rate. For the UF conditions, the model predicts that the soot mass would increase further.

What is interesting to observe is that both conditions seem to converge towards a "stable" soot load, roughly 0.6-0.8 g for the CC case and somewhere between 2.8 and 3 g for the UF case. This condition would represent a balance point, at which the rate of soot oxidation is balanced by the rate at which soot is provided from the engine. This is supported by our experience with filters from a large number of vehicle experiments, which in most cases showed some low level of soot in the order of 0.5-1 g upon receipt in our laboratories.

8.4.3 Examples of Severe Soot Regeneration

While the passive regeneration and decrease in soot load is generally desired, it also should be considered that it can result in extreme exothermic heat release if the reaction is fast and a significant amount of soot is present. This leads to the second main objective of the soot management, protection of the exhaust system from component failure. Extreme conditions with respect to thermomechanical exposure require a combination of conditions to be present at the same time. Two examples of conditions leading to high exothermal reactions inside a GPF are shown in **Figure 8.17**. The diagrams describe examples for a vehicle experiment on real road (left) and an engine bench experiment (right). The vehicle experiments and observations were used to develop the engine bench test procedures,

FIGURE 8.17 Example of deceleration fuel cut-off testing on vehicle (left) and engine bench (right) with GPFs loaded with a large soot mass (2.5-3 g/L).

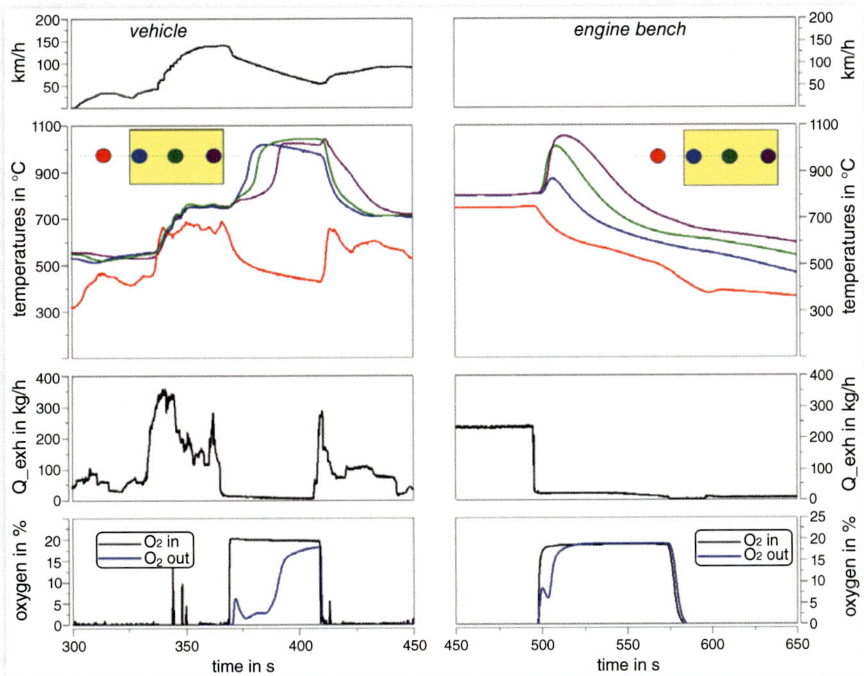

allowing for an easier and better means to test under repeatable conditions. The target of these tests is combining high inlet temperature (requiring high engine load) with a defined amount of soot in the GPF and sufficient supply of oxygen (duration of fuel cut-off). For vehicle testing, this can be achieved by selecting the appropriate route where the vehicle could be accelerated to a high speed, followed by a subsequent deceleration fuel cut-off for several seconds. In the example shown in Figure 8.17 (left side), the GPF temperature at the time of the fuel cut-off was 750°C. During the fuel cut-off, the oxygen concentration at the GPF inlet increases, and as a result of the soot oxidation, the temperature inside the GPF sharply rises. The amount of oxygen that gets consumed by the soot oxidation is seen by the difference between the inlet and outlet oxygen concentration level. The intake-air throttle closes during deceleration, which reduces the exhaust flow rate, limiting the convective heat removal, but also the supply in oxygen.

The high effort to install preloaded GPFs into test vehicles and the difficulty to obtain repeatable conditions motivates the transfer of these tests to an engine bench. Here the desired exhaust gas temperature can be chosen within the entire engine map, the GPF could be preloaded in situ or ex situ of the exhaust system, and the fuel cut-off condition can be repeatedly executed and maintained by motoring (dragging) the engine. An example of such a procedure is shown in the right-hand diagrams of Figure 8.17, performed with a different engine than the one used for the vehicle. For the vehicle, we can observe the severe heat release when the filter with a high soot load is operated at high temperature and then is exposed to oxygen during a fuel cut.

The level of the temperature increase is a function of the soot load (the "fuel" available in the filter) and the inlet temperature level, determining the rate of reaction [21]. In **Figure 8.18**, examples from an engine bench test program with a coated GPF in 300/8 geometry are shown.

FIGURE 8.18 Maximum filter temperature inside the GPF as function of the initial soot load and GPF inlet temperature during fuel cut testing on engine bench.

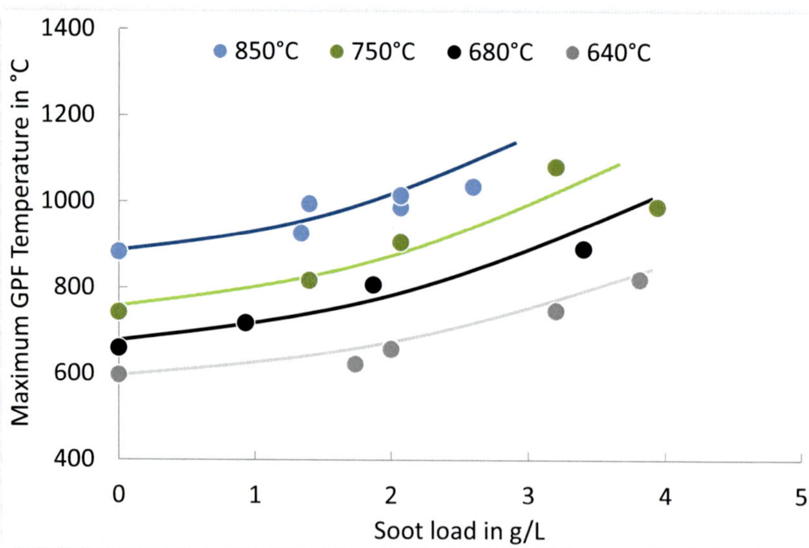

Plotted are the maximum filter temperatures measured inside the GPF, under fuel cut conditions, with different initial soot loads and temperatures. The increase in maximum temperature with soot load, as well as initial temperature prior to the fuel cut is apparent. The base temperature level is set by the temperature prior to the fuel cut, while the increase in temperature with soot load is similar for the four series with different initial temperatures.

The objective of soot management is to identify soot load conditions, which can lead to severe temperatures, and if present, initiate protective measures. The imits for the filter component depend on the technology and the type of exposure. This will be discussed in the next section.

8.5 Thermal Robustness Requirements in Vehicle Applications

In the application, the filter is exposed to different conditions which can result in thermal stress, as well as high temperature exposure. In general, we can separate between the thermal stress applied due to transient changes in the inlet exhaust gas temperature, e.g., the boundary conditions, and stress that results from exothermic heat release due to local soot oxidation or other reactions within the filter. Both conditions will be discussed separately in the following sections.

8.5.1 Thermal Stress Due to Transient Changes in the Boundary Conditions

There are two primary conditions that result in significant transient changes in the boundary conditions at the inlet into the filter. Both are identical to what is well known for TWCs. Changes in engine load due to accelerations or decelerations result in changes in exhaust gas flow rate and temperature. The temporary temperature gradient dT/dt which the filter experiences is higher the closer the component is located to the engine. The actual values for the exhaust gas temperature and gradients vary based on engine technology, specific power, and applicable temperature limits from other engine or after treatment components, for example, a maximum allowed turbine or catalyst temperature. The second condition resulting in significant transient temperature gradients is driven by the need for fast catalyst light-off. To minimize the time after a cold start, until the catalyst reaches its light-off temperature, heating measures are applied within the engine controls, resulting in a rapid increase in exhaust gas temperature [22]. In **Figure 8.19**, the maximum transient temperature change dT/dt and maximum temperatures at the GPF inlet are shown for seven different test vehicles operated within our fleet using our standard thermocouple pattern.

The values shown were recorded during New European Driving Cycle (NEDC) testing (filled symbols) and during normal on-road testing in a mixed drive cycle including full load (open symbols). The most aggressive gradient in the NEDC

FIGURE 8.19 Maximum transient temperature change *dT/dt* and maximum exhaust gas temperature in front of the GPF during New European Driving Cycle (NEDC) testing (filled symbols) and normal on-road driving (open symbols). The legend indicates the engine type and the location of the GPF.

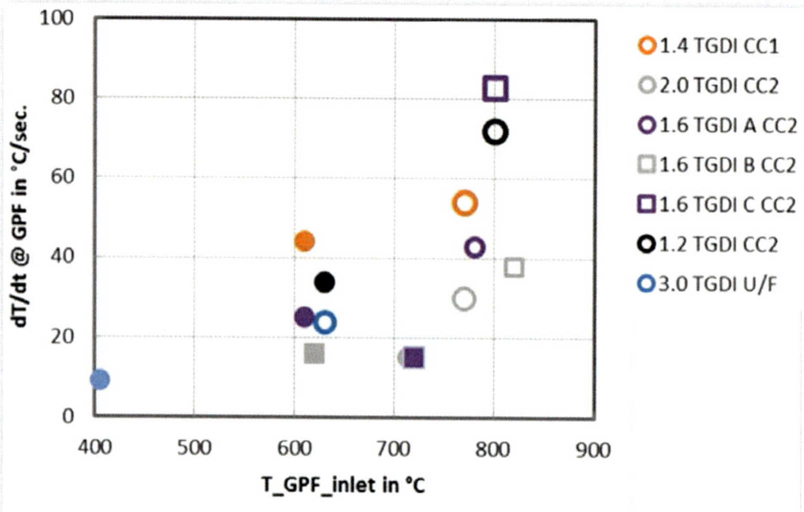

test cycles was observed during the initial catalyst heating phase, with the highest value for the test with the GPF located in the CC1 position. Observed in this case during catalyst heating mode, the maximum heating rate resulted in a transient gradient of 44°C/s. For the CC2 position and the NEDC testing, the maximum temporal gradients were 15 to 34°C/s. Although not shown, it is worth mentioning that we had observed somewhat higher maximum transient gradients during the catalyst heating mode, for applications with naturally aspirated engines. During on-road driving on public roads in Germany, including accelerations at full load, the maximum gradient found for the examples shown in Figure 8.19 were as high as 83°C/s. Measured maximum exhaust gas temperatures in front of the GPF were in the 800°C range.

All of the discussed conditions above are identical to those when the TWC is located in the respective location. Therefore, the typical validation tests, such as natural gas burner test rigs for cyclic exposure, can be carried over for GPF validation. An example for a cyclic product validation test on the gas burner, which we have used in some development activities, is given in **Figure 8.20** [23]. The exhaust gas temperature is repeatedly ramped between 60°C and 960°C, over a predefined number of cycles. The ramp rates are up to 120°C/s measured at the front face inside the GPF, exceeding the values shown in Figure 8.19 for some test vehicles. In these tests, it is important to note that the primary challenge is not the maximum temperature, but rather the difference between the highest and lowest temperature level since this is what causes the thermal stress within the part. A cycle between 50°C and 800°C can be comparable to one between 350°C and 1100°C.

FIGURE 8.20 An example of a cyclic thermal stress test of a GPF performed on a natural gas burner [23]. The different lines represent temperatures recorded at different radial and axial locations within the part tested.

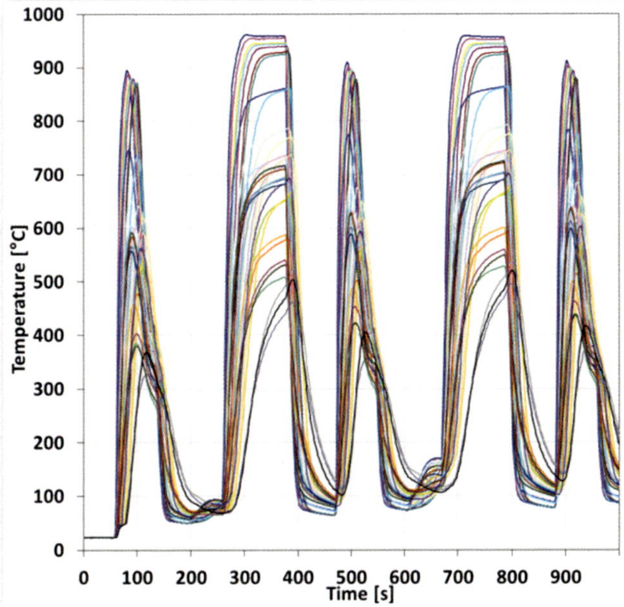

8.5.2 Thermal Stress Due to Severe Soot Oxidation

Further conditions that can result in significant thermomechanical stress within a filter are severe soot oxidations, such as those described above in Figures 8.17 and 8.18. These conditions are unique to filters and no reference exists from TWC experience. They are, however, similar to what is known for DPFs. The soot oxidation usually does not occur uniformly throughout the part. The combustion zone typically moves in the axial direction through the part. It also reaches the colder skin regions later. This results in a highly non-uniform temperature field within the filter, which constantly changes until the oxidation is complete. In **Figure 8.21**, an example of the spatial temperature distribution during a severe fuel cut soot oxidation is shown for one time step, during which high stresses were observed within the part. The maximum temperature in this example was slightly above 1000°C and occurred in the center back of the filter. The radial and axial temperature gradients result in axial, radial, and hoop stresses.

Since the material properties for porous cordierite filters are highly complex, the most reliable method to assess the material robustness is still based on extensive testing, primarily on the engine, as described above. The testing for robustness against soot oxidations can usually be differentiated into survivability tests, during which the filter is exposed to increasing soot and thermal loads, until its filtration function is compromised, and testing for durability, during which the filter is exposed to a representative or a repeated number of severe soot oxidations.

FIGURE 8.21 Spatial temperature distribution within a GPF during a fuel cut test with severe soot oxidation. The data are for a time step with maximum stress level for this specific test. The red color represents the hottest temperatures, and blue represents the coolest temperatures.

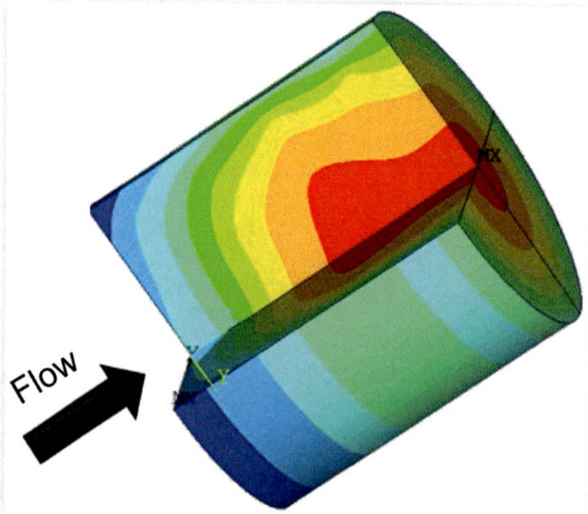

8.5.2.1 Testing for Filter Survivability Under Severe Soot Oxidation Events

The severe soot oxidation during a fuel cut with a soot-loaded GPF has been identified as a potential worst-case scenario that the filter has to survive. In the survivability testing, we expose the filter to increasing soot loads, resulting in increased peak temperatures and thermomechanical stress, similar to what has been shown in Figure 8.18. After each soot oxidation event, the filter is completely regenerated and then tested for its filtration performance using a full-load filtration test. The increase in soot load (i.e., peak exposure temperature) is performed until there is a significant change in filtration efficiency. Since the stress is not related to the soot load, but only the resulting temperature and gradient, it is most instructive to evaluate the temperature and/or temperature gradient until filter failure. In Figure 8.22 data is shown, obtained on a very large number of commercialized Corning® DuraTrap® GC 200/8 (N = 27) and DuraTrap GC HP 300/8 (N = 20) samples, the former uncoated and the latter uncoated and coated with different TWC technologies. The filter sizes ranged from slightly above 100 mm in diameter to almost 170 mm and filter volumes between 1 L and 3.2 L. In these specific tests, no change in filtration efficiency was observed up to a maximum temperature of 1200°C for all 47 samples tested. Above 1200°C, the probability for a loss in filtration increased with temperature. No difference was observed between coated or uncoated samples or the two filter technologies. While the reduction in filtration for these extreme tests was observed only for temperatures in excess of 1200°C, it is important to note that these filters experienced only a small number of tests (about 4-8) with limited hours on the engine, until the test ended, due to the filtration decrease. Therefore, a maximum temperature of 1100°C is recommended for applications, in line with the typical limit for TWCs.

FIGURE 8.22 Change in filtration efficiency (FE) as a function of the maximum temperature observed inside the GPF during a severe fuel cut soot oxidation test [6]. Different symbols represent different filter technology. Different shades of green represent the different individual parts tested. Red symbols indicate conditions where a significant drop in filtration efficiency is observed. Total number of parts shown N = 47.

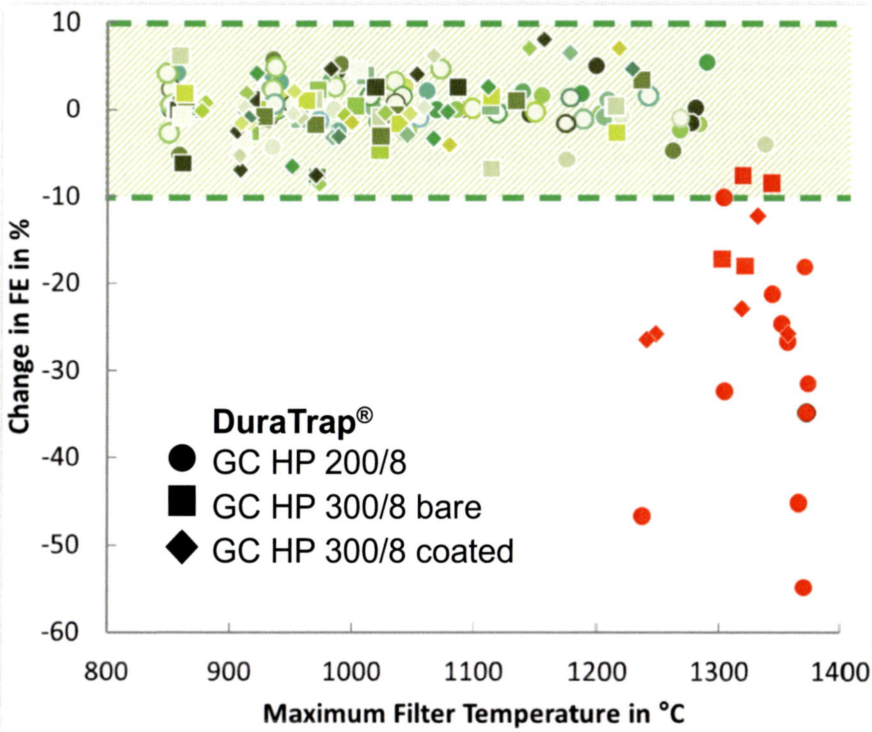

8.5.2.2 Testing for Filter Robustness to Repeated Severe Soot Oxidations

Because GPFs are a new technology, very little is known about the number of the severe soot oxidation events a filter will experience over its useful life. As the engine-out soot generation conditions are typically much lower than for diesel applications, and in many applications passive regenerations can be effective, the number of severe oxidation events is assumed to be lower than DPF applications. In CC applications, where passive regenerations will be even more effective, the probability for a severe event at a high soot load will be even lower. For UF systems, where the risk is higher, protection strategies are developed [19]. In our test program filters, we tested 5 to 20 severe regeneration events with a maximum temperature target of 1100°C, for example, the recommended maximum temperature level. In the test programs, the filters were loaded with soot, then a fuel cut test was performed at high initial temperature (about 850°C), followed by a full clean-out regeneration, and a filtration test at high engine speed and load. Then the test cycle was repeated until the required number of tests was completed. In Figure 8.23, an example for a filter tested over 11 cycles is shown.

FIGURE 8.23 Example of a severe fuel cut durability test over 11 cycles. Testing was done to achieve a T_{max} of about 1100°C. Clean filtration efficiency (FE) was monitored after each fuel cut cycle.

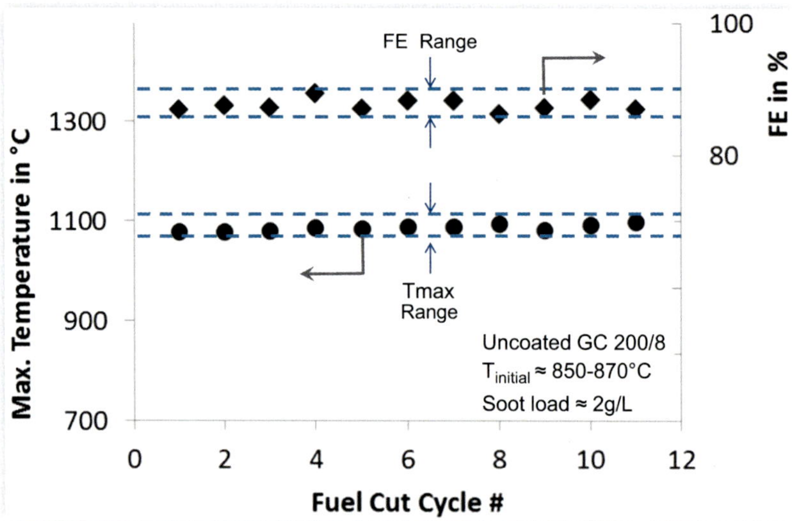

FIGURE 8.24 Robustness testing to repeated severe fuel cut soot oxidations with a target maximum temperature of 1100°C on commercialized Corning DuraTrap® filter technologies [6].

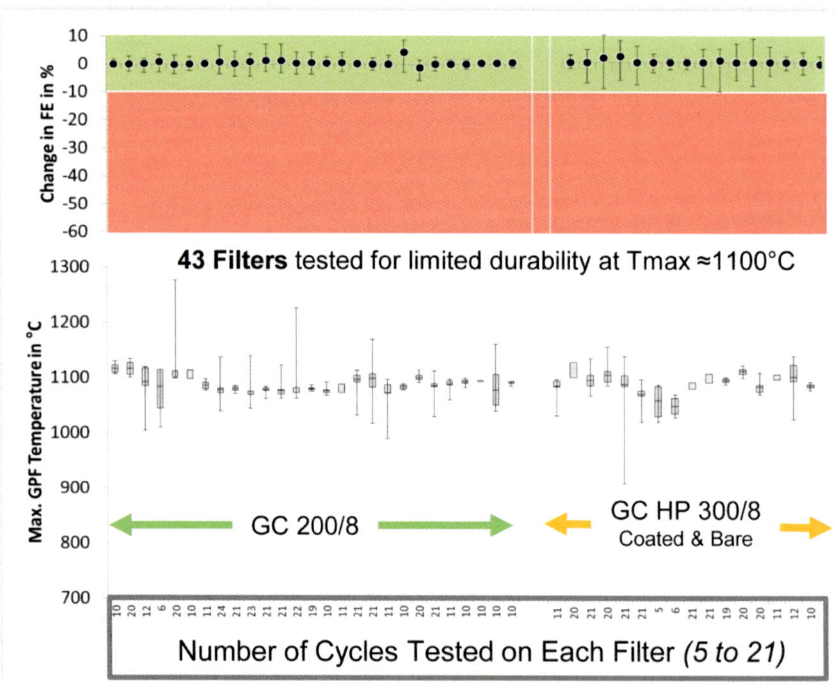

Plotted is the maximum temperature achieved in every cycle together with the filtration efficiency determined just afterwards. As shown, the filtration efficiency remains constant throughout the cycles tested. Since GPF is a new technology, this type of durability testing has been performed on a large number of filters. Examples for this are shown in **Figure 8.24**, where the test results are summarized for a total of 43 filters (Corning® DuraTrap® GC 200/8 and DuraTrap® GC HP 300/8, coated and bare) of different sizes, demonstrating robustness, and within the recommended range of maximum temperatures around 1100°C [6]. Shown in Figure 8.24 are the number of cycles tested for each filter (at the bottom); the maximum temperature range during the cycles (middle), with the error bars showing individual extreme conditions (e.g., one cycle where the target was not met); and the range for the filtration efficiency values for each of the cycles (top). For the filtration efficiency, the criterion described in the previous section, change in filtration efficiency from the original, was used to differentiate test noise from real change in filter performance.

8.6 Summary and Conclusions

GPFs are a new technology and there is still much to learn about their application and optimization. In this chapter, we tried to provide an overview of some of the experiments we have conducted over the course of the last few years, working on the development and implementation of GPF. This new class of particulate filters offer a robust solution that cannot be bypassed, enabling the effective reduction in particulate emissions. Unless the filter gets damaged, filtration typically improves with vehicle operation, due to the formation of an ash layer, acting itself as a filtration medium. Therefore, the emission focus is primarily on the fresh filtration efficiency, which itself is a strong function of the filter, its sizing and placement, as well as the application conditions. While ash helps with filtration, it comes at the price of an increased pressure drop over the useful life of the filter. While the in-service experience is still growing, the results available to date suggest that the filter may accumulate typically about 0.18 ± 0.1 mg/km ash, with one case having had up to 0.4 mg/km of ash. The resulting ash deposition is different than the diesel application, as the majority of the ash is expected to accumulate as a layer onto the channel walls, not primarily in the rear of the filter. The accumulation of ash and its impact on pressure drop has to be considered during filter sizing, providing a secondary design boundary, in addition to the sizing for filtration. To protect the filter and control the increase in component back pressure, the accumulation of soot has to be managed. This requires means to monitor the soot load state and strategies to force soot regeneration, if needed, as well as strategies to protect the filter in case of critical conditions, for example, high load combined with high temperatures and a fuel cut. This task is analogous to DPFs, although the actual approaches taken are different. Under most conditions, an effective passive soot oxidation is possible, but exceptions exist, especially if the engine-out raw emissions are high. The filter can experience significant thermal stress in service as a result of transient changes in exhaust gas temperature or as a result of severe soot oxidations. While thermal stresses are anticipated to be analogous to TWCs today and can be tested using similar methods, the latter condition of severe soot oxidation is new. Examples of methods to test for robustness for severe soot oxidations have been described, with results demonstrating the robustness of commercialized GPF technologies to withstand such thermal exposures.

References

1. Görgen, M., Nijs, M., Lehn, H., Scharf, J. et al., "Current and Future Trends of Gasoline Particulate Filter Technologies, Calibration Strategies and Aging Methods," *Proceedings of 26th Aachen Colloquium Automobile and Engine Technology*, Germany, 2017, 1001-1036.

2. Commission Regulation (EC) No 692/2008.

3. Commission Regulation (EU) 2017/1151 (WLTP).

4. China 6, "Limits and Measurement Methods for Emission from Light-Duty Vehicles," GB18352.6-2016.

5. RDE Act 1, Commission Regulations (EU) 2016/427, RDE Act 2, Commission Regulation (EU) 2016/646, RDE Act 3, Commission Regulation (EU) 2017/1154.

6. Shimizu, M., Park, J., He, S., Lv, J. et al., "Performance Evolution of Gasoline Particulate Filters over Extended Vehicle and Engine Bench Operation," *17th Hyundai-Kia International Powertrain Conference*, South Korea, October 24-25, 2017.

7. Rose, D., Coulet, B., Nicolin, P., Boger, T. et al., "Field-Study and Durability Evaluations on GDI Vehicles Equipped with Various Gasoline Particulate Filter (GPF) Concepts," *Proceedings of 25th Aachen Colloquium Automobile and Engine Technology*, Germany, 2016, 1305-1325.

8. Boger, T., Rose, D., Nicolin, P., and Coulet, B., "Field Experience with Gasoline Particulate Filter Equipped GDI Vehicles," *MTZ Worldwide* (January 2017): 28-34.

9. Rubino, L., Piotr Oles, J., and La Rocca, A., "Evaluating Performance of Uncoated GPF in Real World Driving Using Experimental Results and CFD Modelling," SAE Technical Paper 2017-24-0128, 2017, doi:10.4271/2017-24-0128.

10. Umicore Presentation, "A Study on Gasoline Particulate Filter Systems (GPF) for Upcoming Emission Legislation," *14th Hyundai-Kia International Powertrain Conference*, South Korea, October 28-29, 2014.

11. Ito, Y., Shimoda, T., Aoki, T., Yuuki, K. et al., "Next Generation of Ceramic Wall Flow Gasoline Particulate Filter with Integrated Three Way Catalyst," SAE Technical Paper 2015-01-1073, 2015, doi:10.4271/2015-01-1073.

12. Lambert, C., Bumbaroska, M., Dobson, D., Hangas, J. et al., "Analysis of High Mileage Gasoline Exhaust Particle Filters," *SAE Int. J. Engines* 9, no. 2 (2016): 1296-1304, doi:10.4271/2016-01-0941.

13. Custer, N., Kamp, C., Sappok, A., Pakko, J. et al., "Lubricant-Derived Ash Impact on Gasoline Particulate Filter Performance," *SAE Int. J. Engines* 9, no. 3 (2016): 1604-1614, doi:10.4271/2016-01-0942.

14. Shao, H., Wang, Y., Yang, F., He, S. et al., "Effects of Lubricant Properties on GPF Performance," *SAE International Powertrains, Fuels & Lubricants Meeting*, Beijing, China, October 16-19, 2017.

15. Bernardoff, R., Hennebert, B., Delvigne, T., Courtois, O. et al., "A Study of Ash Accumulation in the After-Treatment System of a Gasoline Direct Injection Engine Equipped with a Gasoline Particulate Filter," SAE Technical Paper 2017-01-0879, 2017, doi:10.4271/2017-01-0879.

16. Rizvi, S.Q.A., *A Comprehensive Review of Lubricant Chemistry, Technology, Selection, and Design*, (West Conshohocken: ASTM International, 2009), ISBN 978-0-8031-7000-1.

17. Heibel, A. and Bhargava, R., "Advanced Diesel Particulate Filter Design for Lifetime Pressure Drop Solution in Light Duty Applications," SAE Technical Paper 2007-01-0042, 2007, doi:10.4271/2007-01-0042.

18. Sappok, A., Govani, I., Kamp, C., Wang, Y. et al., "In-Situ Optical Analysis of Ash Formation and Transport in Diesel Particulate Filters during Active and Passive DPF Regeneration Processes," *SAE Int. J. Fuels Lubr.* 6, no. 2 (2013): 336-349, doi:10.4271/2013-01-0519.

19. Van Nieuwstadt, M. and Ulrey, J., "Control Strategies for Gasoline Particulate Filters," SAE Technical Paper 2017-01-0931, 2017, doi:10.4271/2017-01-0931.

20. Nicolin, P., Rose, D., Kunath, F., and Boger, T., "Modeling of the Soot Oxidation in Gasoline Particulate Filters," SAE Technical Paper 2015-01-1048, 2015, doi:10.4271/2015-01-1048.

21. Boger, T., Rose, D., Nicolin, P., Gunasekaran, N. et al., "Oxidation of Soot (Printex® U) in Particulate Filters Operated on Gasoline Engines," *Emission Control Science and Technology* 1 (2015): 49-63.

22. Zhao, F., *Technologies for Near-Zero-Emission Gasoline-Powered Vehicles*, (Warrendale: SAE International, 2007), ISBN 9780768014617.

23. Cutler, W.A., "Gasoline Particulate Filter (GPF) Performance and Durability," *SAE International Powertrains, Fuels & Lubricants Meeting*, Beijing, China, October 16-19, 2017.

About the Authors

Dominik Rose is currently serving Corning GmbH as manager of systems engineering in Europe. His team is in charge of several engineering activities related to testing and evaluation of Corning products in the appropriate system environment. These activities include vehicle testing, engine bench testing, numerical simulations, and development of surrogate test methods. He joined Corning in 2004 and holds a diploma in mechanical engineering from Frankfurt's University of Applied Sciences.

Dr. Thorsten Boger currently serves as commercial technology director for light-duty products at Corning Environmental Technologies (CET). He is responsible for leading the interface between customer and technology for global Corning customers of gasoline and diesel technologies as applied to passenger cars and chassis-dynamometer-certified light commercial vehicles. In addition, he serves as the regional technology director for Europe, CET. Boger joined Corning in 1997, working in different roles focused on new technologies for mobile emission control as well as industrial applications in the chemical and refining industry. He graduated with a degree in process engineering (Dipl.-Ing.) and received his doctorate (Dr.-Ing.) from the University of Stuttgart.

CHAPTER 9

Three-Way Catalyst Integration into Particulate Filters

Patrick Burk and Thomas Schmitz
BASF

9.1 Introduction

The continued worldwide tightening of automotive emission standards will require ever better performing pollution abatement systems. Incorporation of three-way-conversion (TWC) catalyst functionality into a gasoline filter substrate brings additional emissions catalytic activity to the exhaust system while minimizing the system's footprint.

The European authorities have now introduced real-world driving emission (RDE) into the market [1]. This requirement is an additional test to check particle number and NOx emissions throughout the vehicle life on public roads under real-life conditions. Particle number and NO_x emissions are checked on public roads under real-life conditions using an on-board measurement system (PEMS). As shown in **Figure 9.1**, the long utilized NEDC cycle covers only a small portion of the engine map concerned with RDE emissions conformity. Also shown is the engine map used for the Hannover cycle [2], which takes the RDE requirements and translates them

FIGURE 9.1 A comparison of the operating points found in the NEDC and Hannover test cycles.

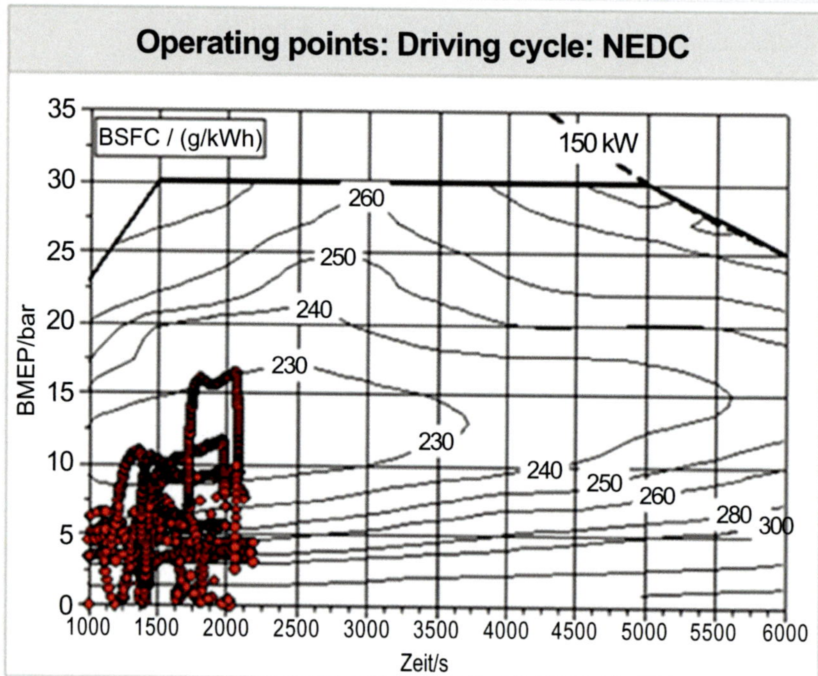

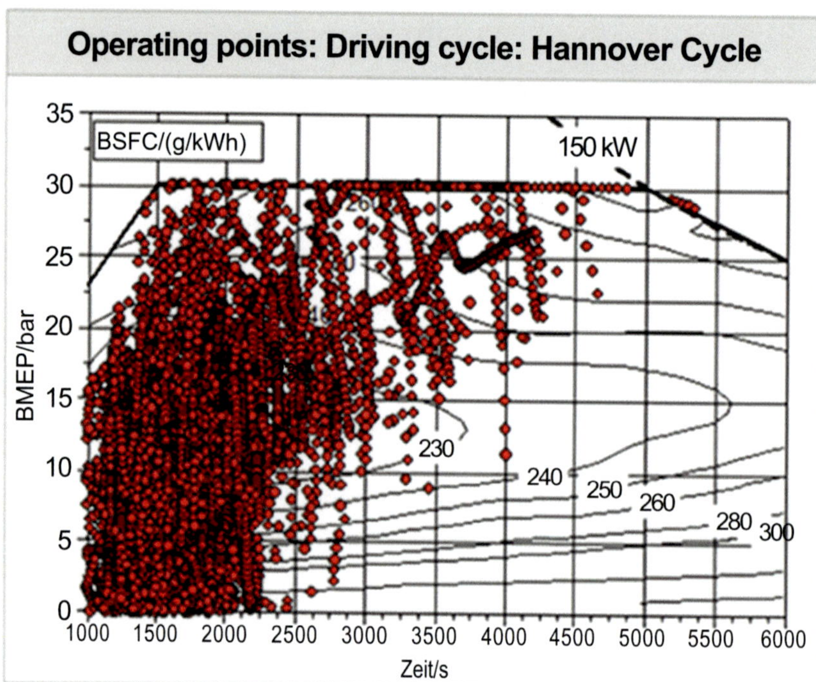

into an on-road driving cycle for a 2.0 L turbocharged gasoline direct injected engine (TGDI) vehicle. BASF utilizes this cycle in developing catalysts that will function over the widely varying conditions found in the RDE.

Major challenges occur during high exhaust volume flows and high NO_x emissions portions of the RDE cycle. Increased catalyst volume is one possible solution to reduce tailpipe emissions significantly under high-load conditions.

In China, in addition to the RDE conditions, the gas-phase standards for China 6b legislation are much tighter than those in Europe [3, 4]. To fulfill these quite stringent conditions, higher overall catalyst volume in the exhaust system will be required to reach emissions safety margins. For both China and Europe, the FWC™ can be an additional catalyst bed with gas-phase and particle filtration functionality.

With regard to the United States, the widespread use of catalyzed filters will be delayed until 2025 when the allowed particulate emission will decrease from 3 to 1 mg/mile over the U.S. FTP [5, 6].

In exhaust systems where filters are used to reduce the particle number and the particle mass emissions, integration of catalytic functionality on the filter substrate is a smart way to further reduce gas-phase emissions. This is already state-of-the-art technology in diesel exhaust gas purification systems, where catalyzed soot filters having either diesel oxidation catalyst (DOC) functionality to oxidize CO and hydrocarbon emissions or selective catalytic reduction (SCR) catalysis for NO_x reduction are in the market.

For gasoline applications, TWC catalyst functionality needs to be integrated on the filter. This combination of the TWC with the filter creates a new exhaust cleaning device, the four-way-conversion (FWC) catalyst. The FWC catalyst removes the gaseous pollutants hydrocarbons (HC), CO, NO_x, and solid particles, for example, soot particles from the exhaust gas at the same time.

The requirements for catalyzed filters include gas-phase activity and filtration efficiency at an acceptable backpressure. Achieving the right balance between these varying factors is related to the mass of the catalyst applied to the filter. High amounts of catalyst can maximize the catalytic activity while driving up backpressure and degrading fresh filtration efficiency. Therefore, it is important to develop FWC technologies which can be tailored to the specific requirements of the application.

A major challenge in the development of FWC catalysts is the integration of the different TWC functionalities on the filter. TWC catalysts are characterized by the reactions (1)–(3) enabling high conversion efficiencies of HC, CO, and NO_x:

$$2CO + 2NO \rightarrow 2CO_2 + N_2 \tag{1}$$

$$2CO + O_2 \rightarrow 2CO_2 \tag{2}$$

$$C_xH_{4x} + 2xO_2 \rightarrow xCO_2 + 2xH_2O \tag{3}$$

Another key characteristic of the TWC catalyst is its capability to store and release oxygen. This oxygen storage function is required so as to keep the reduction and oxidation reactions at peak efficiencies. In addition, the storage and release function of oxygen can be used to monitor the catalyst activity of the FWC as it is used in on-board diagnosis (OBD) of the TWC functionality.

All these properties are correlated with the amount of catalytic material that can be applied on the carrier substrate. However, the catalytic material currently applied on a Euro 6 "state-of-the-art" TWC catalyst cannot be simply transferred to a filter. The backpressure of such FWC would be by far too high.

The catalytic activity of a FWC technology as a function of the catalyst mass was investigated and compared to state-of-the-art TWC technology (**Figure 9.2**). The activity relative to HC and NO_x after full useful life engine bench aging shows a clear dependence on catalyst loading. The same performance of the TWC cannot be reached because of the limitation on the catalyst amount which can be used on the FWC. The catalyst amount is directly correlated to backpressure. The higher the amount of catalyst on the filter, the higher is the resulting backpressure.

Thus, there is a clear trade-off between backpressure and catalytic activity. Complicating the optimization further is the relationship between catalyst loading on the substrate and the filtration efficiency of the FWC. Therefore, the major task for the successful FWC development is the optimization of these trade-offs: achieve the highest possible catalytic activity and filtration efficiency at the lowest possible backpressure (see **Figure 9.3**).

Such a goal can be achieved by opportune modifications of the catalytic materials properties in the coating and optimization of the coating process. The backpressure of a FWC can be improved through the development of specific coating processes

FIGURE 9.2 Aging stability of FWC and TWC catalyst at equal catalyst size and precious metal loading.

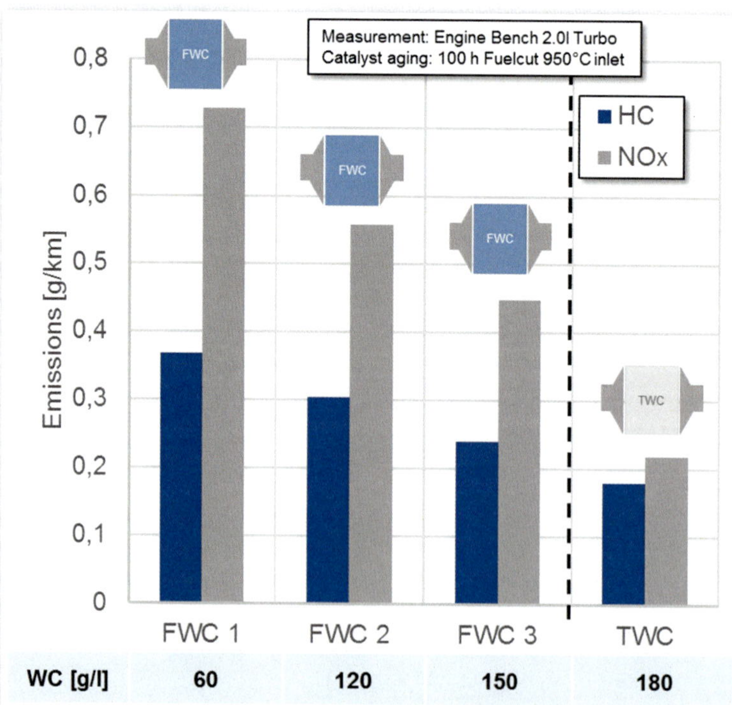

FIGURE 9.3 Trade-offs in development of catalyzed filters.

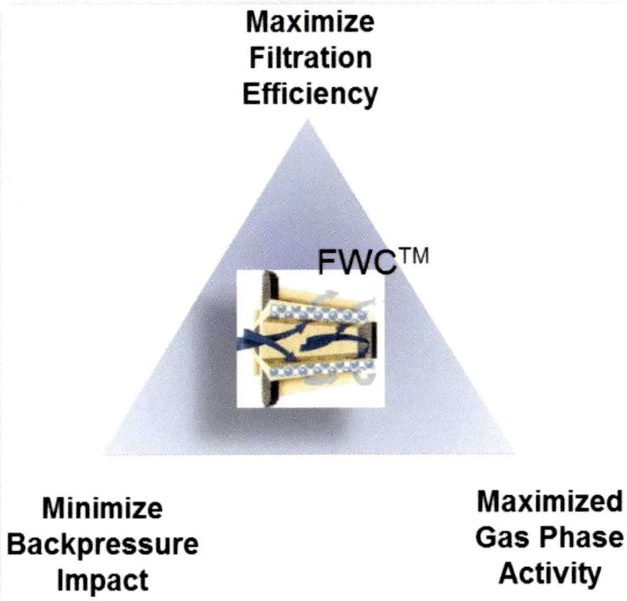

FIGURE 9.4 Illustration of catalyst slurry properties as an important feature for effective FWC coating.

achieving optimal distribution of the catalyst material in the filter. Figure 9.4 illustrates the properties of the catalyst slurry which is applied to the filter. Slurry properties in combination with the process which was developed specifically for coating of porous filter substrates ensure proper distribution of the catalyst.

Filling the pores of the substrate evenly is a method to ensure low backpressure of the FWC. Putting catalyst partially on the wall generates higher backpressure. Figure 9.5 shows a qualitative view on the effect of catalyst in and on the wall of the filter substrate. Proper distribution in the wall can give backpressure levels close to that of a state-of-the-art TWC. Allowing some catalyst to sit on the wall increases filtration efficiency but can result in undesirable backpressure increases.

Figure 9.6 shows an example of a FWC based on an on-wall coating concept in comparison to the in-wall coating approach and the resultant effect on backpressure. For both FWC catalysts, the same amount of catalyst was applied to the filter substrate. The substrate dimensions used were 118.4 mm in diameter, 127 mm in

FIGURE 9.5 Effect of catalyst location on backpressure and filtration efficiency.

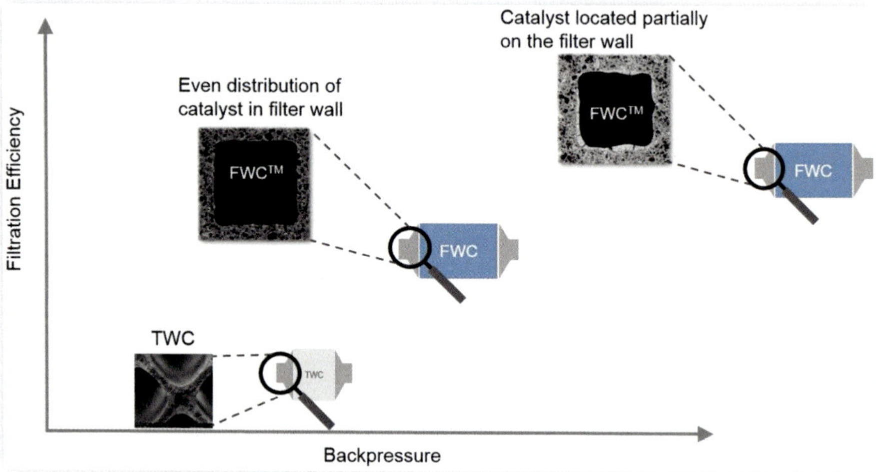

FIGURE 9.6 Backpressure measurement of a FWC with catalyst in the wall and catalyst partially on the substrate walls. Measurement on cold flow equipment with 700 m³/h air flow.

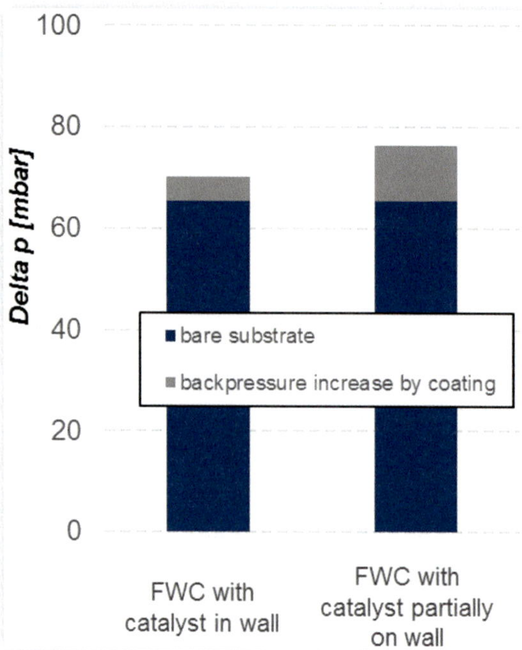

FIGURE 9.7 Effect of catalyst location in the FWC on the particle number emissions measured over the WLTC cycle.

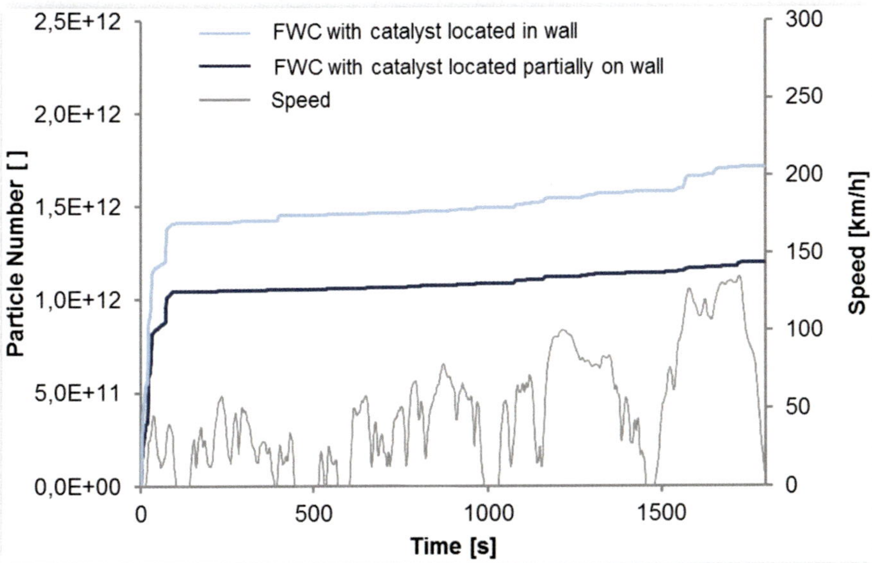

length, with a cell density of 300 cpsi and wall thickness of 8 mil. The porosity of the bare substrate is ~65% with a mean pore size of ~20 μm. In the case using the in-wall coating concept, the contribution of the catalyst to the total backpressure is only to 7%. However, by placing the catalyst on the wall, the contribution to the total backpressure more than doubles to 16%.

The effect of the catalytic coating location in the FWC on the filtration efficiency is shown in Figure 9.7. Here the FWC catalysts have been tested in a fresh state on a dynamic engine bench. The direct injected engine has a displacement of 2.0 L and is turbocharged. The test cycle used is the WLTC, run with a 20°C start temperature and standard market fuel. The effect of the location of the catalyst on the reduction of particle emissions can be clearly seen. The per km emissions value achieved with the FWC with catalyst partially on the wall is 30% lower than with the FWC where catalyst is in the wall.

Catalyst systems that include a FWC will be widely employed and will be found on vehicles with both small and large engines. Depending on the application, different design architectures will be used. Figure 9.8 shows the most common FWC-based system designs. The requirements for the FWC relative to gas-phase activity, back-pressure, and filtration efficiency are mainly driven by the position of the FWC in the exhaust system. The amount of catalyst applied to the filter substrate is an important parameter and needs to be determined based on the following emission requirements:

a. FWC in the close-coupled (CC) position replaces the classical CC TWC catalyst. In this configuration, an additional TWC is typically also found

FIGURE 9.8 Schematic view of Euro 6d gasoline system configurations with FWC—a) close-coupled FWC, b) close-coupled FWC cascade system, c) FWC underfloor system.

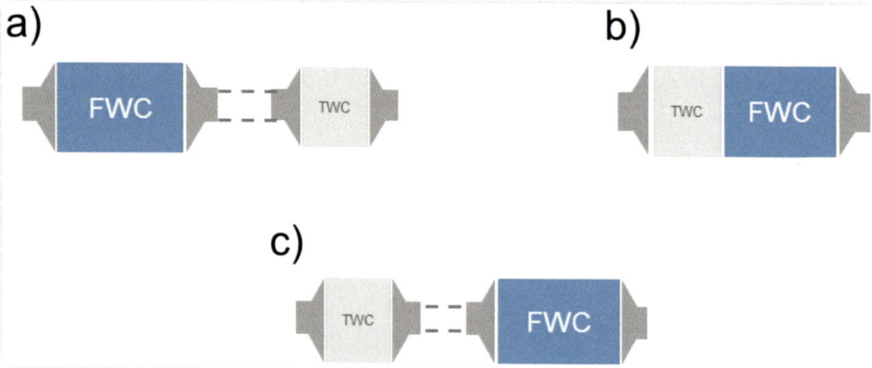

in the underfloor (UF) position for clean-up of CO and NO_x emissions. In this configuration, the demand for hydrothermal stability of the TWC functionality of the FWC is the strongest and requires the highest possible catalyst loading to be applied to the filter.

b. FWC in the CC position behind a TWC is still subject to high thermal stress. If a larger TWC is used, the required contribution of the FWC to gas-phase activity is lower, and therefore a lower catalyst amount can be used. As would be expected, if a small CC TWC catalyst, the washcoat loading on the FWC needs to be raised.

c. FWC in the UF position downstream of a CC TWC is the configuration, where the FWC has lowest thermal resistance requirements. Optimization to the lowest possible catalyst amount to minimize the backpressure impact can be done. Furthermore, the catalyst can be adjusted to deal primarily with NO_x slip from the CC.

Application of a FWC to the CC position is operationally the most challenging because of the need for thermal durability coupled with pressure drop constraints and thus results in a total amount of catalyst mass on a filter being necessarily lower than that of a state-of-the-art TWC. Feasibility studies have shown that boosting the TWC activity can be achieved either by increasing the catalyst volume [this also results in reduced backpressure and higher ash tolerance] or adding more precious metal, that is, Pt, Pd, and/or Rh. While increased precious metal levels can overcome deficiencies in FWC gas-phase activities, higher volumes can overcome a performance deficit and provide more volume for ash accumulation. Figure 9.9 shows an example catalyst layout for a CC application. Relative to the reference TWC, the FWC volume was increased by 30%, and the precious metal content was also increased by 58% for Pd and 121% for Rh. After applying a rapid aging procedure of 100 h fuel cut aging with bed temperatures of 1030°C, emissions were well

FIGURE 9.9 Example of FWC for close-coupled application. Higher volume and higher precious metal are used compared to reference, and after aging emission limits are met safely.

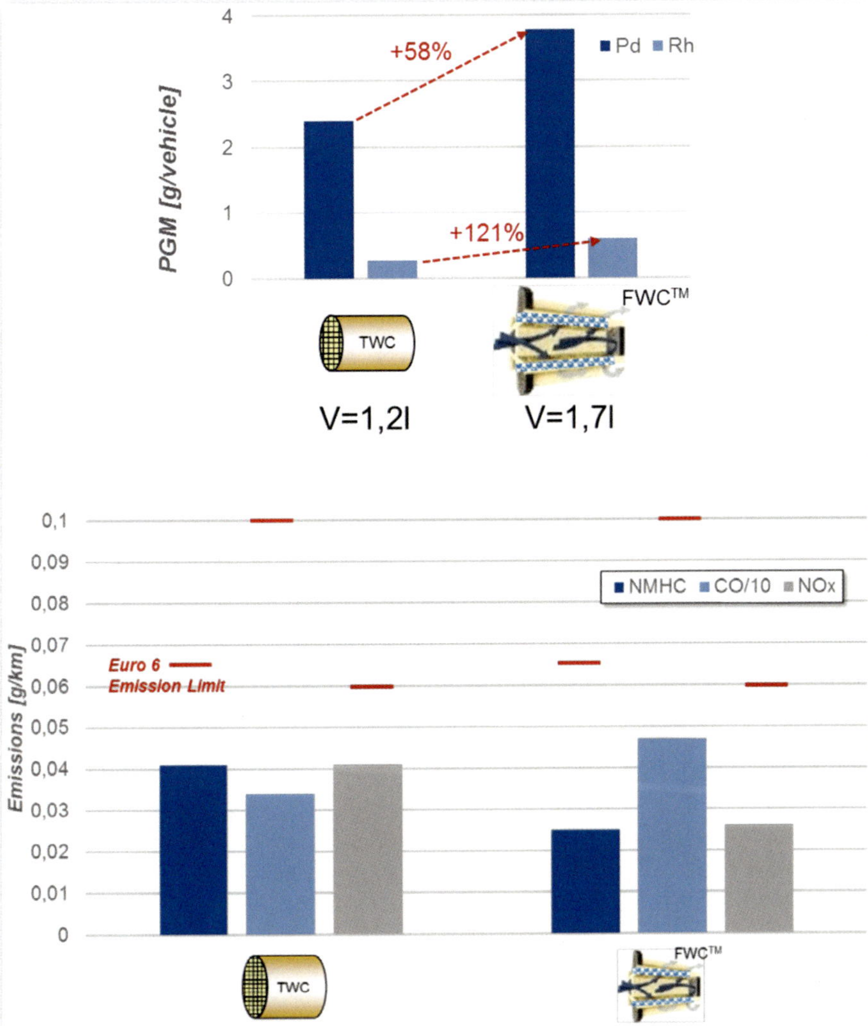

within the Euro 6d limits, and no major differences relative to the TWC reference system were observed.

Crucial for the application of a FWC in the CC position is the ability to meet OBD requirements. To detect the end of life, a certain level of ceria–zirconia oxides (CZO) which serve as oxygen storage capacity (OSC) materials need to be used in the catalyst formulation. The OSC signal from the end-of-life catalyst needs to be well above the minimum detectable level of 100 mg O_2. **Figure 9.10** shows the fresh and end-of-life aged values from the CC FWC. Fresh OSC values of a 1.6 L large FWC are in the range of 800–900 mg O_2. The value after 100 h fuel cut aging

FIGURE 9.10 OSC measurement of fresh and full useful aged FWC—100h fuel cut aging at 1050°C bed temperature.

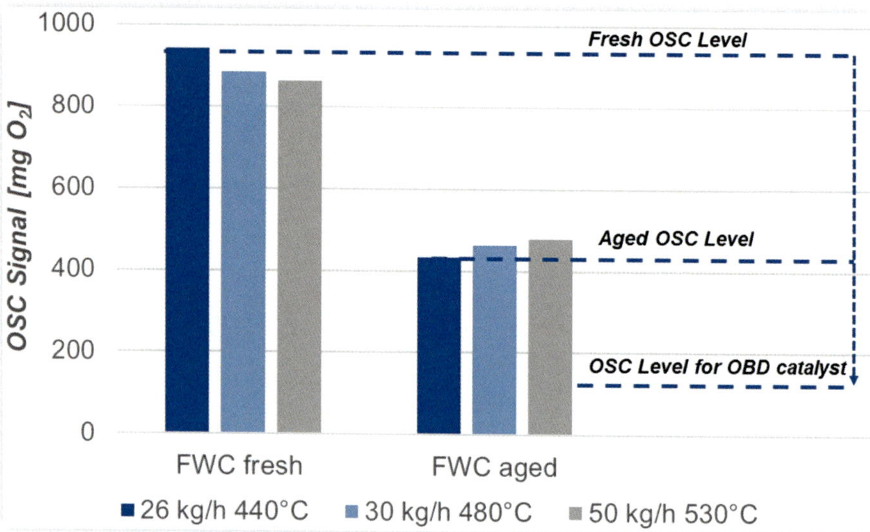

procedure with 1050°C bed temperature is between 400 and 500 mg O_2. The aged OSC value is high enough so that an OBD monitoring system can be readily applied to the system.

For a FWC in the CC position, the maximum gas-phase activity and CZO amount are high enough to ensure the catalyst can be monitored via standard dual oxygen sensor-based OBD methods. Requirements for a FWC in an UF position, however, are different with the focus being on low backpressure. Therefore, a lower catalyst amount can be used for the design of UF applications. For a CC TWC and UF FWC system, **Figure 9.11** shows the relationship between catalyst variation and backpressure. For this experiment, the precious metal content was kept constant for all samples at 2 g/ft^3 Pd and 3.5 g/ft^3 Rh. Three catalyst loadings ranging from ultra-low to medium relative to the total catalyst amount that is possible to put on a highly porous filter were examined. The middle loading (referred as low) has been aged on an engine bench to simulate an end-of-life gas-phase performance. The ash content reached in this aging cycle is not representative of those expected from high mileage trials and is approximately one-tenth what would be expected from on-road aging.

The testing has been carried out on a Euro 6a vehicle with 1.4 L turbo direct injection engine. An end-of-life aged TWC was used in CC position using the same size and precious metal loading as in serial production. Testing was conducted using the RTS 95 test cycle with cold start at 20°C. The RTS 95 cycle is a widely accepted surrogate cycle for RDE testing. It represents high-load and high-speed driving. In combination with a cold start at 20°C, this cycle is challenging for both particulates and high-speed, high-NO_x emissions.

FIGURE 9.11 Cold flow backpressure measurements of FWC catalysts for underfloor application. Catalyst with low loading was aged and re-measured with 4 g ash accumulated.

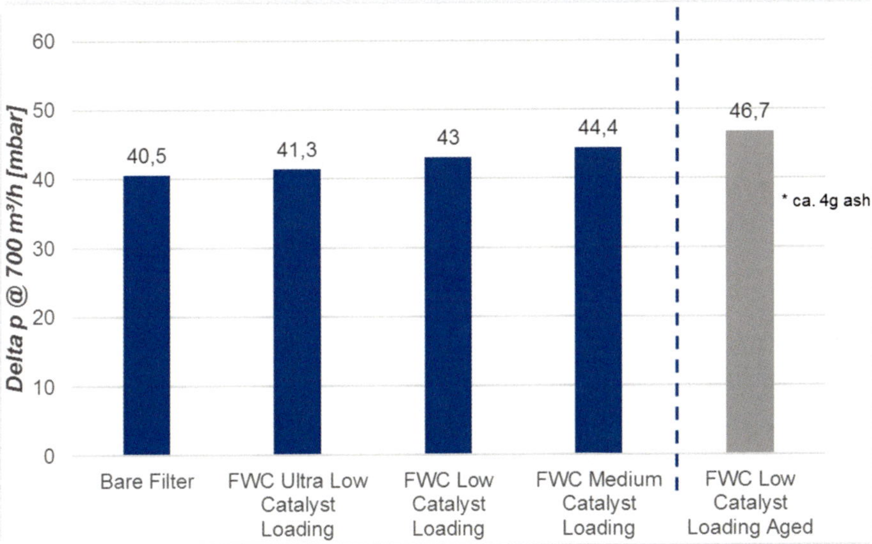

Under the conditions of RTS 95, the FWC catalyst in the UF position is needed to reduce NO_x emissions and filter the particulates. **Figure 9.12** summarizes the emission results relative to HC and NO_x emissions as a function of catalyst amount. The FWCs were tested in the fresh state so as to show the highest contribution that can be made to gas-phase activity. The far left dot represents the emissions as measured behind the CC TWC. To the right, the effect of increasing the catalyst amount from ultra-low to medium can be seen. The contribution of the FWC on hydrocarbon conversion is relatively low and is independent of the catalyst amount.

The FWC with a low catalyst loading was aged using a rapid aging cycle with fuel cut events. The aging duration was 100 h and represents full useful life gas-phase activity. The temperature in the FWC reached 980°C. Two different FWC catalysts with different precious metal contents have been aged: one with 2 g/ft^2 Pd and 3.5 g/ft^3 Rh (can be compared to the fresh catalyst results) and a second one with Rh only at a loading of 4 g/ft^3.

The results are summarized in **Figure 9.13**. Little deterioration of the emission results after aging for CO and NO_x were found. For hydrocarbon, the contribution of the FWC in an UF position is low, and therefore no changes after aging were expected. Comparing the results of the aged Pd/Rh and the Rh-only FWCs, it can be seen that there is basically no difference in emissions. Therefore, it can be concluded that low amounts of Pd in the UF position behind a TWC catalyst have little or no effect on the emission results. In order to optimize the precious metal costs of such system, Rh-only FWC catalysts can be used in the UF position.

FIGURE 9.12 Hydrocarbon and NOx emissions measured in the RTS95 test cycle as function of FWC catalyst amount in underfloor position. TWC in close-coupled position full useful life aged, FWCs in fresh state.

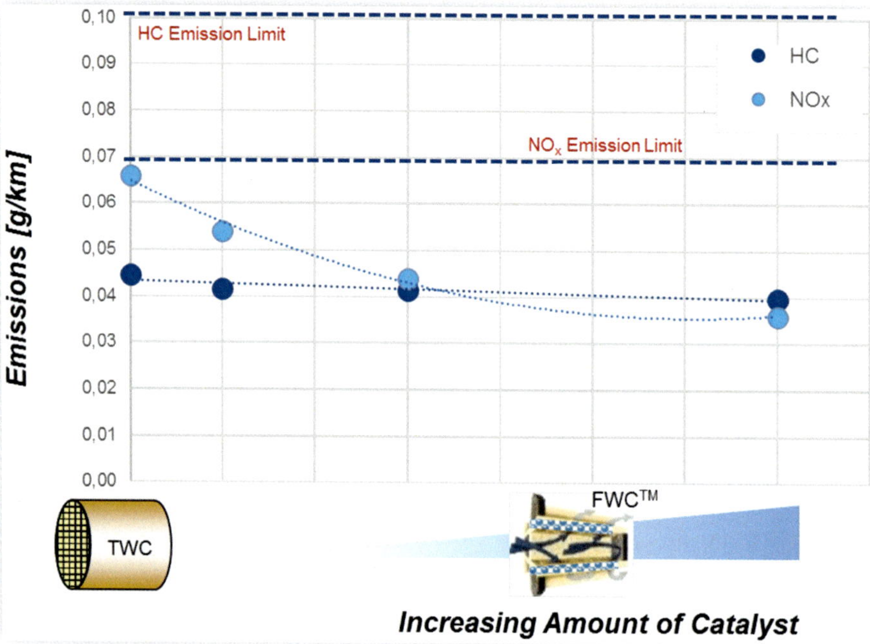

FIGURE 9.13 RTS95 evaluation of FWC with low catalyst amount in fresh and aged state.

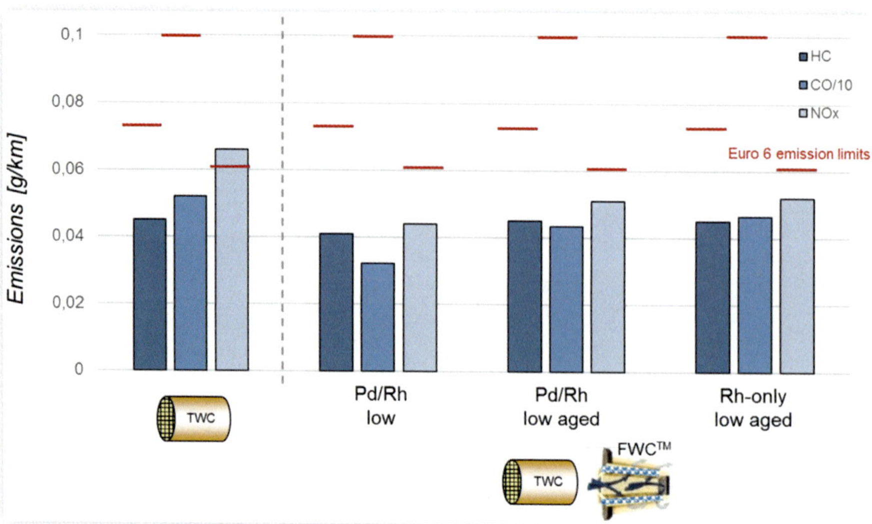

References

1. "2016/2017 Worldwide Emission Standards, Passenger Cars and Light Duty," https://www.delphi.com/sites/default/files/inline-files/delphi-worldwide-emissions-standards-passenger-cars-light-duty-2016-7.pdf.

2. Harth, K., Caudle, M., Li, Y., Punke, A. et al., "Emission Control for Diesel Passenger Cars to Meet Euro 6c," *35th International Vienna Motor Symposium*, Vienna, Austria, 2014.

3. He, H. and Yang, L., "China's Stage 6 Emission Standard for New Light-Duty Vehicles (Final Rule)," International Council on Clean Transportation, March 2017.

4. Johnson, T. and Joshi, A., "Review of Vehicle Engine Efficiency and Emissions," SAE Technical Paper 2017-01-0907, 2017, doi:10.4271/2017-01-0907.

5. "Attachment A-1, Final Regulation Order, Amendments to Sections 1900, 1956.8, 1960.1, 1961, 1961.1, 1965, 1968.2, 1968.5, 1976, 1978, 2037, 2038, 2062, 2112, 2139, 2140, 2145, 2147, 2235, and 2317, and Adoption of new Sections 1961.2 and 1961.3, Title 13, California Code of Regulations," https://www.arb.ca.gov/msprog/levprog/leviii/attacha1.pdf.

6. "An Update on the Measurement of PM Emissions at Lev III Levels," https://www.arb.ca.gov/msprog/levprog/leviii/lev_iii_pm_measurement_feasibility_tsd_20151008.pdf.

About the Authors

Dr. Patrick Burk is an R&D director in environmental catalysis at BASF Corporation currently responsible for Light Duty Gasoline Exhaust Abatement Technologies located in Iselin, NJ.

Patrick began work on environmental catalysis in 1988 at Engelhard/BASF and has successfully managed the development of catalysts for both gasoline and diesel exhaust abatement applications. He has more than 20 patents in this area. Prior to joining BASF, he worked at Smith Kline and French Laboratories and Exxon Research and Engineering Company developing processes for the selective the transformation of organic moieties to higher value-added products.

Patrick earned his Ph.D. in organic chemistry with the Nobel Laureate Robert Grubbs at Michigan State University followed by post-doctorals with Prof K. Barry Sharpless at M.I.T. and Prof. John Osborn at U. Louis Pasteur, Strasbourg, France.

Dr. Thomas Schmitz joined the automotive catalyst business in 2001. He held several positions in R&D, Application Engineering, and Sales. In 2011, he took over a project manager position at Gasoline System Development Europe located in Hannover, Germany. He and his team are developing gasoline exhaust system solutions with special focus on European customer requirements, meeting most stringent emission standards. Special emphasis is given to the Four-Way Conversion Catalyst, which is now being introduced to the market.

He received his PhD in physical chemistry from the University of Wuppertal, Germany. In his postdoc, he worked at the research center in Jülich at the Institute for Tropospheric Ozone Research on the characterization of vehicle emissions under real-world driving conditions. His focus was the speciation of hydrocarbon emissions in order to evaluate the contribution of traffic-related emissions to ozone formation.

CHAPTER 10

Three-Way Catalyst Gaseous Emissions: Gasoline Particulate Filter vs. Flow-Through Substrates

Chris Morgan and Jonathan Cooper
Johnson Matthey

10.1 Introduction

Tightening emissions legislation for gasoline vehicles around the world will require the removal of particulates as well as the control of gas-phase CO, HC, and NOx emissions. In Europe and China, the focus is on gasoline particulate number, while regulations in North America refer to particulate mass. Control of particulate number or mass can be achieved through the addition of a wall-flow gasoline particulate filter (GPF) in series with a conventional flow-through three-way catalyst (TWC) system. As a further step, coatings with TWC functionality can be incorporated into the GPF to convert gaseous pollutants, in addition to particulate filtration.

Such a coated GPF with TWC activity can be used to replace some of the existing TWC volume. This is an attractive option for packaging reasons, reducing the overall volume of the aftertreatment system and enabling more compact designs, especially beneficial for smaller vehicles. Alternatively, it can be used to supplement the existing TWC system to meet more demanding legislative requirements, such as real-world driving emissions (RDE), where higher conversion efficiency is required. In either case, if the coated GPF is contributing to the overall gaseous emissions performance of the aftertreatment system, then it is critical to understand the filter design parameters available and how they affect the overall conversion of gaseous emissions from the system.

10.2 Typical System Architectures

Many different system architectures are being considered, with simplified schematics shown in **Table 10.1**.

Some systems have retained a full-size TWC component, with an additional uncoated GPF downstream to remove particulates. An option is to coat the downstream GPF, to provide additional TWC volume to clean up emissions which slip through the conventional TWC component. This approach can be used, for example, where the original catalyst system has insufficient activity to control emissions at high load and/or high space velocity in the highway section of an RDE cycle. As the

TABLE 10.1 Examples of GPF system architectures

Architecture type	System schematic (Exhaust flow from left to right)
Full TWC system plus uncoated GPF	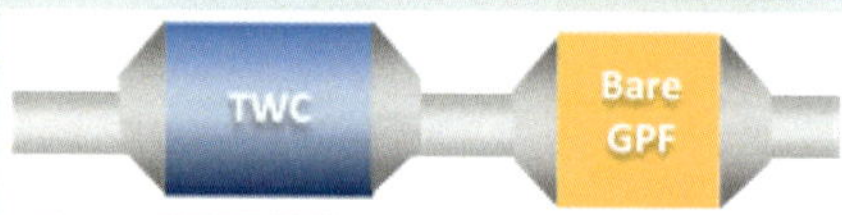
Full TWC system plus coated GPF	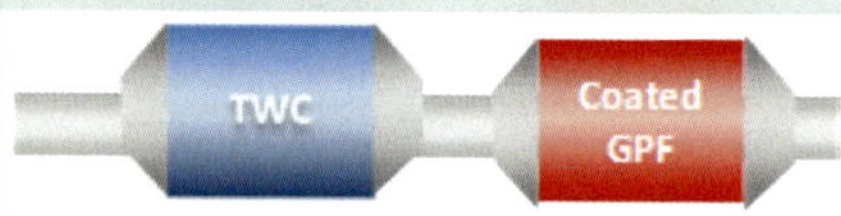
Reduced TWC system plus coated GPF	 Or 
Close-coupled coated GPF plus optional TWC	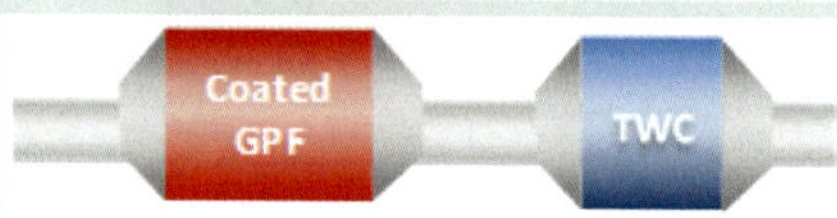

purpose of the coated filter is to supplement the TWC component, it is likely that a low to medium level of conversion activity will be required, relative to a standard TWC.

In other systems, the volume of the TWC component is reduced and the coated GPF is required to replace the TWC activity of the removed catalyst volume. Typically, the filter volume will be larger than the removed catalyst volume, in order to limit its backpressure impact on the system, and it is likely that a medium to high level of conversion activity will be required, relative to a standard TWC.

A third approach, most applicable on smaller vehicles where packaging space is at a premium, is to completely replace the close-coupled TWC with a coated filter. This places the greatest demands on the coated filter technology, requiring an equivalent high level of conversion activity to a standard TWC, as well as the need to satisfy demands for on-board diagnostics (OBD) monitoring.

Therefore, the choice of system architecture leads to a broad spectrum of requirements for TWC activity from the coated filter. This is further diversified by legislative requirements; emission test cycles with higher vehicle speed sections, such as the highway component of an RDE test or the US06 cycle, will generally require a higher level of gaseous emissions conversion.

10.3 Design of Filter Coating and Effect on Gaseous Conversion

The material components of a TWC coating have been optimized for over 40 years, with a good understanding developed of the physical and chemical properties required to meet both activity and durability requirements. This understanding can be transferred into the design of coatings for GPFs, and similar alumina and cerium-zirconium mixed oxide raw materials are typically employed, although the particle size distribution of each material may be modified for filter coating. Parameters such as washcoat loading, washcoat composition, and washcoat location can be optimized to tailor the level of gaseous emissions conversion efficiency, but also impact the particulate number filtration efficiency and the filter backpressure. Therefore, a critical design feature is to achieve the best balance between these features in the vehicle application.

A simple parameter to demonstrate is the effect of washcoat loading. As shown in **Figure 10.1**, the oxygen storage content (OSC), which corresponds well to CO and NOx conversion efficiency, and particle number filtration efficiency (over the World-Harmonized Light-Duty Test Cycle, or WLTC) both increased with washcoat loading. However, the backpressure measured across the filter also increased with washcoat loading. Therefore, the preferred washcoat loading depends on the emissions conversion and filtration efficiency requirement of the coated GPF and the acceptable backpressure increase.

In the example architectures referred to in Table 10.1, a coated GPF designed to supplement a full TWC system would typically need to provide a low to medium level of conversion activity to clean up gaseous emissions that slip through the TWC system under higher space velocity or higher load conditions. This can be achieved with a lower washcoat loading of, for example, 25–50 g/L, which also limits the additional backpressure from the coated GPF. When the size of the TWC system is reduced,

FIGURE 10.1 Effect of GPF coating loading on NOx conversion, filtration efficiency, and backpressure (all samples coated on 4.66 × 4.5″, 300/8, cordierite GPFs and tested on a 1.4L vehicle). Flow-through and filter substrate properties will be referred to using the convention of cell density (cells per square inch)/wall thickness (milli-inches).

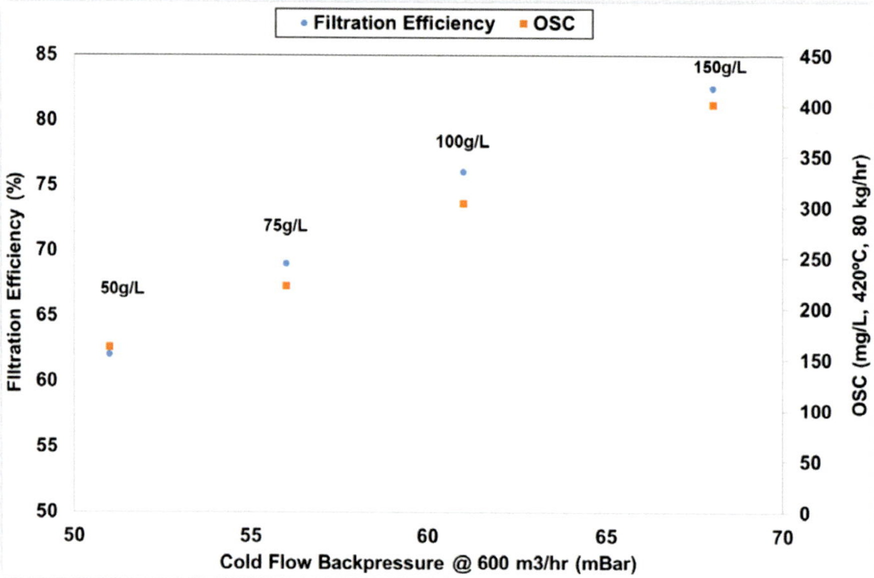

the demand for conversion efficiency from the downstream coated GPF increases, and higher washcoat loadings of 50–125 g/L are typically required to compensate for the missing TWC volume, at the expense of increased backpressure. If a coated GPF replaces the first (normally close-coupled) TWC, then to achieve an equivalent level of conversion activity even higher washcoat loadings of at least 100–150 g/L may be required.

Another degree of freedom is the precious metal content of the coating, although the improved gaseous conversion efficiency achievable through increased precious metal loading must be considered against the increase in system cost. For a coated GPF downstream of a TWC component, it is most likely that the gaseous emissions conversion requirement will be to reduce NOx, as this is generally a slower reaction than CO or HC oxidation in the upstream TWC. An effective way to improve NOx conversion is to increase the rhodium content of the coating, from, for example, 2 to 5 g/ft^3, as shown in **Figure 10.2**. In this experiment, the increased Rh content on the 75 g/L coated GPF yielded a 30% reduction in tailpipe NOx emissions, with particular benefit observed during the high-speed transient sections of the cycle.

The location of the coating within the filter will also affect the gaseous emissions conversion activity. The interaction between gas and catalyst in a flow-through substrate is well understood. The substrate can be considered as a system of parallel coated pipes, with laminar and turbulent gas flow within the open diameter, diffusion of gas into and within the coating layer, and reactions occurring when pollutant molecules come into contact with active sites on the surface of particles within the coating.

FIGURE 10.2 Effect of Rh content on NOx conversion over a coated GPF. TWC + GPF system vehicle tested over the WLTC (warm start).

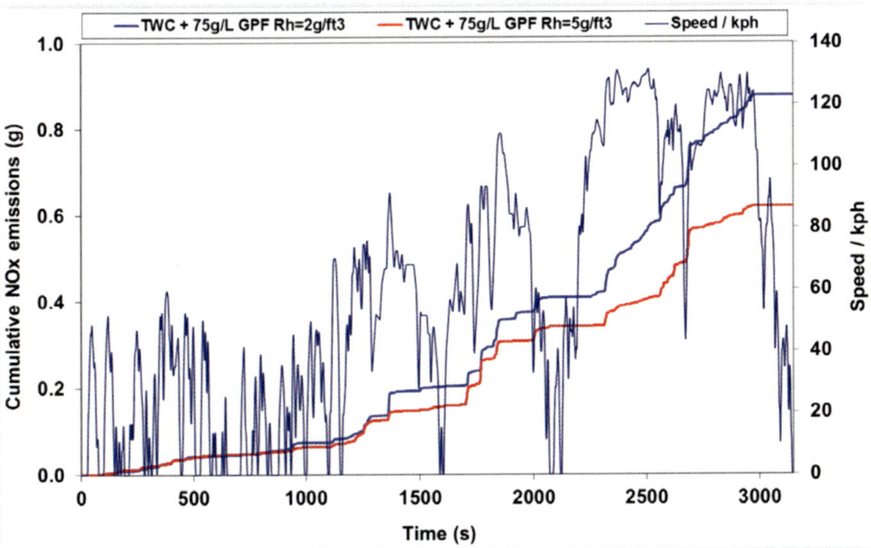

However, there are a number of additional complexities when considering gas flow through a particle filter. During its passage through the filter, the gas flow must transfer from the inlet to the outlet channels, and the relative flow along each channel varies along the filter length, as illustrated in **Figure 10.3** [1]. Magnetic resonance imaging of gas flow at 10 locations along the length of a diesel particulate filter, with similar cell density and pore diameter to a typical GPF, demonstrated that at the filter entrance the gas velocity was high in the inlet channels and low in the outlet channels. At the exit of the filter, gas velocity was high in the outlet channels and low in the inlet channels. The gas velocity was equal in inlet and outlet channels just over halfway along the filter.

FIGURE 10.3 Illustration of measured gas flow within a particulate filter. Figure reproduced with kind permission from the authors.

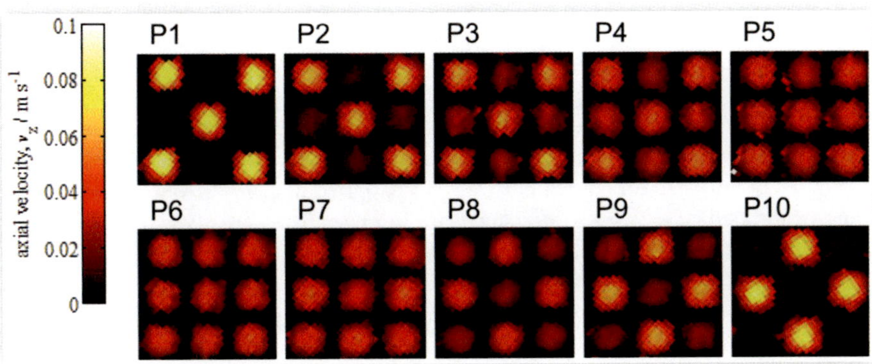

The gas flow through the filter wall was calculated and found to be non-uniform. It was high near the inlet due to the significant backpressure differential between the incoming gas being driven by the engine and the gas in the outlet channel. It was also high near the outlet, where the absolute pressure in the outlet channel decreases progressively faster. The gas flow through the filter wall was lowest in the region where gas velocities in the inlet and outlet channels were similar. Also, as incoming gas flow rates were increased, the crossover point moved downstream within the filter, and the flow rate through the filter wall increased and became less uniform along the length of the filter. Therefore, the distribution of gas flow within a particulate filter will vary significantly under different operating conditions, which has implications for both particulate deposition (which generally follows gas flow) and the design of a coating to maximize interaction with pollutants in the exhaust gas.

When a coating is applied to the filter, it can be located within the filter wall, on the surface of the inlet channels, on the surface of the outlet channels, or as a combination of in-wall and surface coating. A surface coating can lead to good contact with pollutants when there is significant gas flow within the channel. However, as the gas passes through the wall from inlet to outlet channel, it is inevitable that in a portion of either an inlet or outlet channel, there will be reduced interaction between a surface coating and the gas phase, relative to a flow-through catalyst.

If the coating is located primarily within the filter wall, the diameter of the pores is significantly smaller than the diameter of the filter channels, leading to improved contact between pollutant gas molecules and the coating, and the active sites contained therein. However, as the gas passes through the wall from inlet to outlet channel, the contact time is also reduced. Diffusion back into the filter wall is possible, but also varies along the length of the filter depending on the lateral flow rates. These factors make it significantly more complex to design the ideal coating location for a coated GPF and to predict its emissions performance than for a flow-through TWC.

Finally, differences between the physical properties of a filter substrate and a flow-through substrate can also influence the gaseous emissions, especially if the coated filter is the first emissions control component in the aftertreatment system. As filter substrates designed for coating typically have higher porosity (*ca.* 60–65%) than the flow-through substrates they may replace, they also tend to have lower thermal masses. This can enable good heating rates and quicker achievement of the coating light-off temperature. However, this effect is offset by the plugging of half of the channels at the inlet face and the effective reduction in geometric surface area in the front section of the filter, as well as the tendency to use lower cell densities than flow-through TWCs, typically 300 cells per square inch.

Therefore, experiments can be carried out where equivalent washcoat and precious metal quantities are applied to typical flow-through and GPF substrates of equal volumes and tested under the same exhaust flow conditions. For example, **Figure 10.4** compares an equivalent washcoat applied at 150 g/L to a conventional flow-through substrate and a GPF substrate in an engine bench lambda scan test. A good level of TWC performance was obtained from the coated GPF, although it was slightly inferior to that of the flow-through TWC due to a combination of the factors discussed above. Note that a conventional TWC would typically have a coating loading of >200 g/L.

FIGURE 10.4 Coated flow-through vs. filter at equal washcoat loading and PGM content. Lambda scan test at 450°C, ±4% lambda perturbations at 1 Hz, mass air flow 125 kg/h.

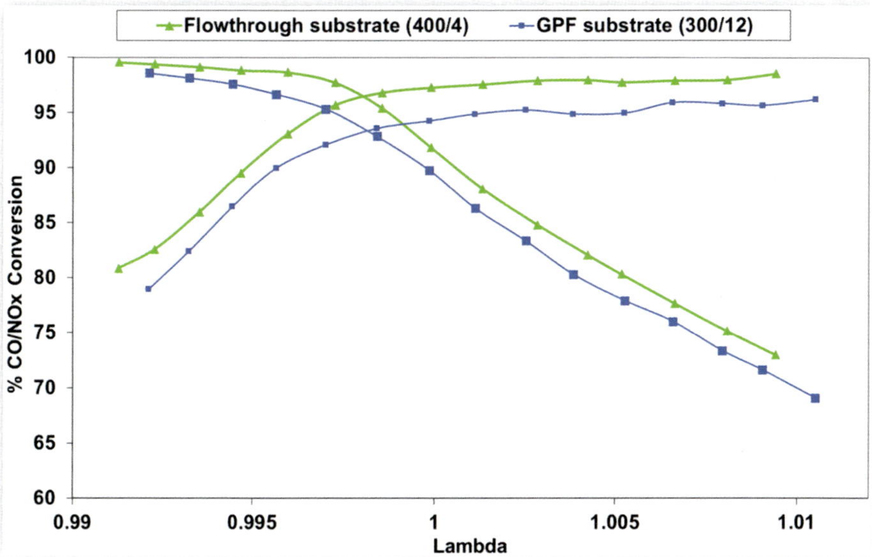

10.4 Examples of the Gaseous Emissions Conversion Performance of Coated GPFs

This chapter has outlined a number of ways in which the design of the coated filter and the overall gaseous emissions performance of the aftertreatment system is influenced by the system architecture and the legislative requirements. To better illustrate these effects, a number of examples will be given from development TWC plus coated GPF systems tested on vehicles in the laboratory over a range of drive cycles.

The first example is on a vehicle designed to meet the Euro 5 and Euro 6b gaseous emissions limits with a single TWC over the MVEG-B drive cycle (also known as the New European Drive Cycle, NEDC). To meet the latest European PN (particle number) emissions limit of 6×10^{11}/km, a GPF will be added downstream of the TWC. By coating the GPF, the conversion of gaseous pollutants can be improved, making the system more robust to the introduction of the new WLTC and RDE testing in the Euro 6d legislation. In such applications, most of the conversion of gaseous emissions will take place over the upstream TWC, and it is expected that the concentration of pollutants reaching the GPF will be low. Consequently, the GPF can be designed with low levels of both washcoat and PGM—in this case a washcoat loading of 50 g/L and a Rh-only precious metal loading of just 2 g/ft^3.

FIGURE 10.5 Comparison of gaseous emissions over the WLTC for TWC only (1.2L, 50/0:46:4) vs. TWC + GPF system (1.4L, 50 g/L washcoat, 2 g/ft^3 Rh) on a 1.0L vehicle. All catalysts engine bench aged to 160K km equivalent. Precious metal loadings are referred using the convention x/a:b:c, where x is the total precious metal loading in g/ft^3 and a:b:c is the ratio of platinum:palladium:rhodium.

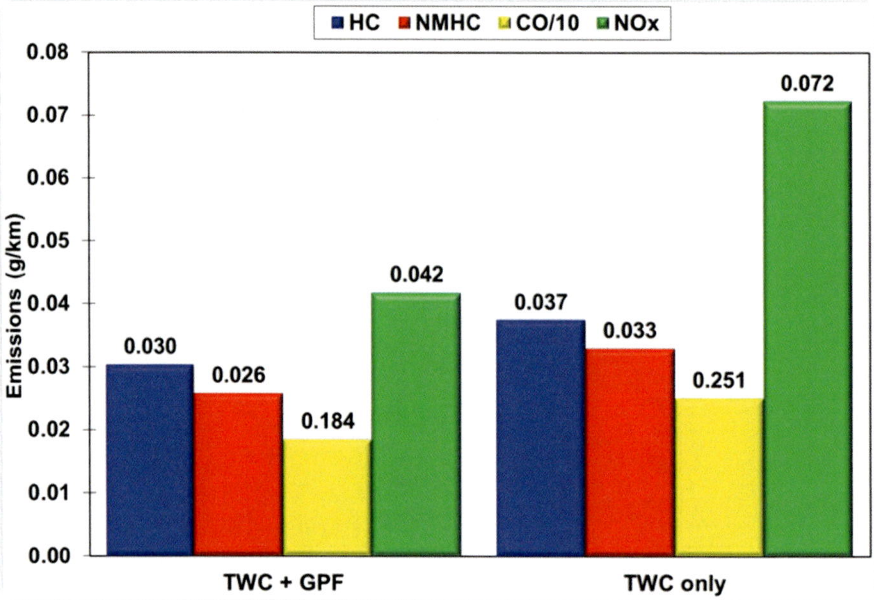

As shown in Figure 10.5, over the more transient WLTC, the TWC-only system gives NOx emissions outside the 60 mg/km limit, although this could be improved through modifications to the engine calibration. HC and CO emissions are well inside the limits, suggesting that there is a slight lean bias. However, adding the extra coated volume of the GPF leads to a higher conversion efficiency of all pollutants on the WLTC cycle, despite the low precious metal and washcoat loadings. HC and CO emissions are reduced by 20–25%, while NOx emissions are over 40% lower. Closer inspection of the modal data shows that most of the benefit is seen in the final, higher-speed section of the WLTC test. This is as expected, given the front TWC was designed to fully comply with emissions legislation as measured over the MVEG cycle; driving conditions outside this window cause some pollutants to break through the front TWC, which can then be converted by the coated GPF. Thus, in aftertreatment systems where a coated GPF is added to a flow-through TWC system of significant size, the additional coated GPF volume provides extra conversion efficiency, for example, at high-load, high-speed conditions.

A second example illustrates how a coated GPF can be used to replace part of the TWC volume of an aftertreatment system and how the coating loading affects the gaseous emissions performance. A twin TWC system designed to meet Euro 5 emissions standards was modified such that the second TWC (representing nearly 60% of the total system volume) was replaced by coated GPFs of the same dimensions and precious metal content.

FIGURE 10.6 Effect of replacing a second TWC with coated GPFs at various washcoat loadings (MVEG cycle, 1.6L vehicle, 0.66L + 0.84L twin close-coupled catalyst system. PGM loadings 75/0:14:1 + 18/0:5:1). All catalysts engine aged to 160K km equivalent.

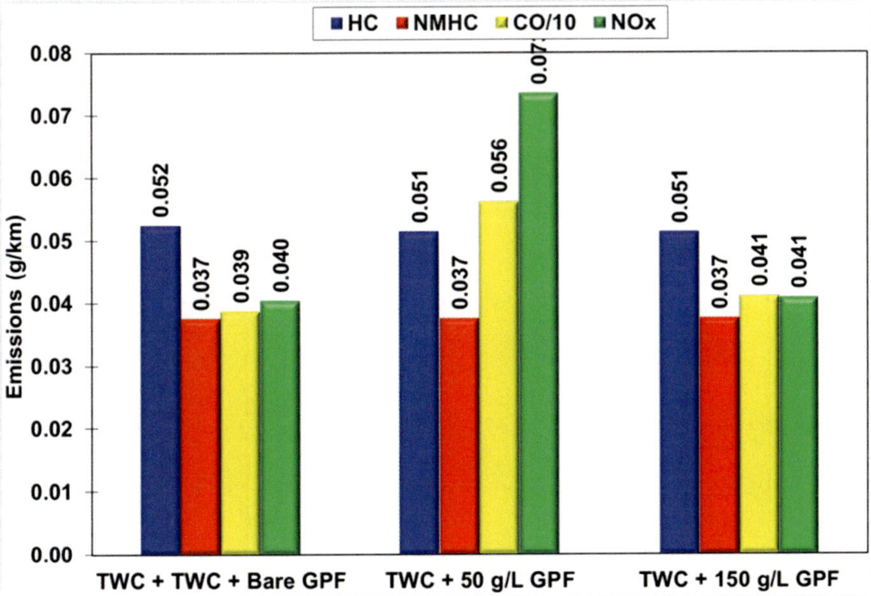

Figure 10.6 compares GPFs with a coating loading of 50 g/L and 150 g/L against the reference system with a second TWC and an uncoated GPF added in the underfloor position for PN control. In all cases, the HC emissions were equivalent. It is well known that the main challenge for HC emissions is light-off performance, and, as the first flow-through substrate in the system, the initial TWC will chiefly be responsible for this activity. In this application, where the GPF volume is replacing TWC volume, a washcoat loading of 50 g/L was not sufficient to give equivalent activity, and there was a significant increase in CO and NOx emissions at the lower filter coating loading. However, when the coating loading was increased to 150 g/L, the additional catalyst support and oxygen storage capacity enabled performance equivalent to the original TWC (within measurement error).

Therefore, it is possible to design coated GPFs which give equivalent conversion activity to a TWC, and the TWC + 150 g/L GPF is a more compact system than the TWC + TWC + uncoated GPF system it replaces. However, it should be noted that while the contribution to total system backpressure was similar for the twin TWC + uncoated GPF and TWC + 50 g/L coated GPF, the backpressure was approximately 40% higher for the TWC + 150 g/L coated GPF system.

This example demonstrates that coated GPFs can be used to replace TWCs and achieve either equivalent emissions or equivalent backpressure. An obvious question is whether a compromise can be achieved which gives an acceptable balance between these properties. Therefore, in a third example, another twin TWC system was compared with a system where the second TWC was replaced with a GPF at a

FIGURE 10.7 Effect of replacing a second TWC with a coated GPF at 100 g/L washcoat loadings (MVEG cycle, twin close-coupled catalyst system, 1.6L vehicle, 0.88L + 1.26L catalyst volumes, PGM 150/0:29:1 + 22/0:20:2). All catalysts engine aged to 160K km equivalent.

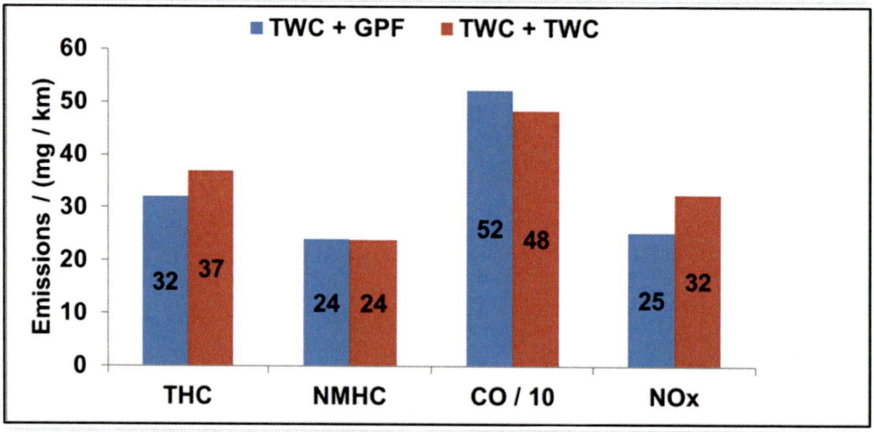

coating loading of 100 g/L. The systems were tested over both the MVEG-B cycle and a significantly more demanding "random aggressive" cycle with a high proportion of rapid transients and high-load, high-speed operation, which can be considered to represent the upper end of the spectrum of driving conditions for RDE development.

Over the MVEG-B cycle, there was no clear difference between the two systems (**Figure 10.7**), with the TWC + GPF system giving similar HC and CO emissions performance and possibly offering a small benefit for NOx emissions, although this is close to being within test error. Both systems comfortably passed the European legislative limits for gaseous emissions. The TWC + GPF system also comfortably passed the particulate number limit over the MVEG-B cycle, with emissions of 4×10^{10}/km for the fresh system and lower emissions after aging due to accumulation of ash residues in the filter.

With the more aggressive cycle, the GPF-containing system gave equivalent HC performance to the twin TWC system but ~10% higher CO emissions and ~15% higher NOx emissions. As the vehicle used for testing was designed to meet the Euro 5 emissions limits, it is very likely that improvements in calibration could enable the TWC + coated GPF system to meet the future Euro 6d full RDE legislation where a conformity factor of 1.5x is required for NOx emissions. If necessary, a small increase in precious metal content could boost the conversion activity, as illustrated in Figure 10.2.

10.5 Summary

To summarize, coated GPFs can be used within exhaust aftertreatment systems to control gaseous emissions in a number of ways, from supplementing the performance of the existing TWC components to providing equivalent performance when replacing some or all of the TWC volume. This leads to a diverse list of requirements from existing and planned exhaust system applications and a wide range of coated GPF solutions and design tools to address them. Key design parameters include the GPF size and location, the coating loading and composition, and the precious metal content.

With knowledge of the impact of these parameters, and the benefit of over 40 years of materials development from TWC formulations, coated GPFs have been demonstrated to achieve a good level of TWC activity, after full useful life aging and over a range of driving conditions. With careful design, it is possible to optimize the GPF technology to give equivalent TWC gaseous conversion activity to a flow-through TWC within an aftertreatment system. However, it must be recognized that there is a trade-off between improved gaseous emissions conversion activity and filtration efficiency and increased system backpressure and component cost. The gas flow through the coated filter and its interaction with the coating is also more complex than for a flow-through TWC, which makes predicting gaseous emissions conversion more difficult.

Vehicle testing has confirmed the contribution of coated GPFs to gaseous emissions control, both meeting the gaseous emissions targets when replacing part of the existing TWC system and enhancing the emissions control when supplementing the existing TWC system. Therefore, coated GPFs are an important tool for the vehicle manufacturer to meet current and future legislative limits for gaseous pollutants, as well as controlling the particulate number or mass emissions.

A system design approach can enable benefits by combining the TWC coating onto a filter substrate in a more compact aftertreatment system. Within the system, it may also be possible to rebalance the performance between the upstream TWC and the coated GPF, for example, by increasing the volume or PGM content of the TWC, so as to reduce the demand on the coated GPF and enable lower gaseous conversion activity (and, hence, lower backpressure) designs.

Reference

1. Ramskill, N.P., Sederman, A.J., Gladden, L.F., and York A.P.E., "Magnetic Resonance Velocity Imaging of Gas Flow in a Diesel Particulate Filter," *Chemical Engineering Science* 158 (2017): 490.

About the Authors

Dr. Chris Morgan has worked for Johnson Matthey since 1997, and in his current role as Technology Director, Clean Air Europe is responsible for the development, scale-up, and testing of autocatalyst coatings for both gasoline and diesel applications. Chris previously managed the Gasoline Product Development team, developing new families of three-way catalysts and leading JM's early work on coatings for gasoline particulate filters, as well as the Catalyst Characterisation and Prototype Samples groups. Before joining Johnson Matthey, he completed a DPhil at Oxford University on the preparation and application of high-temperature ceramic superconductors.

Dr. Jonathan Cooper has worked for Johnson Matthey since 1999. Having worked in gasoline aftertreatment in various roles throughout his career, including being part of the team to begin studies into coated GPFs, he is currently Gasoline Development Manager for Clean Air Europe, responsible for developing new TWC and GPF products. Prior to joining Johnson Matthey he completed a DPhil at Oxford University looking at photo-activated electroanalysis using platinum electrodes in flowing solutions.

CHAPTER 11

Uncoated Gasoline Particulate Filter Integration and Examples

Christophe Colignon
Groupe PSA

Summary

Particulate mass and number emissions have become mandatory with Euro 6 legislation and the implementation of Euro 6c legislation reduces the limit to Diesel-like levels. In order to comply with these emissions standards, gasoline direct injection (GDI) engines could require the use of a gasoline particulate filter (GPF). Various GPF concepts can be used, and this paper will present the design and integration of an uncoated GPF system, and particular attention is paid to the validation tests of the system. The impact of the use of a GPF on the vehicle attributes is also addressed.

11.1 Introduction

September 2017 coincides with the introduction of additional requirements for regulated pollutants by the EU commission. In addition to the change in the certification cycle from the "New European Driving Cycle" (NEDC) to the "Worldwide Harmonized Light Duty Test Cycle" (WLTC), emission testing in real-life conditions in a non-predictable drive schedule (unlike tests performed on a chassis dynamometer) will be introduced. The real driving emissions (RDE), as it is called, will be measured on public roads with a portable emission measurement system (PEMS), and two pollutants will be particularly scrutinized, particulate number (PN) and NOx. **Figure 11.1** gives an overview of the timeframe of the introduction of RDE, with two different steps, corresponding to two different values of the conformity factors (CF).

Gasoline direct injection (GDI) engines have higher particulates raw emissions than port-fuel injection (PFI) or multi-port injection (MPI) gasoline engines, especially in cold-start conditions when the engine is running rich and for some specific vehicle applications (heavily loaded) there is a risk of not fulfilling the limits with engine-out emissions. The use of a wall-flow particulate filter, called GPF (gasoline particulate filter), in order to differentiate it from the Diesel particulate filter (DPF) system, could help or even be compulsory to respect the future emissions standards. The objective of this chapter is to describe the integration of such a system on a vehicle, which differs a lot from Diesel (substrate characteristics and regeneration mode).

11.2 GPF Design/Integration

There are currently two different options for GPF systems: either an uncoated GPF (also called bare GPF, chosen by PSA Group for its future GDI engines) or a coated GPF (with a three-way catalyst (TWC) function added on or in the filter walls), also called FWC (four-way catalyst) or cGPF (coated GPF). The choice between these two technologies largely depends upon vehicle integration constraints and vehicle applications (targeted emissions standards, GPF position in the exhaust line (close coupled or underfloor), vehicle raw emissions and temperature levels in front of the GPF, etc.). As will be described later in this chapter, the substrate choice closely relies upon the functional requirements of the system, and the targeted filtration efficiency plays

FIGURE 11.1 RDE introduction calendar and limits.

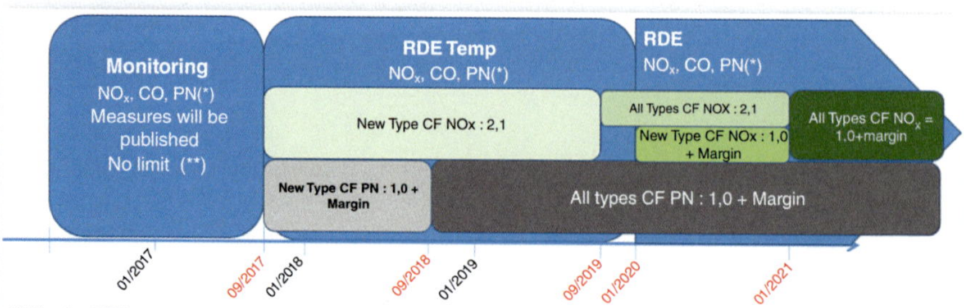

(*) Diesel and GDI
(**) local certification authorities could however introduce more severe constraints than the legislation

a key role. For instance, selecting a filter to respect only Euro 6c legislation or selecting it in order to fulfill both Euro 6c and China6 standards (with different fuel characteristics, leading to modified raw emissions) will not necessarily lead to the same substrate definition, as the minimum filtration efficiency required is not the same.

Once the filter size is frozen (mainly due to packaging constraints), the only design freedom to meet project targets relies upon the filter choice, that is to say, the substrate material, its microstructure (mean pore size and porosity), and its geometry definition (symmetrical or asymmetrical cells, cell density, and wall thickness).

11.2.1 Functional Requirements

The filter has to fulfill a certain number of targets and functions. The functional requirements can cover a very large range as described below:

1. The primary function of the GPF is to filter soot particles contained in the exhaust gas (one can include the PM size of the official measurement means)—linked to the GPF filtration mechanisms.
2. The filter should be resistant to failure during soot combustion (maximum amount of soot to be clearly specified)—linked to the material thermomechanical resistance.
3. The filter should exhibit an exhaust backpressure lower than a maximum backpressure value (for a given mass flow rate and a given exhaust gas temperature)—linked to the filter permeability (material microstructure), geometry, and placement in the exhaust system.
4. The filter should remain functional during all the vehicle life (duration may vary from a passenger car application to a light commercial vehicle application)—durability aspect, including ash accumulation (coming from the fuel and the lubricant).
5. The filter should resist to any potential chemical reaction with the lubricant and fuel ash components—thermochemical resistance.
6. The filtration efficiency should be compatible with specific PN/PM emission standards (the target efficiency will depend upon raw emissions/vehicle application, targeted geographical area and fuel quality). The target will also depend upon the margin taken compared to the official limits (taking into account production discrepancies).
7. The GPF system should also be completely transparent for the driver, compared to vehicles not using a GPF with a similar engine, whatever the filter soot loading—linked to the engine control unit (ECU) strategies.

Additional requirements can be added, increasing the complexity of the filter choice, which will necessarily be a compromise of various assets:

a. NVH targets
b. Filter (or global system) cost

c. Production system capability and quality
d. Other system and supply chain constraints

11.2.2 GPF System Operation

The GPF system is mainly composed of the following items:

a. A filter
b. A differential pressure sensor in order to measure the pressure drop of the filter
c. A large number of strategies in the ECU in order to manage and monitor the GPF

Figure 11.2 shows the exhaust line layout of the PSA engines using a bare GPF system.

The GPF operates in a very different way than the DPF, its equivalent for Diesel engines, as it is based upon a passive regeneration of the filter. This is due to the much higher exhaust gas temperature level of gasoline applications (see **Figure 11.3**) compared to Diesel where the natural temperature of soot combustion is barely obtained naturally.

It is also due to the fact that engine-out particulate emissions are much lower on gasoline engines, reducing the soot accumulation speed. The GPF relies on a passive regeneration process, that is, the soot accumulated in the filter burns automatically when its self-ignition conditions are obtained. The regeneration opportunities depend on two parameters, the GPF inlet temperature and the presence of oxygen in the exhaust gas. As most GDI engines are running at the stoichiometry, oxygen is only available during fuel cut operations, which reduces the possibilities and durations of soot combustion. **Figure 11.4** shows how the regeneration opportunities can vary on a given profile, depending on the hypothesis taken as the minimum values to obtain sufficiently fast combustion (combination of a minimum temperature and a minimum oxygen content). The figure also shows that in real life driving, there is an accumulation of short periods (a few seconds or minutes) of soot accumulation in a GPF, followed by short periods of soot combustion ($T > 600°C$, $[O_2] > 2\%$). This differs significantly from

FIGURE 11.2 Exhaust line architecture for PSA Group engines equipped with a GPF.

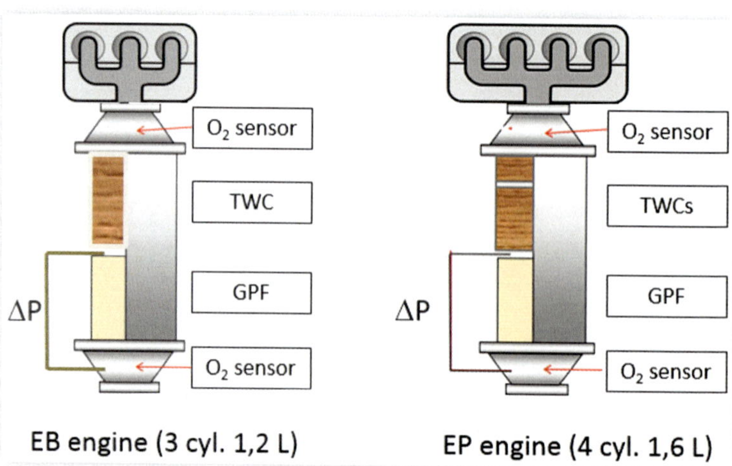

FIGURE 11.3 Temperature profile comparison between Diesel and gasoline applications.

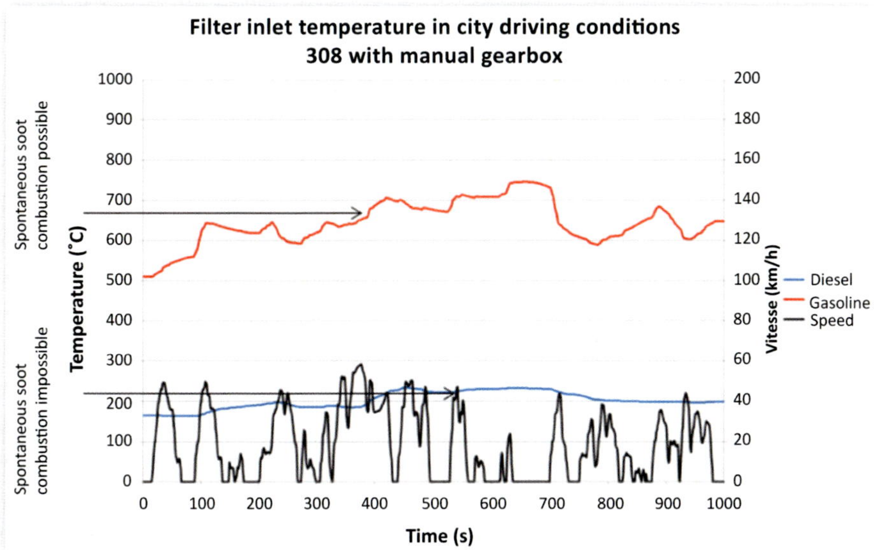

FIGURE 11.4 Regeneration opportunities on high speed driving profile.

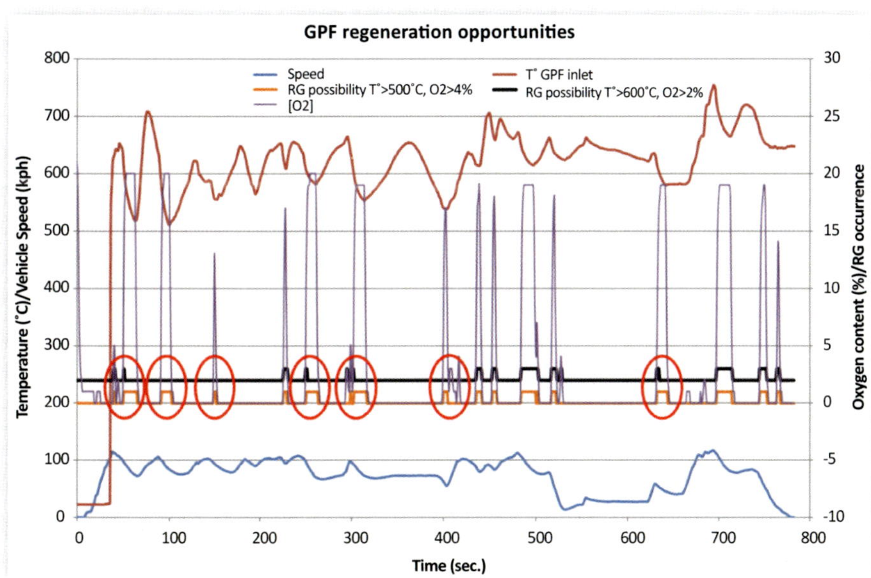

Diesel applications where soot accumulates continuously in the filter during hundreds of kilometers and soot combustion lasts several consecutive minutes.

It is clear that there are plenty of opportunities for regeneration, but each one is very short, so if there was a large amount of soot accumulated in the filter, the whole amount of soot would only be fully eliminated after many fuel cuts.

11.2.3 Associated Test Procedures

In order to assess the GPF system performance in line with the described requirements, a series of specific tests can be applied. The first tests that are carried out to evaluate a GPF filter technology are fresh filtration efficiency, pressure drop measurement, and thermomechanical resistance evaluation. Initial filtration performance is evaluated on a chassis roller with a fresh (non-stabilized) GPF on various profiles (WLTC + RDE constraints) using the official PN/PM measurement tools, so there is no need to further describe this test.

11.2.3.1 Pressure Drop Measurement

Pressure drop is preliminary tested on a cold flow bench (see **Figure 11.5**) with un-canned substrates, without soot, in order to have a first positioning of the technology compared to competitors and, more important, the project target for global exhaust line backpressure. Some candidates are sometimes already eliminated after the cold flow bench tests.

The engine bench is then used for a comprehensive pressure drop characterization, first without soot and then with various soot loadings, the loading being performed either on a steady state operating point or using transient cycles, which requires a longer test duration. This time the filter is tested in a canning, if possible the target definition in the targeted exhaust line position. The exhaust gas volumetric flow rate varies from the minimum (idle) to the highest values (rated power) in order to obtain the full $\Delta P = f(Q_{vol})$ curve, as can be seen on **Figure 11.6** below. The range of soot loadings depends on the strategies used by the manufacturer, especially for diagnostics purposes. It is vital, for pressure drop characterization with soot, to ensure that there is no oxygen in the exhaust gas during the whole test, in order to prevent

FIGURE 11.5 Cold flow bench pressure drop measurement.

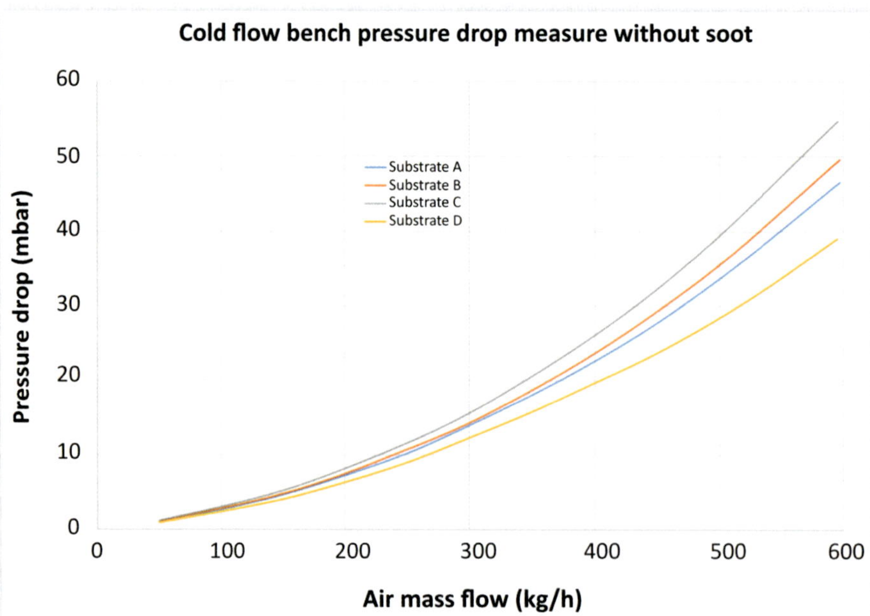

FIGURE 11.6 Engine bench pressure drop characterization.

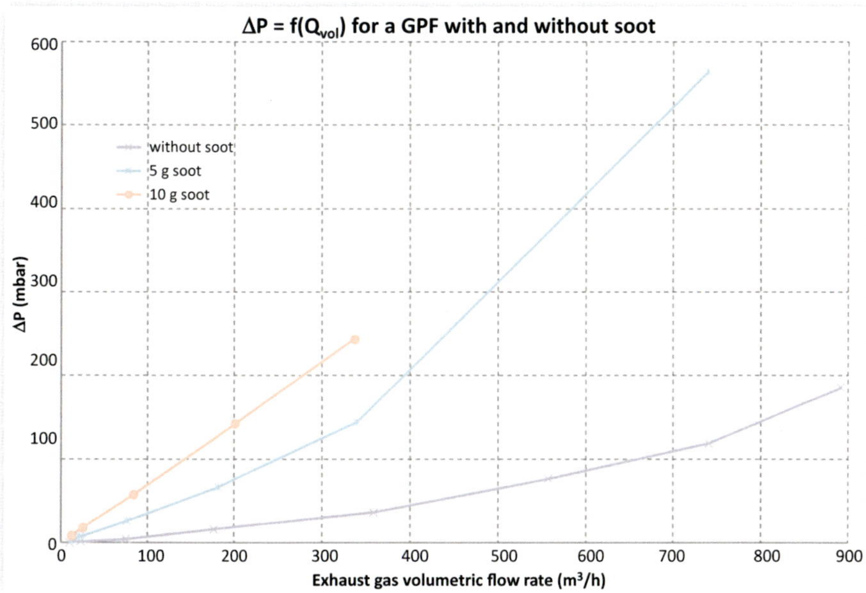

from any soot combustion that would ruin the test result. That means that fuel cuts are disabled and that there should not be any leak in any portion of the exhaust line upstream the GPF.

11.2.3.2 Severe Regeneration Test

Like for Diesel filters, thermomechanical resistance behavior is assessed on engine bench, using a severe regeneration test. This test consists in loading a filter with soot up to a target value and then initiating soot oxidation with high combustion rates in order to create the highest possible temperature levels and thermal gradients inside the filter. The filter and its canning are equipped with a large number of thermocouples in order to measure temperature levels upstream and downstream of the filter as well as inside the substrate. **Figure 11.7** shows that the filter is first heated up to 800°C at lambda 1 (no oxygen in the exhaust gas to avoid any soot combustion) and once the conditions are stabilized, a fuel cut is operated, bringing suddenly a large amount of oxygen (20–21% of the exhaust gas content) in the soot environment. This oxygen presence, combined with high GPF inlet temperature levels, results in a very fast regeneration, generating high internal temperatures and thermal gradients in the substrate.

The objective of this test is to mimic what could happen to "critical" customer driving profiles. For instance, a driver with very low exhaust gas temperature (e.g., very short distances in city driving, which could even be worsened by cold-start conditions; see **Figure 11.8**) would accumulate a large amount of soot, without ever creating the conditions for their combustion. Soot will burn very quickly and intensely if the driving profile is changed, resulting in much more favorable exhaust gas temperature levels (e.g., motorway or highway driving), coupled with a fuel cut (e.g., when exiting the highway or pulling down to a traffic light) that will bring a sufficient amount of oxygen

FIGURE 11.7 Severe regeneration test.

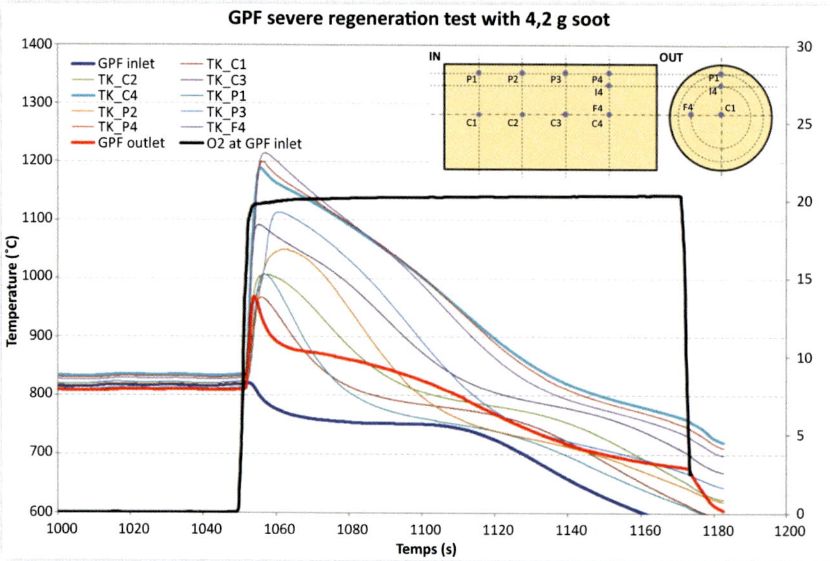

FIGURE 11.8 Influence of coolant temperature at engine start on raw emissions.

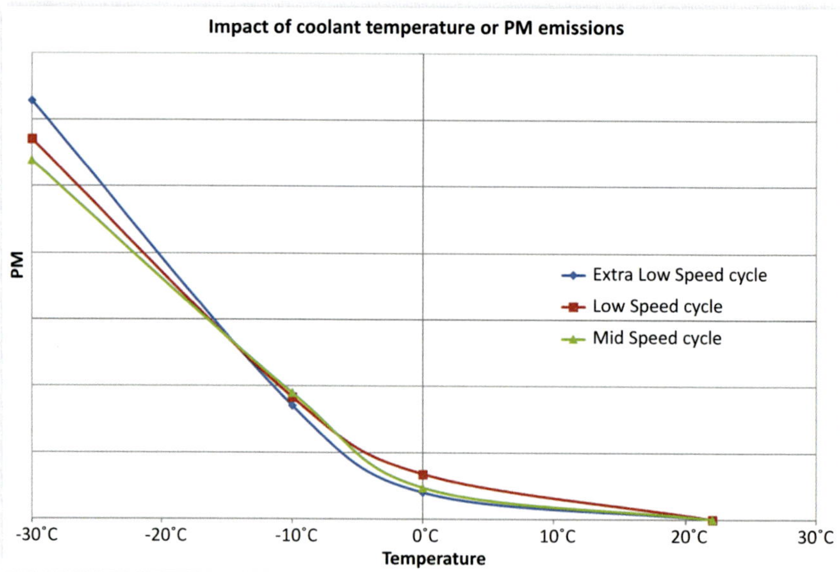

to ignite the soot accumulated in the filter. This would produce a large exotherm inside the filter and high thermal gradients.

Once this severe regeneration is performed, the filtration efficiency of the GPF is assessed on WLTC and RDE profiles, in view of the in-service conformity (ISC) evaluations. We will see later in this paper other tests procedures that allow one to comprehensively assess the performances of the GPF system.

11.3 Tuning Impact of the GPF

First of all, the most common impact of a GPF system on a vehicle comes from the pressure drop of the filter. The backpressure of the uncoated filter depends on four main factors:

- Material microstructure, that is, porosity and mean pore size (MPS)
- Filter geometry, that is, length, entry surface area (diameter for a cylindrical substrate), cell density, and wall thickness
- Engine parameters, such as exhaust gas flow rate, exhaust gas temperature, and exhaust gas composition
- The amount of soot and ash accumulated in the filter

Under the same boundary conditions, the differential pressure of the filter decreases with an increasing porosity or MPS. The increase in the cell density and in the wall thickness will result in the reduction of the open frontal area (OFA), causing pressure drop to increase. For a given filter volume, a low pressure drop level can be achieved using a large diameter. However, the substrate diameter choice is sometimes not flexible and driven by the vehicle packaging constraints.

The amount of soot stored in the filter depends upon several parameters such as the engine-out emissions and the regeneration ability. Both of these parameters depend on the driving conditions (engine load and speed as well as ambient temperature, for cold-start operation, and pressure), as the amount of soot burned by passive regeneration is directly linked to these parameters.

The amount of ash accumulated in the GPF depends primarily on the engine oil consumption (and the oil ash content), so the driving profile impacts the ash collected. The actual amount of ash accumulating in the filter can significantly differ from the amount calculated based on the oil consumption and oil ash content. Indeed, some of the ash can go directly in the oil sump while some of them are stored along the exhaust line before the GPF (in the TWC). Oil quality (grade and ash content), driving profiles (gas flow rate), TWC design (cell density), and number of flow-through substrates therefore influence the ash finding rate.

11.3.1 Engine Performance

The pressure drop generated by soot and ash negatively influences the engine running:

- Power/torque reduction as the engine volumetric efficiency decreases
- Fuel consumption increase
- Risk of valve reopening (potentially leading to a contact with the piston in the cylinder) in extreme cases

The way ash is accumulated in the filter largely depends upon the driving profile and the flow rate. Unlike Diesel applications (especially using a fuel-borne catalyst to promote soot combustion speed) where ash is accumulated at the back of the filter (due to a sintering and migration process during soot regeneration events), gasoline ash is

FIGURE 11.9 Ash accumulation on the walls throughout the filter length.

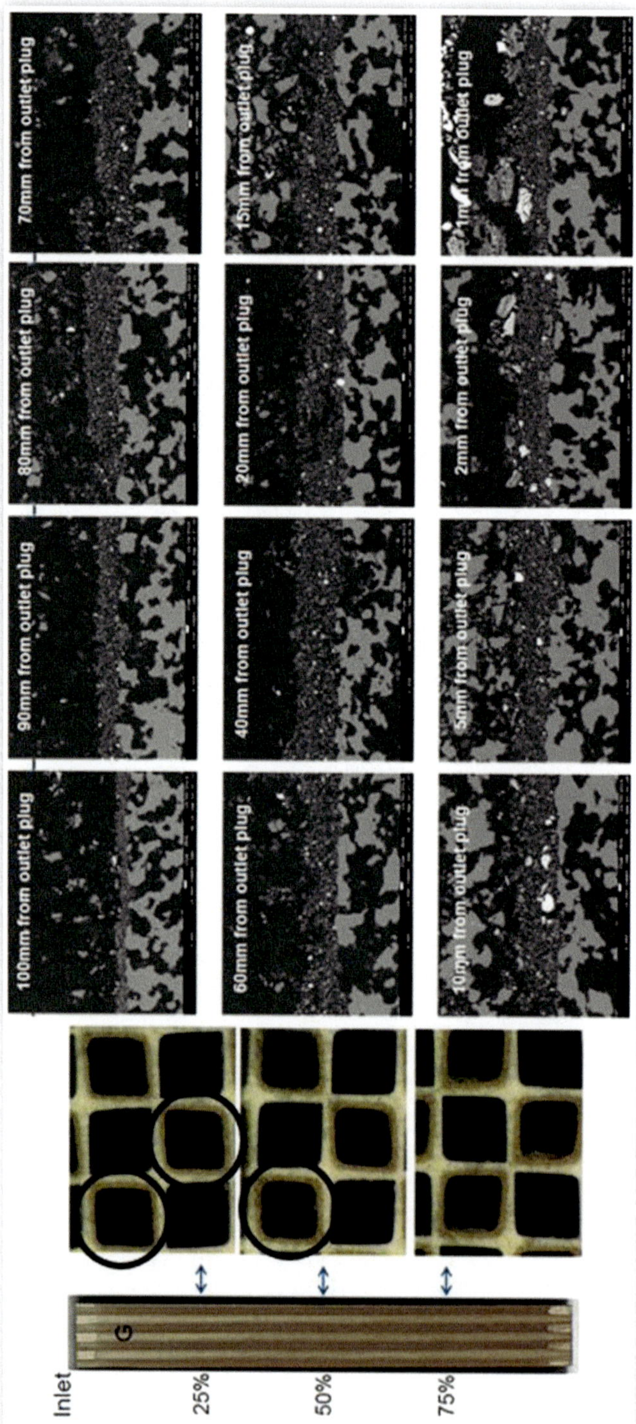

FIGURE 11.10 Ash accumulation in the back of the filter.

not only accumulated as plugs in the rear of the filter. Due to the passive regeneration (whereas the regeneration is mainly active for DPFs), a large amount of ash is accumulated as a layer on the whole length of the inlet channels, as we can see it on **Figure 11.9**. The ash plugs/ash layer ratio depends upon the driving profile, with an increasing plug ratio for high speed. There are two reasons for that. First of all, the amount of soot present in the filter is very low in these conditions, and this absence of soot promotes ash build-up as a layer. Secondly, the high flow rates of this driving pattern tend to push ash to the bottom of the filter, as shown on **Figure 11.10**. Globally, for an average driver, 65–70% of the ash is deposited as a layer and 30–35% as plugs located in the back of the filter. We'll explain later how pressure drop is impacted by ash accumulation.

11.3.2 GPF Monitoring Strategies

The differential pressure sensor is used for the continuous monitoring of the filter, in order to assess the amount of soot accumulated in the filter. Depending on the driving conditions and the exhaust gas flow rate, a direct or an indirect use of the measured pressure drop (ΔP) is used. As for Diesel vehicles, there are some thresholds to diagnose the system and detect if the amount of soot continues to increase, which means that the driving conditions are not favorable for a passive regeneration. Such a "severe" driving pattern could lead to a filter overload and generate a filter damage once the regeneration spontaneously takes place. In order to avoid the risk of a "too severe" filter degradation, if the soot ignition conditions are fulfilled, a strategy of "active" regeneration could be used. Such a strategy can be triggered when reaching a pressure drop threshold. The active regeneration would consist in modifying the engine running, without any customer impact, in order to simultaneously increase exhaust temperature and oxygen content to reach the soot self-ignition conditions. This strategy means running the engine slightly in lean mode in order to limit the oxygen content. Indeed, if the oxygen content is too high, as the filter soot loading is already high, soot combustion would be fast; limiting the oxygen content in the exhaust gas allows controlling regeneration speed and reducing the risks of filter damage. This "active" regeneration strategy would only be used by a very small percentage of customers, as it is a dysfunctional mode. For instance, with 20°C ambient temperature, it can be necessary to accumulate soot during around 7500 km on a severe driving cycle to reach the amount of soot that triggers the "active" regeneration. The impact is more important for short-distance journeys, shorter than 10 min.

In addition, as previously mentioned, cold ambient temperatures can also further increase the severity of the driving pattern as raw emissions are widely influenced by air temperature, especially below 0°C (see Figure 11.8).

11.4 GPF System Validation

11.4.1 Filtration Efficiency Tests

As described earlier, the engine bench is used to make severe regeneration tests, in order to assess the filter thermomechanical resistance. The filter that went through these tests is then evaluated on a chassis roller in order to measure the filtration efficiency of the damaged filter. This allows to set the maximum soot loading (MSL) value, the mass of soot that the system shall not exceed in order to guarantee that the PN/PM emissions standards will always be respected. The chassis roller is used to assess the filtration performance of the GPF in two different conditions:

- Fresh filter in order to simulate COP (conformity of production) tests
- Damaged filter after a severe regeneration to simulate ISC tests

In order to validate the system in the worst-case scenarios, the tests take into account the discrepancies that can be observed in the production, especially for engine-out emissions:

- For COP tests, the highest tailpipe PN results will be obtained with the highest raw emissions levels.
- For ISC evaluations, we test the vehicle in the worst possible conditions, that is, the highest raw emissions and a completely empty filter (no soot, no ash). In real life, the vehicle is tested in its current state, potentially with a small soot cake and a slight or medium accumulation of ash, both of them being favorable for an improved filtration performance.

For each configuration, the filtration performance is evaluated both on the future homologation cycle (WLTC) and on driving profiles representative of the RDE test conditions.

To select the best filter candidate, we first took into account what we considered as the most important parameters: initial filtration efficiency, pressure drop without soot and with soot, and MSL. Figure 11.11 shows an example of the candidate comparison with two parameters taken into account, fresh filtration efficiency and initial pressure

FIGURE 11.11 Filter choice matrix.

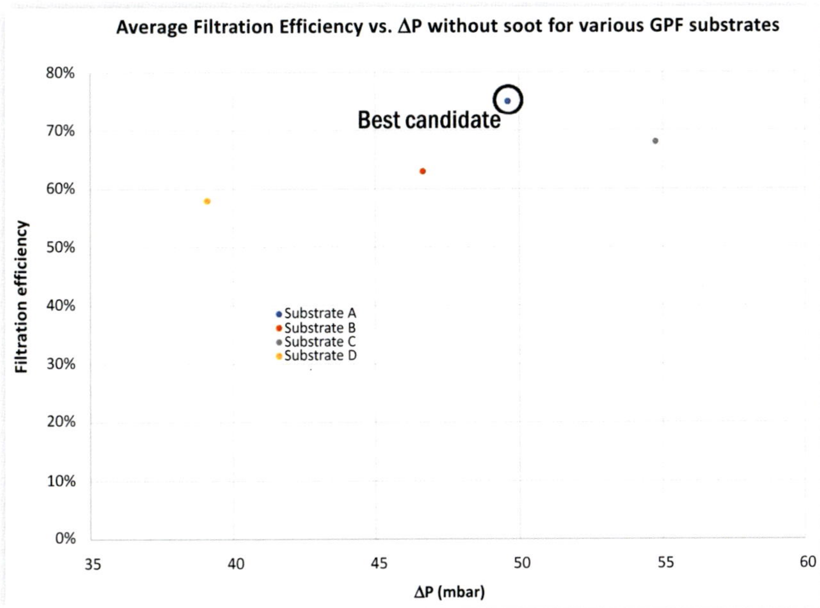

drop level. This helped reducing the number of competitors, prior to the final choice. Overall, considering all parameters (including filter price, thermochemical resistance, etc.), the best candidate proved to be a cordierite Corning DuraTrap® GC 200/8 filter.

11.4.2 Vehicle Road Tests

In addition to all the tests that have already been described, some road tests are carried out for an assessment of the whole system (including the GPF monitoring strategies) or for a durability test of the filter. These vehicle road tests are the final validation sequences for a project.

Two kinds of tests are performed using real-life driving conditions. On the one hand, there are some short-distance tests carried out on various driving profiles, using a comprehensive data acquisition in order to monitor the filter loading and the way the strategies are behaving. These tests require the regular weighing of the filter, so a dismountable canning is used, instead of the serial one that includes TWC and GPF. The filter is also equipped with internal thermocouples for temperature monitoring and regeneration severity assessment. On the other hand, in order to validate the filter durability and investigate the pressure drop behavior with ash, a long-run endurance is performed. A 240,000 km endurance with Europe type oil and fuel was organized in collaboration with Corning GmbH, with regular filter checks (filtration performance on chassis roller, then lab tests such as cold flow bench pressure drop, weight, CT scan, and filter cleaning to remove soot and quantify the actual soot loading) at service intervals. In order to have a driving pattern representative of a majority of our customers, we used a mix of city, highway, and "mountain-like" driving conditions, as shown in **Figure 11.12**, resulting in profiles illustrated in **Figure 11.13**.

FIGURE 11.12 Driving profile statistics of the real-life endurance test.

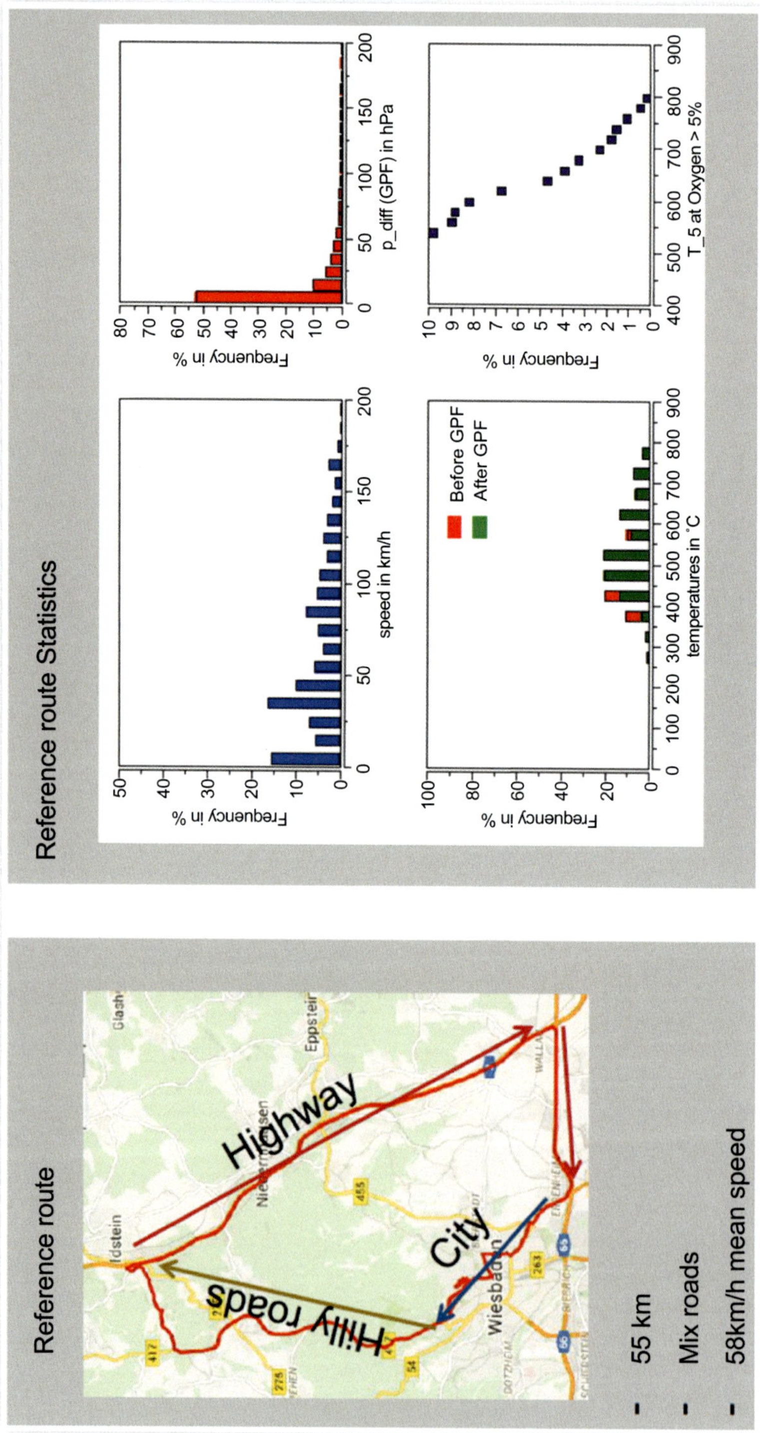

FIGURE 11.13 Data acquisitions of the real-life endurance test.

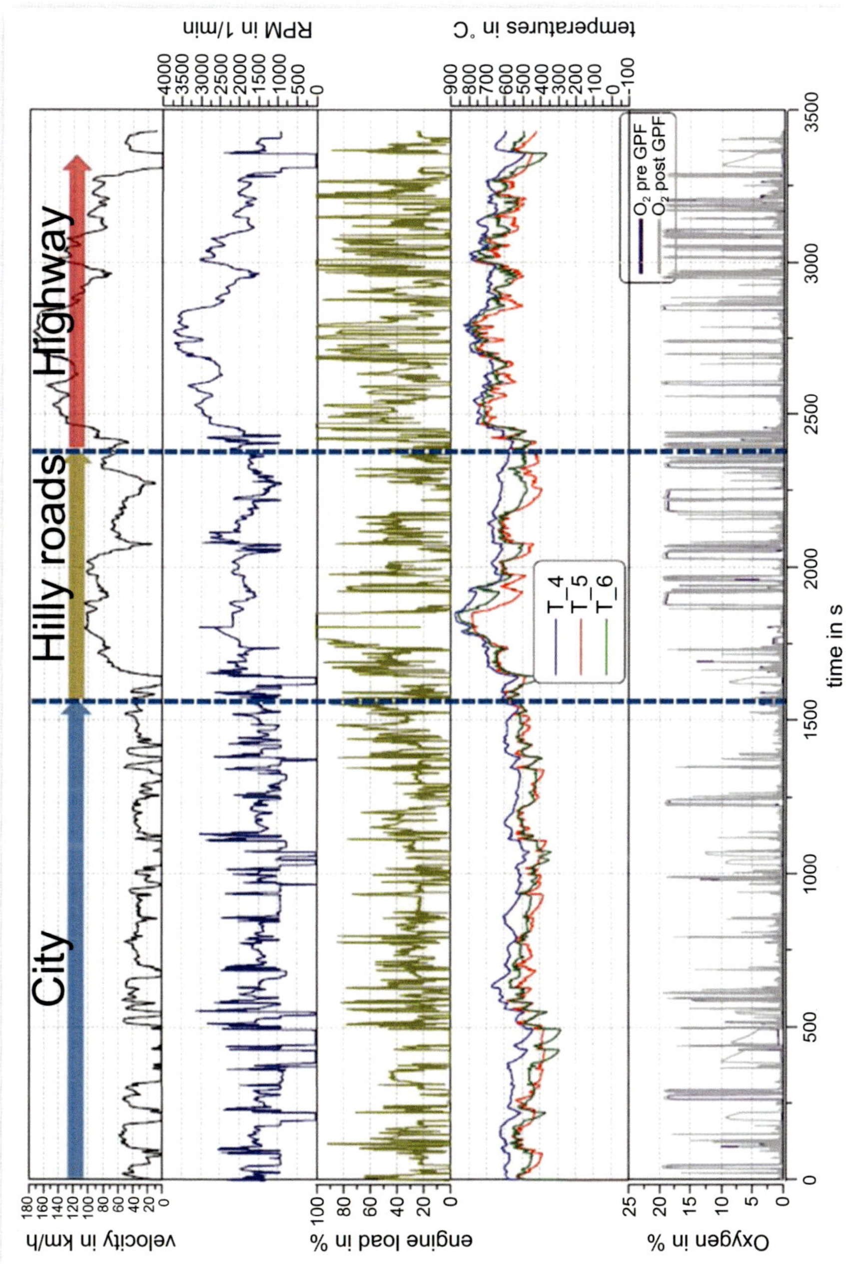

FIGURE 11.14 Filtration performance results on various driving profiles.

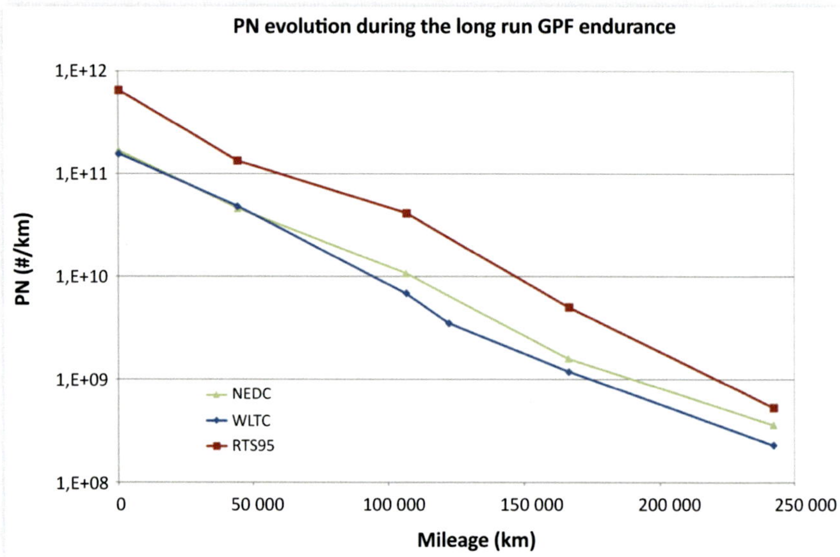

FIGURE 11.15 Cold flow bench pressure drop evolution during the endurance.

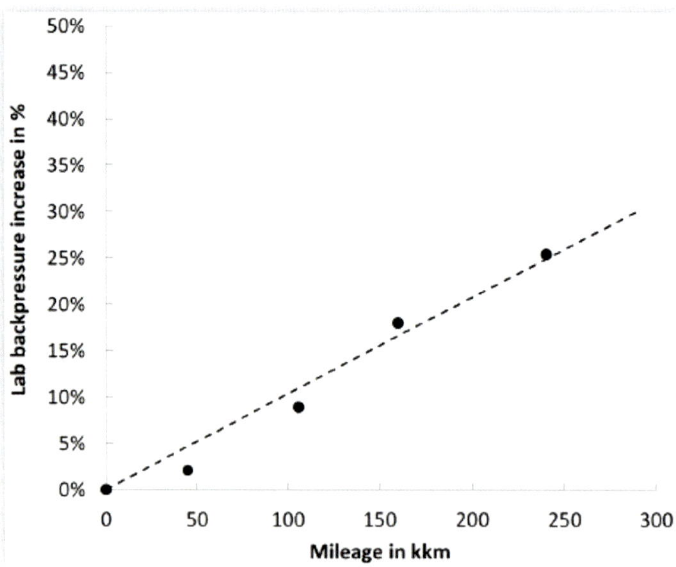

The exhaust gas temperature upstream and downstream the GPF was monitored in order to detect any potential exothermic regeneration taking place with a large amount of soot. The regular filter weighing before and after soot regeneration during the checks, the filtering performance (shown on **Figure 11.14**), and the fact that the GPF outlet temperature never exceeded 900°C showed that during the whole endurance the MSL remained at a low level (estimated below 3–4 g).

FIGURE 11.16 ΔP evolution for a honeycomb particulate filter.

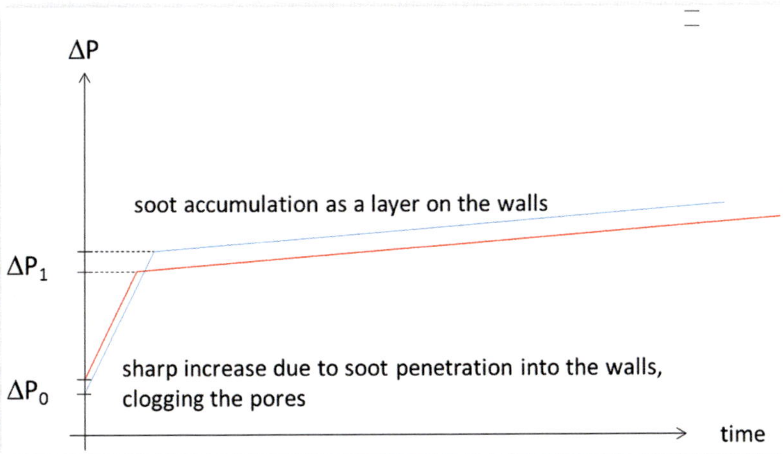

The regular cold flow bench measurements showed that pressure drop without soot was increasing linearly with the progressive ash build-up, as can be observed on **Figure 11.15**. With higher oil consumption values, the ΔP penalty would obviously increase and start to impact fuel consumption and engine performance.

We confirmed it with engine test bench ΔP characterizations after 105,000 km and at the end of the endurance. As shown on **Figure 11.17**, once soot builds up in the filter, there is no more penalty due to the ash. This can be explained by the fact that the ash layer prevents soot from penetrating into the walls, which generates a very steep ΔP increase when there is no ash in the filter, as can be seen on **Figure 11.16**.

FIGURE 11.17 Engine bench pressure drop characterization with and without ash.

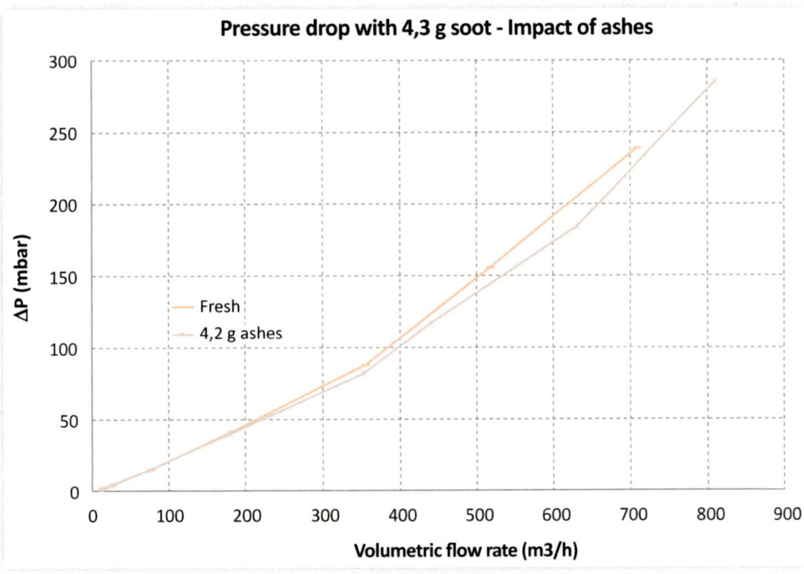

11.5 Conclusion

The implementation of a GPF in the exhaust line of GDI engines will become compulsory with the introduction of more stringent emissions standards throughout the world. The use of a particulate filter impacts the engine performance, but it can be reduced by specific control strategies in the ECU. When vehicle integration constraints allow a minimum flexibility, the filter design (diameter and length only, as cell density, wall thickness and microstructure can be optimized, whatever the filter size) can play a major role in reducing the ΔP level. Comprehensive test programs on engine bench and on the vehicle showed that for an uncoated cordierite GPF application, Corning DuraTrap® GC 200/8 filters was our best candidate, with a good level of performance in all conditions, from fresh to end of life. Indeed, a long-run real-life endurance demonstrated that its filtration performance was improved thanks to the build-up of an ash layer on the whole length of the filter channels, with a relatively limited impact on the exhaust line backpressure.

About the Author

Christophe Colignon studied mechanical engineering and graduated from IFP-School in 1997, with an engines engineering degree. He worked 16 months in the technical Department of Renault UK After-Sales Services before joining Groupe PSA (PSA Peugeot Citroën at that time) in April 1999. He has worked for reducing emissions of gasoline and Diesel engines in all the positions he occupied in the Company, becoming a specialist for particulate filter systems. He also lectures on emissions control systems at several universities.

CHAPTER 12

System Integration and Application for a Three-Way Catalyst-Coated Gasoline Particulate Filter

Jung-min Seo and Chi-bum In

12.1 Design of TWC-Coated GPF

A TWC-coated GPF needs to exhibit both a gaseous emissions reduction function and a soot-filtering function simultaneously. Therefore it needs to be optimized for both particulate matter (PM) removal and gaseous emissions reduction performance, considering the system configuration, which may be located in close-couple (CC) or in the underflow (UF) positions [1, 2].

Since TWC-coated GPFs can replace all or a portion of a conventional TWC substrate, gaseous emission performance needs to be comparable, back pressure needs to be minimized, and vehicle PM/PN emissions need to be taken into account. Design of TWC-coated GPF can be divided into two categories. The first category includes the GPF contour, volume, and position, which are external considerations. The second category is about GPF properties (cell density, web thickness) and catalyst properties (coating method, loading amount), which are internal considerations. The many design parameters which need to be considered simultaneously and comprehensively include gaseous emission reduction performance, PM/PN filtration efficiency (FE), and engine power loss due to back pressure. Generally 200 or 300 cells per square inch (cpsi) GPFs are selected considering back pressure, but web thickness needs to optimize between high particle number (PN) FE and high engine power loss due to elevated back pressure with thicker web [3].

- Design parameters for a TWC-coated GPF
  - PN FE
  - Gaseous emissions
  - Back pressure
  - Active regeneration (soot mass limit (SML))
  - Ash storage

12.1.1 PN FE

PN FE targets for TWC-coated GPF need to be set considering raw PN level of vehicle and the variables below in the fresh (soot- and ash-free) state.

- Variables which can impact distribution of PN FE include (see Figure 12.1):
  - GPF variability (GPF properties, wash coat loading (WCL) amount, distribution, etc.)
  - Fuel specification
  - Test variability (equipment, driver, etc.)

FIGURE 12.1 Target PN values.

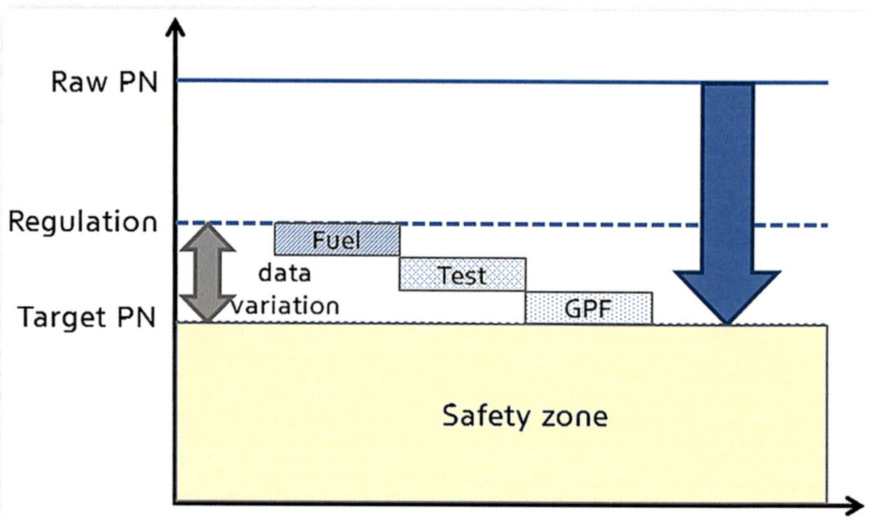

Since PN FE is impacted by GPF properties (porosity, median pore size, etc.) variability and wash coat properties (loading amount, distribution, etc.) variability, the variability needs to be taken into account for the GPF design.

Furthermore fuel specification has significant impact on the emission results and especially on PN. Specifications need to be compared carefully between the fuel used for development and the fuel in vehicle sales area. Evaporation is difficult for high boiling point fuels due to strong van der Waals force, and thermal decomposition is difficult for stable molecular structures, such as for benzene ring (aroma) compared to linear chain structure. It can be confirmed by a simplified PM index from the research of JAMA (Equations 12.1 and 12.2) [4]:

$$\text{PM index} = \sum_{i=1}^{n}\left(\frac{\text{DB}+1}{\text{V.P}(443\,\text{K})_{\text{i}}}\times Wt_i\right) \tag{12.1}$$

$$\begin{gathered}\text{Simplified index} = \alpha\times \text{E170}+\beta\times\text{E130}+\delta\\ (\alpha=-0.0647,\quad \beta=0.0324,\quad \delta=9.92405)\end{gathered} \tag{12.2}$$

where

DBE is the number of double bonds

V.P(443 K) is vapor pressure at 170°C [kPa]

Wt is mass%

Therefore, the variability of fuel specifications (max. Simplified Index) in vehicle sales area needs to be considered during GPF development.

12.1.2 Optimization of PN FE

Since high PN FE is necessary for high raw PN levels for a vehicle, the methods for PN FE improvement are as follows.

- PN FE improvement methods for a given GPF size
  - WCL amount increase and wash coat distribution
  - Web thickness increase
  - Pore size distribution modification and control
  - Porosity control

Optimization of GPF coating amount needs to consider GPF characteristics which change during mileage accumulation. Since PN FE increases with respect to mileage accumulation, regulation should be met at fresh GPF (soot- and ash-free) state just out of the production line. On the other hand, gaseous emission should meet government regulation at the end of vehicle life. Furthermore, SML reduction is possible as the effective volume is reduced over lifetime from ash accumulation (mostly from engine oil). Generally, fresh GPF PN FE of commercial products is around 50~80%. A higher PN FE GPF is selected for higher raw PN vehicle.

WCL amount should be balanced between emission performance reduction and back pressure increase, which result in vehicle power loss. Conventional TWC substrates use around 250~300 g/L WCL with 2~3 layers, but GPFs use relatively low WCL, around 25~150 g/L, with single layer coating, to have comparable catalytic conversion efficiency. The back pressure increase of the coated GPF is marginal, up to around 50 g/L of WCL, but it will increase significantly with higher WCL. So the WCL amount needs to be optimized considering catalytic conversion efficiency. Although PN FE will be improved with thicker web, the back pressure will be increased as well, resulting in engine power loss.

PN FE can be improved with managing the pore size distribution of GPF by narrowing the distribution width to decrease large pores (e.g., over 30 μm) and shifting median pore size to a smaller region. However, this shift to smaller median pores for FE needs to be balanced with the ability to apply the wash coat inside the porous wall. Also PN FE can be improved slightly by lowering porosity, but it should be optimized because it can impact engine power loss due to high back pressure. Furthermore co-work with catalyzers is necessary to optimize on-wall and in-wall coating ratio to GPF, which influences both FE and back pressure. Increasing the GPF size is also an effective lever in improving both FE and back pressure.

12.2 Soot Regeneration

Since GPFs are made of cordierite, which is the same base material as TWC substrates, the thermal threshold temperature should not be higher than 1400°C. The target maximum temperature should be lower, due to the potential for local heat generation from soot loading and regeneration. Also in case of the TWC-coated GPF, the thermal threshold temperature should be considered in TWC performance perspective, as well as thermal structural integrity of GPF itself. Therefore 1000~1100°C may be reasonable thermal threshold temperature for TWC-coated GPF. The regeneration strategy and corresponding SML will be influenced by the GPF properties, the wash coat properties, and thermal durability limits of each.

12.2.1 Soot Regeneration with Respect to Oxygen Concentration

Since emission temperatures of gasoline engine are relatively high, the soot collected inside the GPF can be managed by passive regeneration with frequent fuel cuts. In the event active regeneration is needed, the rapid soot oxidation reaction, which can cause high temperatures, can be controlled by active regeneration parameters. Soot regeneration in filter is possible by reactions with NO_2 and O_2. However, low temperature, 200~400°C, reaction ($C + 2NO_2 \rightarrow CO_2 + N_2$) with NO_2 and soot is limited because most of NO and NO_2 are eliminated from TWC prior to GPF. Therefore the majority soot regeneration comes from O_2 reaction ($C + O_2 \rightarrow CO_2$) around 600°C. There are several methods to provide oxygen, such as fuel cut at deceleration, lean control (e.g., $\lambda = 1.05$), and secondary air injection from outside of the system.

To avoid thermal damage, it is necessary to control temperature of GPF under the thermal threshold. If the cold operation of engine is repeated with cold starts and without passive regeneration, the SML should be set considering active regeneration. Soot accumulation needs to be managed so not to exceed thermal threshold temperature of the coated-GPF, due to rapid soot regeneration. The amount of soot can be calculated continuously from soot generation in engine vs. soot regeneration in the GPF. The soot regeneration rate is the function of temperature in fuel cut condition, with extra oxygen. The SML will be determined from GPF inlet temperature. Usually the peak temperature occurs at the rear end of the filter since soot regeneration begins from the front end and heat transfers to the rear end. The location of peak temperature during soot regeneration varies with respect to mileage accumulation because there is no soot in the area with ash, so the peak temperate needs to be monitored considering accumulated ash in the GPF with mileage.

Experimentally, the GPF peak temperature was measured with respect to inlet temperature after soot loading. To record the exothermic reaction, the GPF is fitted axially and radially with a total of around 20 thermocouples across the length of the GPF. **Figure 12.2** shows soot regeneration characteristic inside the GPF with and without fuel cut.

Peak temperatures were monitored for the cases with and without fuel cut at the inlet temperature of 700°C. In the case of fuel cut condition with oxygen (a), a temperature increase of GPF is found due to soot regeneration. It is confirmed that soot was completely regenerated by no weight change during three repeated fuel cuts. On the other hand, there is no soot regeneration if there is no fuel cut at 700°C, due to the lack of oxygen (see Figure 12.2b). However, there is similar temperature profile with (a) for soot regeneration during the second cycle with fuel cut as shown in Figure 12.2b. It is confirmed that there will be no soot regeneration at a relatively high temperature, 700°C, without oxygen.

FIGURE 12.2 Peak temperature of GPF with or without fuel cut (inlet temperature: 700°C).

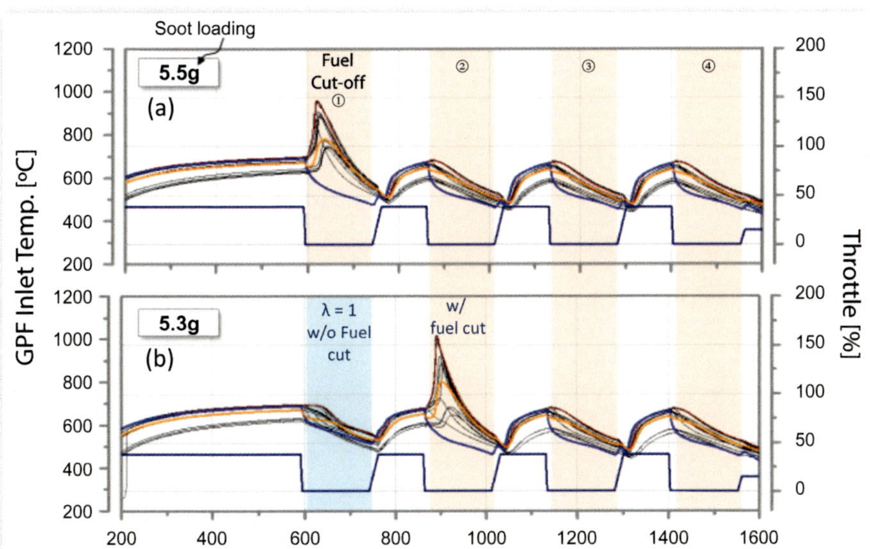

12.2.2 Soot Mass Limit

Figure 12.3 shows peak temperatures within a GPF with respect to soot loading amount with various inlet temperatures. The peak temperatures are measured using a sufficient amount of oxygen under fuel cut conditions.

If thermal threshold temperature of GPF is assumed to be 1100°C, considering deterioration of TWC, the GPF operation window can be determined from inlet temperature vs. SML (see **Figure 12.4**), and varies upon engine and GPF specifications. If soot loading is less than A, it is in a stable region, without GPF failure. If soot loading is between A and B, it is a limited fuel cut region, in which fuel cut in the region should be limited by soot loadings and inlet gas temperature. On the other hand, if soot loading is more than B, peak temperature should be controlled by active regeneration. It can be also achieved by controlling oxygen concentration result in relatively slow

FIGURE 12.3 Peak temperatures in GPF vs. SML at various inlet temperatures.

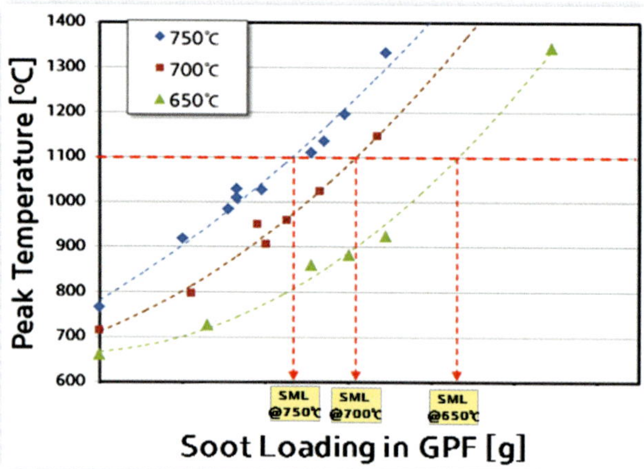

FIGURE 12.4 GPF operating window.

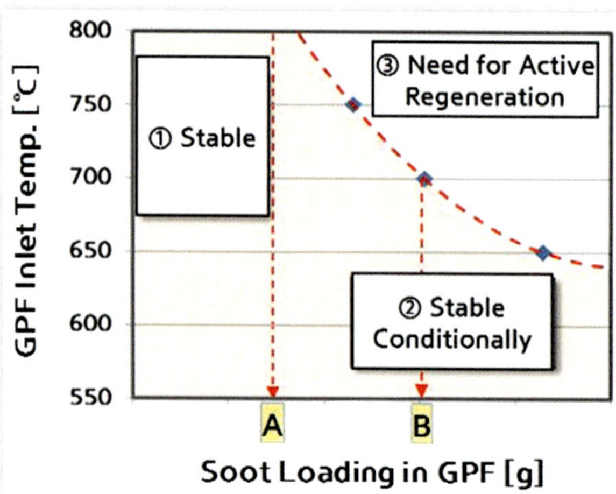

soot regeneration. Soot regeneration is limited at GPF inlet during engine operation under theoretical air-fuel ratio because oxygen concentration is close to zero.

12.2.3 Soot Regeneration Rate

Soot regeneration efficiency can be calculated from the equation below (Eq. 12.3):

$$-\frac{dm_c}{dt} = A \times \exp\left(-\frac{E_a}{RT}\right) \times m_c \times Po_2 \tag{12.3}$$

where

T is temperature of emission gas

m_c is soot mass

P_{O2} is oxygen partial pressure

It can be found from the equation that soot regeneration rate increases with higher emission gas temperature and higher oxygen concentrations. **Figure 12.5** shows the soot regeneration efficiency with respect to oxygen concentration and inlet temperature. Soot regeneration efficiency linearly increases at 600°C with respect to time and reaches saturation after a certain period of time. Therefore, the regeneration duration is set to 300 s to secure linearity of soot regeneration efficiency vs. time.

Soot regeneration efficiency is relatively high when oxygen concentration is higher than 1%, but it rapidly decreases with less than 1% oxygen concentration at a GPF inlet temperature of 700°C. Therefore, it can be concluded that rapid soot regeneration can be controlled through oxygen concentration control (Lambda control). At a GPF inlet temperature of 600°C, the soot regeneration efficiency gradually increases with the oxygen concentration, but the overall efficiency is in a relatively low level. **Figures 12.6** and **12.7** show for soot regeneration in a coated GPF the change in regeneration rate vs. initial soot loading and vs. GPF inlet temperature. It is observed that there is elevated regeneration rate if initial soot loading is high, due to a large exotherm during generation at 700°C. As shown in Figure 12.7, the temperature

FIGURE 12.5 Soot regeneration rate as a function of O_2 concentration and inlet temperature.

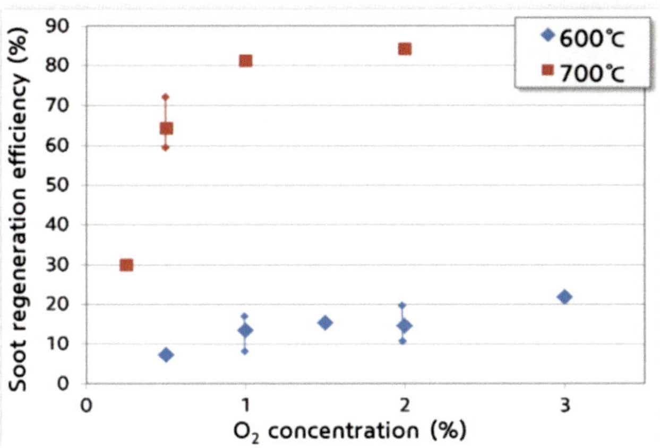

FIGURE 12.6 Soot regeneration vs. initial soot loading (at 700°C).

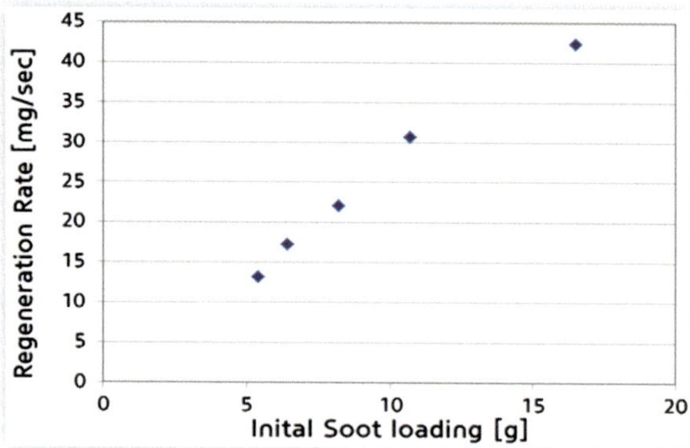

FIGURE 12.7 Soot regeneration vs. GPF inlet temperature (at O_2 1%).

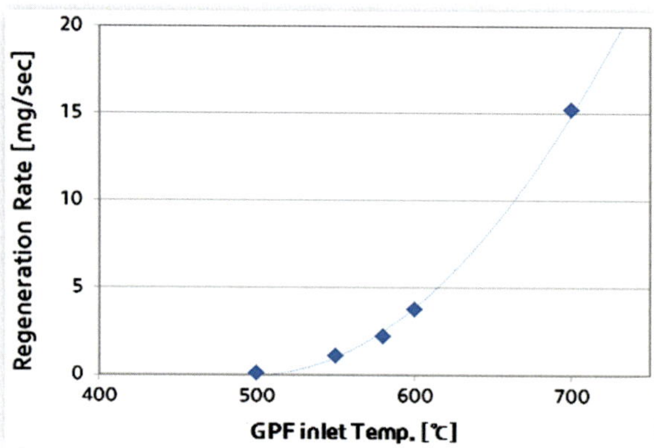

should be higher than 550°C for actual soot regeneration. At less than 1% oxygen concentration, the soot regeneration rate is ~0 mg/s at 500°C and 5 mg/s at 600°C, which requires 1000 s to regenerate 5 g of soot. However, the soot regeneration rate is ~15.2 mg/s at 700°C which is three times faster than that of 600°C.

12.3 Durability Test

12.3.1 PN FE

PN FE, catalytic conversion efficiency, engine power performance, and thermal characteristics need to be considered for GPF durability. Figure 12.8 shows computed tomography (CT) and optical images of GPF after 160,000 km fleet durability test.

As shown in CT image, Figure 12.8, ash is usually deposited at the rear side of GPF in a cone shape. Since emission gas flow in center line is relatively faster than that of GPF periphery, ash is also concentrated in the center flow. As shown in the microscopy image in Figure 12.8, ash is deposited on the entire wall and some at the rear end, where the ash shows higher density. Ash does not penetrate appreciably into the porous walls

FIGURE 12.8 CT and optical microscopy images of GPF after 160,000 km fleet durability.

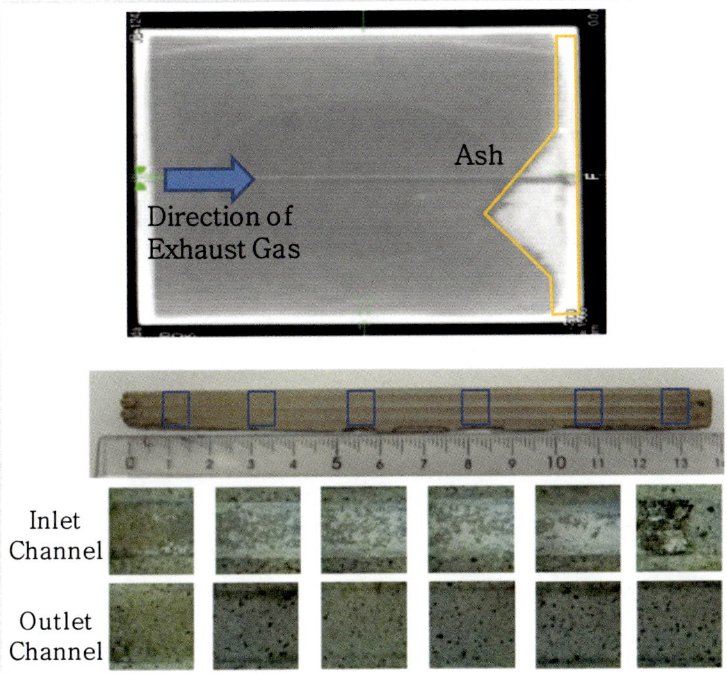

and is not observed on pores near outlet channels. Differentiated from DPF (most of the ash is deposited in the rear side due to active regeneration), the major portion of ash is deposited on porous channel walls and only some ash is located in the rear side. Ash on the wall is one of the reasons for increasing back pressure and increasing FE of the GPF.

As shown in Figure 12.9, major contents of ash are P and Ca from XRF (X-ray Fluorescence). Based on the ash content, it can be inferred that most of the ash in GPF comes from engine oil. The back pressure increase of GPF, 10~30% of increase after 160,000 km of fleet durability test, is impacted by ash amount and distribution. PN FE is increased drastically and catalytic conversion efficiency of GPF is comparable to that of conventional UCC substrate.

12.3.2 Ash Effect on SML

After the fleet durability test, the GPF peak temperature characteristics, with respect to soot loading, are studied to investigate the ash effect. As shown in Figure 12.10, SML decreases for the GPF with ash compared to the fresh GPF, which can be attributed to the ~40% effective volume decrease, due to ash accumulation. Since the accumulated ash near the plug is less than 10% of GPF volume, the rest of the effective volume decreases by 30%, due to ash deposited on the wall. Typically, 20~40% of ash is deposited on the rear plugs and 60~80% of ash is deposited on the wall. Since the accumulated ash disturbs the soot flow in GPF and decreases effective GPF volume, higher peak temperatures are observed with the same amount of soot. Therefore, the SML during active regeneration is determined based on the effective volume decrease,

FIGURE 12.9 Ash compositions of GPF after 160,000 km fleet durability.

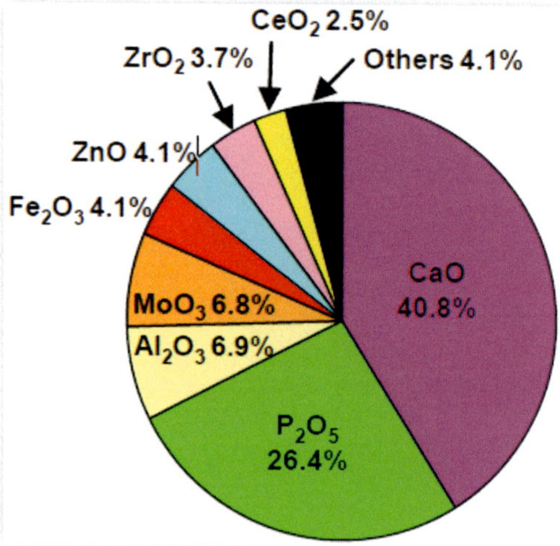

FIGURE 12.10 Ash effect on SML.

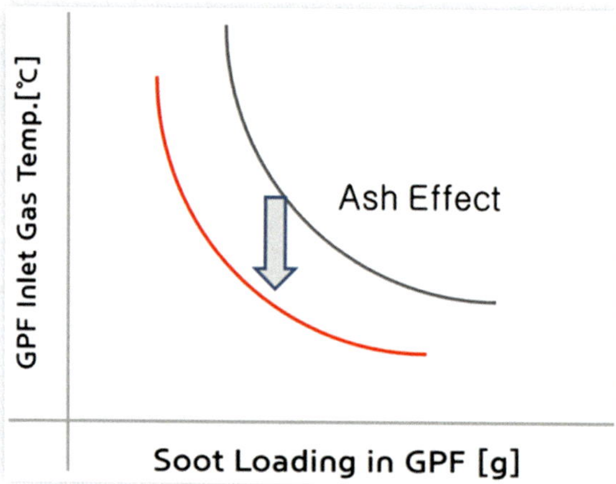

due to ash accumulation. Since the ash amount is impacted by engine oil content, consumption amount, driving pattern, etc., multiple variables should be taken into account to predict the ash influence.

To reduce ash amount, engine oil selection is important, in addition to reducing engine oil consumption. Since most of diesel vehicles are now equipped with DPFs, engine oils with less ash contents are typically used. It is possible to reduce ash generation with low contents of sulfated ash, phosphorus, and sulfur (SAPS). Generally, high SAPS content gasoline engine oils have around 30% higher ash content than that of low SAPS (ACEA, C2/C3 grade) engine oils. Therefore, the ash accumulation amount during a fleet durability test can be decreased by 30%, using low SAPS oils under the same oil consumption conditions. The ash deposit amount can be predicted based on engine oil consumption, since GPF weight increases linearly vs. mileage accumulation. Based on the prediction, GPF volume can be optimized and SML can be set during active regeneration.

FIGURE 12.11 Schematic of underflow GPF.

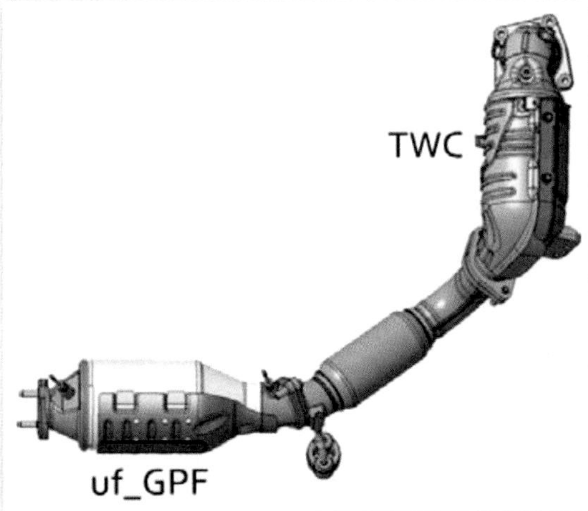

12.4 Implementation of GPF in Mass Production Vehicles

Since early 2000, Hyundai Motor Group (HMG) has been developing various technologies related to performance, durability, and logic for GPF implementation. HMG is planning to implement GPFs in all GDI (gasoline direct engine) and Turbo-GDI platforms starting from 2018 in the EU market, and proliferation to other areas will follow. **Figure 12.11** shows the schematic of under-floor GPF which is under development in HMG. HMG continuously puts effort in improving GPF system performance and durability while reducing raw PN from engines for eco-friendly vehicle development.

References

1. Craig, A., Warkins, J., Aravelli, K., Moser, D. et al., "Low Coat LEV-III, Tier III Emission Solutions with Particulate Control Using Advanced Catalysts and Substrates," SAE Technical Paper 2016-01-0925, 2016, doi:10.4271/2016-01-0925.

2. Nieuwstadt, M.V. and Ulrey, J., "Control Strategies for Gasoline Particulate Filters," SAE Technical Paper 2017-01-0931, 2017, doi:10.4271/2017-01-0931.

3. Rose, D., Boger, T., Nicolin, P., Ingram-Ogunwumi, R. et al., "Advanced Filter Technologies to Reduce Particulate Emissions of GDI Engines, Implications on Exhaust Gas Aftertreatment Systems and Functionalities," *22nd Aachen Colloquium Automobile and Engine Technology*, Germany, 2013.

4. JAMA-ACEA, "Executive Summary – RDE Heavy Fuel Impact on PM Emission," 2016.

About the Authors

Mr. Jung-min Seo is the leader in the development of aftertreatment system of gasoline, diesel, and hybrid engine in the R&D headquarters of Hyundai Motor Group. He joined Kia Motor Co as project manager in 1991 and was the design engineer of diesel engine from 1995 to 2003. Then he developed the diesel aftertreatment system by 2014. His major was mechanical engineering.

Dr. Chi-bum In earned both his master's degree in Material Engineering, in 1991, and PhD in Electronic Material Engineering, in 1994, from Korea Advanced Institute of Science and Technology, Korea. His background is in ceramic material and thin film engineering. He joined Hyundai Motor Group in 1991 and is currently working in the area of aftertreatment system development. His current responsibilities include developing the advanced GPF system for the upcoming global emission legislation.

Section 4

Measurement, Modeling and Control

CHAPTER 13

Measurement of Gasoline Particle Emissions: Laboratory and On-Board Vehicle

Les Hill
Horiba

13.1 Background to the Development of Particle Number (PN) Counting

In the late 1990s, as particulate matter (PM) emissions continued to be an increasing issue for public health, concerns were raised with respect to:

- New vehicles were reducing PM mass emissions, but the increasing use of higher injection pump pressures and also more effective injector spray patterns could be producing higher numbers of smaller particles that could be more injurious to public health. These smaller particles contribute little to the mass of PM measured.

- The ability of the filter mass-based gravimetric method to accurately measure the PM emission as it continued to be reduced in the future.

Consequently a working group, the Particle Measurement Programme (PMP) was formed under the Group De Rapporteurs Sur La Pollution Et L'Energie (GRPE; English name—Working Party on Pollution and Energy) under the World Forum for Harmonization of Vehicle Regulations (WP.29) under the United Nations Economic Commission for Europe (UNECE).

The aim of the PMP was to develop a new system of PM measurement (instrument, sampling, and procedures) to complement or replace the existing gravimetric filter method. PMP is open to governments or industries who are members of GRPE, and all contributors funded their own research.

Following a literature study, the process of the PMP was organized into various phases:

- PMP Phase 1: Measurement systems addressing several key particle properties including mass, number, active surface, and chemistry were evaluated along with appropriate dilution methods, sample conditioning, and consideration of cost and logistical aspects.
- PMP Phase 2: The best performing systems from Phase 1 were subjected to more rigorous evaluations in order to determine fundamental levels of repeatability within a single laboratory during a variety of steady-state and transient tests on both engine-out and post-DPF exhausts.

One major exercise of candidate system evaluation was made by EMPA (Swiss Federal Laboratories For Material Science and Technology) using a heavy-duty (HD) diesel engine fitted with a diesel particulate filter (DPF) with a bypass to simulate an engine with higher emissions.

This program involved around twenty instruments including the use of the "enhanced" gravimetric PM mass measurement system similar to which had been described by Environmental Protection Agency (EPA) to measure the lower PM mass emissions from HD vehicles for 2007 (and which is now the basis of what is specified in the EPA CFR 1065 regulations for PM mass measurement).

The PMP working group, having reviewed all of the data provided by the various programs including the EMPA trial, concluded with two recommended methods:

- PM measurements applying the EPA-2007 style gravimetric sampling technique for particle mass
- PN counting technology using the condensation particle counter (CPC) which had demonstrated very high sensitivity for the low-level particle emissions from vehicles fitted with high-efficiency particle filters. The CPC also had the advantage of not being proprietary to a single supplier since the patent for its operating principle had expired.

13.2 CPC Design (Figure 13.1)

Exhaust particles which have been diluted by a Constant Volume Sampling (CVS) (or a Partial Flow Dilution System in the case of HD engines) are sampled into the CPC inlet. The aerosol passes into a heated section where a working fluid vapor

FIGURE 13.1 Outline of a condensation particle counter (courtesy of TSI).

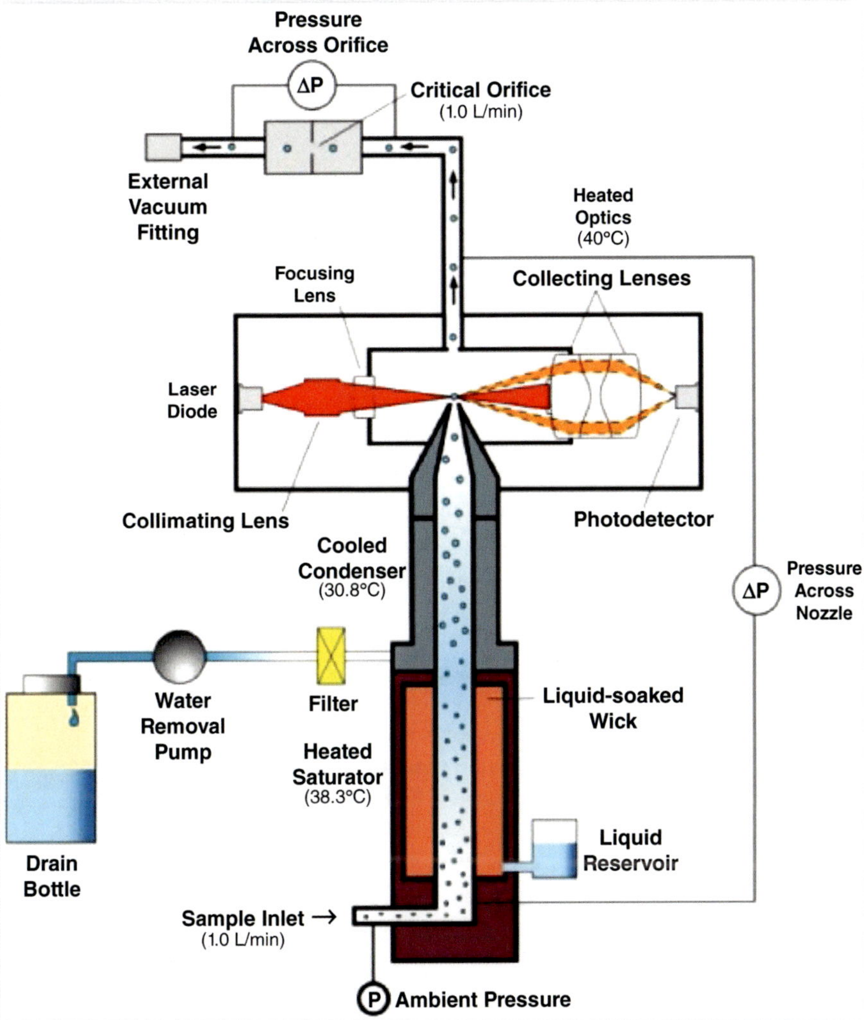

(typically but not exclusively butanol) is entrained into the sample. The aerosol/butanol mixture is then drawn into the a cooler section, where the butanol vapor condenses around the particles in the aerosol and grows them considerably from their initial size to that where each particle can be detected by the scattering of laser light which is beamed across the condenser outlet. The actual flow rate of the sample into the CPC is measured, in the above case, by a flow orifice. The CPC therefore provides the number of particles per unit of aerosol volume, typically per cubic centimeter at the reference conditions (p/cc or p/cm^3).

One aspect of the PN concentration measurement in a CPC is the case where two or more particles break the laser beam simultaneously: such a case can be determined empirically from the fundamental CPC calibration procedure and a coincidence correction factor generated and applied.

13.3 PN Counting System Design

One conclusion from the various programs was that volatile particles could not be measured reproducibly as their formation was highly dependent on a number of factors such as exhaust temperature, dilution ratio, dilution air temperature, and humidity.

Consequently, it was recommended by the PMP that the new PN parameter would be based on the counting of solid particles—volatile particles being eliminated by the design of the PN measurement sampling system (**Figure 13.2**).

In addition, there must be sufficient dilution to maintain the CPC in single particle counting mode—defined as being below the number at which the coincidence correction factor is less than 10%.

Exhaust gas from the vehicle or engine is diluted with high-efficiency particulate air (HEPA)-filtered air and mixed homogenously within the CVS dilution tunnel section. A sample is then taken and optionally passed through a size classifier, such as a cyclone, which is designed to have a 50% cut point for particles between 2.5 and 10 μm at its specified sample flow rate, such particles being considered as artifacts of the measurement system, for example, re-entrained from the walls of the exhaust transfer pipe or dilution tunnel sections.

The diluted exhaust is then further diluted by hot HEPA-filtered air at a temperature of 150 to 350 (±10) degrees C inhibiting the formation of volatile particles. The minimum dilution ratio for this stage is specified to be at least 10, but higher dilution ratios are typically applied, dependent on the PN concentration of the sample gas, in order to maintain the CPC in single particle counting mode.

The hot double-diluted exhaust is then passed into an evaporator unit or tube typically maintained at 350°C, the purpose of which is to vaporize any volatile particles that may have survived the initial hot dilution stage.

Finally, the sample is diluted again in order to bring the temperature below 35°C, at which point it can be supplied to the CPC without affecting the CPC operation. Such room temperature dilution to reduce the sample gas temperature minimizes the thermophoretic losses of particles that would occur should the hot diluted sample after the evaporator tube be simply allowed to cool down to room temperature. The

FIGURE 13.2 Schematic for the PMP sampling system.

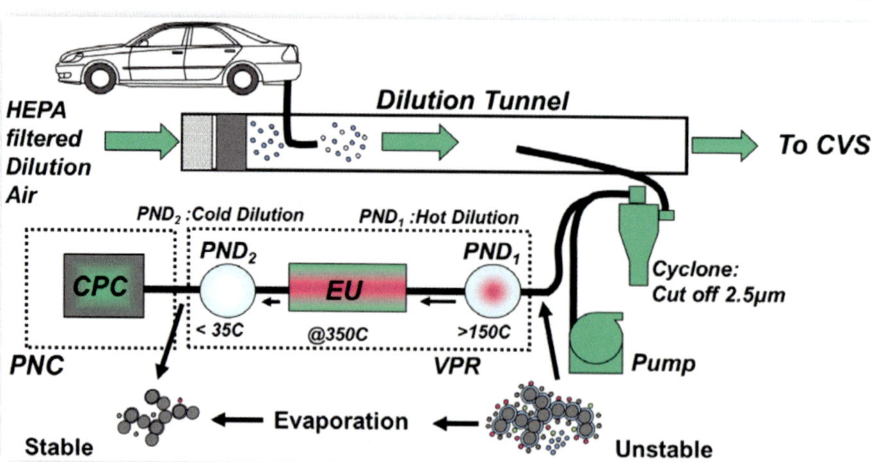

entire diluted exhaust handling from the dilution tunnel to the CPC is defined as the volatile particle removal (VPR) system.

In order to further inhibit any measurement of volatile particles that might remain after the VPR, the design of the CPC was defined by the PMP to have a reduced sensitivity to particles less than 30 nm which are predominantly volatile: specifically, the CPC itself should have a detection efficiency of 50% (±12) for particles at 23 (±1) nm with an efficiency of more than 90% for particles greater than 41 (±1) nm.

13.4 Performance Criteria for Particle Counting Systems

The primary purpose of the VPR is to remove volatile particles before counting: the performance specification of which is that the VPR must remove more than 99% of 30 nm tetracontane particles at an inlet concentration of at least 10,000 particles per cm^3.

The other purpose of the VPR is to reduce the concentration of the particles sufficiently to maintain the CPC in single count mode. The reduction in the particle concentration through the VPR comes from two sources: the overall dilution applied to the exhaust sample and also the solid particle losses within the VPR. These two elements are combined within the calibration process of the VPR by determining the particle concentration reduction factor (PCRF) which is determined for each overall dilution ratio used for testing.

The PCRF is determined by generation of mono-size dispersed aerosols of 30, 50, and 100 nm. The PN concentration is measured at the inlet of the VPR and the outlet of the VPR for each particle size, and then the individual PCRFs are averaged to form the PCRF for that overall dilution ratio as follows:

$$\overline{f_r} = \frac{f_r(30\ \text{nm}) + f_r(50\ \text{nm}) + f_r(100\ \text{nm})}{3}$$

The criteria for PCRF require that the penetration of 100 nm particles through the VPR is at least 70% and that the ratio of loss of 50 nm particles should be less than 20% higher, or 5% lower than that of the 100 nm particles. For 30 nm particles, the ratio should be less than 30% higher or 5% lower.

13.5 Periodic Calibration and Validation

The PCRF determination must be made by the instrument supplier for all the possible dilution ratios during the original manufacturing.

Thereafter, the CPC calibration and particle detection efficiencies should be checked on an annual basis. Similarly, the PCRF values must also be verified on the dilution ratio settings that are used for Type Approval and be within ±10% of the original PCRF values.

In addition the efficiency of the VPR to remove volatile particles must also be confirmed annually using the tetracontane volatile particle generator.

The CPC flow rate must be checked on a monthly basis and both the CPC and complete PN counting system must be leak checked using HEPA-filtered air on a daily/test basis.

Subsequent to reports of some PN counting systems suffering from the contamination of the butanol wicks that are used to saturate the sample aerosol affecting the CPC particle counting efficiency, it became mandatory to either replace the wick or validate the accuracy of the PN counting system 6 months after the annual PCRF check.

13.6 PN Limit Specified for EURO 5 Compression Ignition Vehicles

Typically limits for criteria pollutants such as NOx, CO, PM, etc. are set in order to maintain or reduce atmospheric concentrations to below the levels that are known to affect human health. The setting of the PN limit was defined by a different consideration in the absence of data on atmospheric PN levels and their effects on health.

The limit of $6 * 10^{11}$ particles per kilometer specified for EURO 5 Phase 2 was defined following an empirical program of measurements (the PMP Inter-Laboratory Correlation Exercise) from a wide range of vehicles including light-duty diesels fitted with high-efficiency (>99%) wall-end type particle filters, plus a level of allowance for the uncertainty of the PMP method.

Subsequently, the same limit was applied to gasoline direct injection (or direct injection spark ignition) light-duty vehicles for EURO 6 (in force September 2014), although an initial derogation was allowed for 3 years to a level of $6 * 10^{12}$ particles per kilometer.

13.7 Other Legislative Applications of PN Limits

Since the adoption of PN in Europe for LD vehicles, it has been extended for other automotive applications such as on-road trucks and buses and most recently for some categories of Non-Road Mobile Machinery (NRMM) applicable for EU Stage V in 2019.

Similarly, it has been adopted by other countries such as China, Japan, and India for light-duty and HD vehicles. Additionally, a Global Technical Regulation (GTR) working group was formed in July 2018 to develop a globally harmonised test procedure and equipment specifications for the measurement of vehicle real driving emissions (RDE) using PEMS.

13.8 PN Counting Systems for Real Driving Emissions

As early as 2012, during the development of the application of portable emissions measurement systems (PEMS) for gaseous components to light-duty vehicles as part of the move towards measuring real driving emissions (RDE), the EU Directorate-General for Enterprise and Industry (DG-ENTR) issued a public call for the development and application of a portable PN measurement system for RDE.

The call resulted in a program of testing of candidate instruments by the EU Joint Research Centre (JRC) located in Ispra, Italy, and also the formation of a PEMS-PN Working Group consisting of instrument suppliers, light-duty vehicle OEMs, technical authorities, and member states.

The successful application of a PEMS-PN system required key targets to be met: reduction of size, weight, and especially power, as the RDE legislation requires that the PEMS are powered entirely independently of the vehicle electrical system, for example, by battery. In addition, the system must be largely unaffected by vibration, shock, road gradient, temperature changes, and altitude variation as well as being sufficiently sensitive, accurate, and reliable: ideally with a good and consistent correlation to the laboratory PMP method used for Type Approval.

The program included instruments with a variety of measurement principles including CPCs, diffusion chargers (both simple and advanced), and also a multistage particle mobility type analyzer.

As a conclusion of the PEMS-PN program phases, two instruments (one CPC based, one an advanced diffusion charger design) were selected to proceed on to the next phase in the form of a round robin program. This involved a test vehicle, along with the two chosen PEMS-PN instruments, being circulated to a number of test facilities. The purpose was to obtain PEMS-PN test data from the vehicle and instruments in different laboratories, different RDE routes, and ambient conditions. In addition, both instruments were compared, sampling from the vehicle tailpipe, against each laboratories own PMP-based PN systems sampling from the CVS.

The conclusions from the round robin not only confirmed the viability of the PEMS-PN systems for RDE testing but also resulted in the preparation of the PEMS-PN specifications, performance criteria, calibration method, and test procedure that would eventually be included in the RDE package 3 regulation (EU Directive 2017-1154).

The RDE package 3 also specified that the PN Conformity Factor of the RDE test to Type Approval testing would be 1.0 with an additional 0.5 to allow for the higher uncertainty of the PEMS-PN instrumentation. It is notable that there is generally a systematic difference between PN measured at the tailpipe and PN measured from the CVS: a value of typically up to 30% has been observed, with the tailpipe measurement reading higher than CVS. It remains the intention of the EU to review the PN Conformity Factor on an annual basis with the aim of reducing the value related to PEMS-PN uncertainty eventually to zero with the expected continuing improvement of the instrument and test method over time and with increasing experience.

13.9 Example: A PEMS-PN System Based on a CPC (HORIBA OBS-ONE-PN) (Figure 13.3)

The exhaust is sampled using a probe from the center line of the vehicle exhaust tailpipe. It is immediately diluted (1) in order to reduce particle losses from raw exhaust gas transfer and also from thermophoresis. In order to reduce the dimensions and power consumption, the diluting gas is created from recirculated exhaust gas which has been HEPA filtered and dehumidified by a desiccant cartridge (2): the flow rate and water content of which is monitored by sensors (3).

The diluted exhaust then passes from the tailpipe via a heated line into a cyclone to remove any large artifact particles before entering the heated catalytic stripper (4) to remove volatile particles. The capacity of the heated catalytic stripper is sufficient

FIGURE 13.3 CPC-based PEMS-PN schematic.

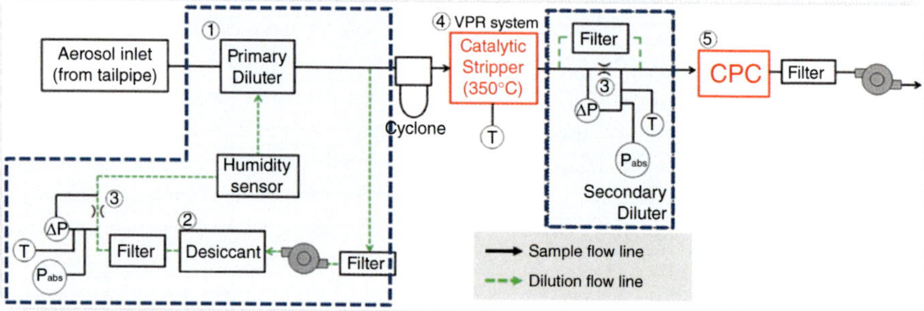

not only to meet the requirements of the PMP method (>99% of tetracontane particles at 30 nm) but also the more stringent specification for RDE which specifies a VPR efficiency of >99% for emery oil with a median particle diameter of >50 nm and a mass of >1 mg/m^3.

The diluted exhaust with purely solid particles then is diluted a second time by means of a exhaust splitter using an orifice and a HEPA filter, with the secondary sample flow rate being monitored and controlled (3). Finally, the sample aerosol is then measured by a CPC, operating in this case with isopropyl alcohol as the working fluid, adsorbed into an easy-to-use cartridge type wick that is replenished after the on-road test is completed, thus eliminating the liquid reservoir that is commonly used in the CPCs for the laboratory PN counting systems.

The use of the CPC provides a direct correlation to the PMP method and is not sensitive to the specified range of particle size and also offers higher sensitivity to the alternative systems using Diffusion Charge type detection.

FIGURE 13.4 Example PEMS-PN using CPC detector.

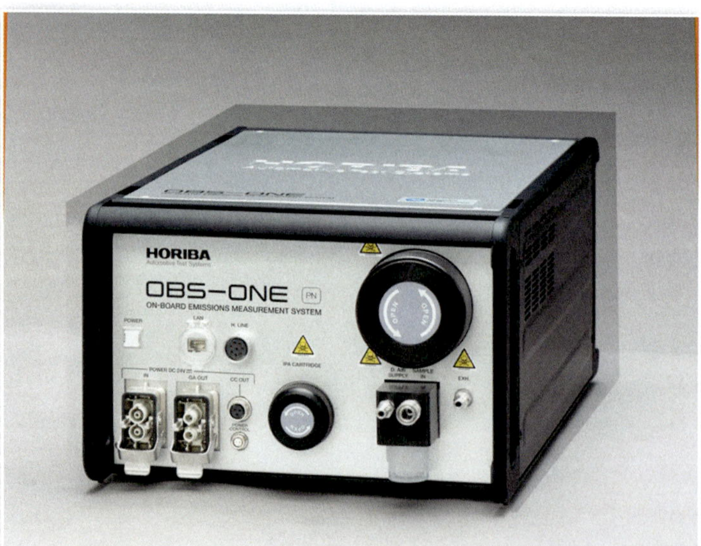

The system is compact and light (18 kg) and has a maximum consumption of 250 W (during warm-up) and has a measurement range of up to $5 * 10^7$ p/cm^3 at the tailpipe (**Figure 13.4**).

The instrument described was a "Golden" system in the PEMS-PN Inter-Laboratory Correlation Exercise organized by the EU JRC.

For the particle counting application, the Diffusion Charger type detector suffers from an inherent sensitivity to the size of the particles, as well as their number. Therefore the specifications for the PEMS-PN specified the use of "advanced" designs where this limitation is largely overcome.

One such design is described.

13.10 Example: A PEMS-PN System Based on an Advanced Diffusion Charger Design (TESTO NANOMET-3)

Instruments based on unipolar diffusion chargers have the potential to be sensitive, robust, and portable. They are based on the electrical charging of particles by a corona discharge. The Diffusion Size Classifier (DiSC) sensor (see **Figure 13.5**) charges the aerosol in a unipolar diffusion charger. After charging, excess ions are removed in an ion trap. The charged aerosol passes through a stage where particles are deposited by diffusion and detected as an electrical current $\boldsymbol{D = I_{diff}}$.

The remaining particles continue to a second stage, the filter stage, and the current is also measured $\boldsymbol{F = I_{filt}}$.

The ratio of these two currents is a measure of the average particle size and is determined during instrument calibration $\frac{F}{D} = I_{filt}/I_{diff} \xrightarrow{yields} d_{gm}$.

Additionally, the diffusion charger DC signal $(F+D) = (I_{filt} + I_{diff})$ correlates well with lung-deposited (alveolar or tracheobronchial) surface area: $\underline{LDSA} = k_0 \cdot (F+D)$. Because the charge per particle is a function of particle diameter $\overline{q}_p \propto d_{gm}^X$, once this

FIGURE 13.5 Schematic for a PEMS-PN using an advanced diffusion charger design.

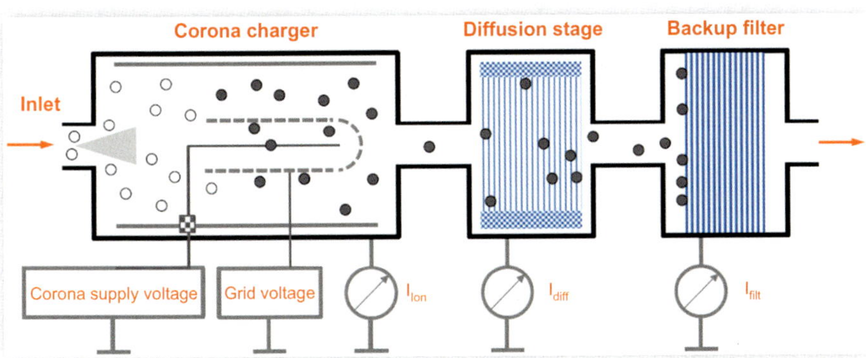

FIGURE 13.6 TESTO NANOMET-3 installed in a vehicle.

is known, the PN concentration can be computed from the total current and flow rate $\overline{q_p}\left(d_{gm}\right)$, $LDSA$, $\varnothing \rightarrow N$.

The sample is taken from the exhaust using the same probe as described for the CPC-based system. In the case of this advanced diffusion charger system, the exhaust is diluted by a miniaturized PMP type stage by means of a rotating disk with an adjustable dilution factor from 10 to 300, minimizing thermophoretic losses. An evaporation tube is then used to remove any volatile particles, and no recondensation occurs as long as the gas has been sufficiently diluted.

This system was a "Golden" system in the PEMS-PN Inter-Laboratory Correlation Exercise organized by the EU JRC (**Figure 13.6**).

13.11 Application of PEMS-PN to Other Automotive Exhaust Applications

The PEMS-PN as specified under the EU 2017-1154 Directive is under consideration for application to HD on-road trucks and buses for In-Service Conformity (ISC) testing as part of their Type Approval. At the time of publication, the EU JRC had concluded a Round Robin exercise with the HD OEMs: the results of which were used to specify the Conformity Factor (CF) that will be applied for PN measurements when the updated EU Directive on HD On-Road ISC is published in early 2019.

Further into the future, the application of the PEMS-PN for NRMM In-Service Monitoring as required under EU Stage V will need to be investigated.

13.12 Future of PN Measurement Within the EU

Following a number of reports on significant numbers of solid particles below 23 nm by JRC and others, the EU has expressed interest in reducing the PN counting size threshold from 23 nm down to at least 10 nm. EU financial support was offered to projects for this initiative under Horizon 2020 that called for programs for high-efficiency diesel and gasoline engine development, with aftertreatment, with special attention to reducing particle emissions below 23 nm. The requirement also included the design and validation of PN systems with the ability to count particle emissions at least down to the 10 nm size for both laboratory and on-board use. These projects are scheduled to conclude and report by the end of 2019.

In addition to the above, GRPE has mandated the PMP to request that the EU JRC investigate the potential of modifying existing PMP compliant systems to count particles from 10 nm upwards. This program is currently underway and involves the circulation of a GDI vehicle (with gasoline fuel to eliminate potential PN emission variation due to fuel composition/volatility differences) and two commercial PMP compliant PN systems that have been simply modified for 10 nm particle counting, along with an additional 10 nm reference PN counting system supplied by JRC. The program involves the participants testing the vehicle with the supplied instrumentation along with their own standard PMP compliant PN systems. The results and conclusions for the program are expected in early 2019.

13.13 Future Applications of PN Counting for Non-exhaust Emissions

It has been reported by a number of sources that a large proportion of atmospheric particle pollution comes from non-exhaust automotive sources such as brakes and tires.

Consequently, the GRPE has mandated the PMP as part of their activities to investigate test methods including measurement devices and procedures. This initiative has now passed the theoretical and literature research stage, and a task force has been formed to move into an actual measurement program, the results of which may be used to define the future standard test and measurement technical regulation.

Acknowledgments

The author of the chapter would like to acknowledge the contributions of

- Dr L. Cachon, Testo SE & Co. KGaA: For schematics, photographs and operating details of the NANOMET-3
- TSI Incorporated: for use of CPC schematic
- Y. Otsuki, HORIBA Ltd: For proof reading and corrections

References

1. EMPA Research-Report No. 202799, Comparison Study of Particle Measurement Systems for Future Type Approval Applications.

2. Particle Measurement Programme (PMP), Light-Duty Inter-Laboratory Correlation Exercise (ILCE_LD), Final Report: Informal document No. GRPE-54-08 Rev.1.

3. Giechaskiel, B., Riccobono, F., Vlachos, T., Mendoza-Villafuerte, P. et al., Vehicle Emission Factors of Solid Nanoparticles in the Laboratory and on the Road Using Portable Emission Measurement Systems (PEMS). *Frontiers in Environmental Science* 3 (2015): 82, doi:10.3389/fenvs.2015.00082.

4. Giechaskiel, B., Riccobono, F., and Bonnel, P., *Chassis Dynamometer Evaluation of Portable Emission Measurement Systems (PEMS) to Measure Particle Number (PN) Concentration: Phase II*, (Publications Office of the European Union, 2015), ISBN 978-92-79-44651-1.

5. Maricq, M., "Monitoring Motor Vehicle PM Emissions: An Evaluation of Three Portable Low-Cost Aerosol Instruments," *Aerosol Science and Technology* 47, no. 5 (2013): 564-573, doi:10.1080/02786826.2013.773394.

6. Bémer, D. and Subra, I., "Monitoring Particle Emission for Non-Road Diesel Machineries Equipped with Particulate Filters," *Journal of Aerosol Science* 113 (2017): 153–165, doi:http://dx.doi.org/10.1016/j.jaerosci.2017.07.019.

7. Fierz, M., Meier, D., Steigmeier, P., and Burtscher, H., "Design, Calibration, and Field Performance of a Miniature Diffusion Size Classifier," *Aerosol Science and Technology* 45, no. 1 (2011): 1-10, doi:10.1080/02786826.2010.516283.

8. Gallus, J., Kirchner, U., and Vogt, R. et al., "On-Road Particle Number Measurements Using a Portable Emission Measurement System (PEMS)," *Atmospheric Environment* 124 (2016): 37-45. doi:doi.org/10.1016/j.atmosenv.2015.11.012.

About the Author

Les Hill has worked on automotive emissions measurement for more than 40 years. Originally trained as an analytical chemist at Lanchester Polytechnic, his introduction to exhaust emissions measurement was at Triumph Motor Company in Coventry, England. In 1982 he joined HORIBA Instruments Limited in the United Kingdom and was fortunate to work in many emissions testing facilities on a wide variety of projects that allowed him to increase his knowledge and experience in all aspects of exhaust emission measurement. In recent years, he has been responsible for emissions measurement product planning activities within HORIBA.

In addition to organizing Emissions Measurement and Testing sessions for the SAE Congress, he has contributed to several industry, legislative, and technical working groups on exhaust emissions measurement. This includes the PMP working group responsible for the development and introduction of the particle number measurement method. He also actively works for global coordination for emissions measurement technology and procedures. He is very grateful to HORIBA Limited who actively encourage their employees to support and contribute to the technical organizations of the automotive industry.

He was presented with the Forest R. McFarland Award from SAE in 2007 in recognition of his contribution to SAE activities.

CHAPTER 14

On-Board Diagnostics for Gasoline Particulate Filters

Jim Daley, Joseph Dekar, Jordan Easter, Nithin Baradwaj, and Brian Terwedo
FCA US LLC

> *By providing technicians with timely and accurate emissions data and helping vehicle owners better maintain their vehicles, OBD plays a key role in helping states meet national air quality standards.*
>
> —**Environmental Protection Agency**

14.1 Introduction

In order for vehicle-focused regulatory requirements to affect change in air quality, it is imperative that OEM's emissions components and systems are monitored over the life of the vehicle. Gasoline particulate filters (GPFs) may become an integral part of the aftertreatment systems for the reduction of fine particulate matter. If employed as a method of soot reduction for the emissions control system of a gasoline engine, the health of this component during the vehicle's life must be held to the same on-board diagnostic (OBD) monitoring standards as other emissions components. Throughout this chapter, the development of GPF diagnostics focused on monitoring filtration and system emissions as they relate to the GPF will be introduced through the following topics: regulatory requirements, failure modes, sensing technologies, algorithm development, and validation.

14.2 Regulation Overview

This section is to be used as an overview for GPF diagnostics as it relates to light-duty gasoline applications and not as a list of design requirements for an OBD-compliant solution. For specific details about GPF regional compliance, the regulations and regulators for that specific market should be consulted. As regulations are subject to change, it is noted that this document reviews regulations as they stood in early 2018.

On-board diagnostic (OBD) regulations typically parallel tailpipe emissions standards for new technologies. As the tailpipe emissions standards for particulate mass (PM) and particulate number (PN) are made more stringent, and in some cases introduced, GPFs are being added to the aftertreatment system for PM and/or PN reduction. The application of GPFs in light-duty automobiles drives the need for corresponding GPF diagnostics. Currently there are three governing bodies that have implemented OBD requirements for PM/PN and GPFs: California Air Resources Board (CARB), the European Union (EU), and China. It is important to mention that currently there are no PN, only PM, OBD emissions thresholds mandated by these regulatory entities. Though CARB is not a government agency, the regulatory agency for the United States, the Environmental Protection Agency (EPA), accepts CARB OBD and does not have any unique requirements for GPF diagnostics.

The CARB OBD requirements for GPFs are derived from the CARB requirements for diesel particulate filters (DPF). Though the OBD requirements are similar for GPFs and DPFs, there are significant operating conditions and porosity differences, discussed in other sections, which drive different compliance solutions for the two technologies [1]. The CARB OBD II regulation, section 1968.2 of title 13, California Code of Regulations, is split into two major sections. One section is "Monitoring Requirements for Gasoline/Spark Ignited Engines" and the other section is "Monitoring Requirements for Diesel/Compression Ignition Engines." Each section contains regulations for components typically equipped on gasoline and diesel engines. However, the push for increased fuel economy and lower PM/PN emissions has led to the application of some traditionally diesel technologies on gasoline engines. For these cases, CARB's guidance has been to use the component part of the regulation that best matches what is being installed on the vehicle, no matter what section it is in [2]. In the case of GPFs, the corresponding regulation is in Section F "Monitoring Requirements for Diesel/Compression Ignition Engines" under section 9 "Particulate Matter (PM) Filter Monitoring" [1].

Additionally, CARB has issued "Proposed LEV III Thresholds for Gasoline" that include specific PM monitoring thresholds for most OBD monitors including GPFs. The CARB mandated gasoline OBD emissions threshold for the malfunction indicator lamp (MIL) is a PM of 17.5 mg/mile on a Federal Test Procedure (FTP) Cycle.

Specifically for GPFs, the PM emissions level for a particular package is determined by running a failed GPF on a FTP cycle. As will be discussed in the "GPF Failure Modes" section to follow, any failure mode that would result in PM over 17.5 mg/mile must illuminate the MIL.

Though Section F(9) on "Monitoring Requirements for Diesel/Compression Ignition Engines/Particulate Matter (PM) Filter Monitoring" includes regulations that are specific to diesel engines [1], this chapter will focus on the parts most relevant to GPFs and will summarize why some sections may only apply to DPFs. The first part of the Section F(9) regulation is on filtering performance degradation and has the same requirements for DPFs and GPFs, except for the 17.5 mg/mile OBD emissions threshold mentioned above. There is also a section in the regulation on "missing filters," which typical diagnostics detect as an extreme case of low filter performance. If the standard filtering performance monitor does not detect the missing substrate, a separate diagnostic would need to be created specifically to detect the missing GPF. Regeneration frequency monitoring is a requirement for DPFs due to the need for active exhaust temperature and lambda control. Regarding GPFs, the temperature and lambda conditions to promote regeneration may be met during normal driving given higher exhaust temperatures and lower engine-out soot levels. Accordingly, a regeneration frequency monitor may not be a requirement for a typical GPF application. If a GPF is coated with a washcoat containing platinum group metals (PGM), then the conversion of carbon monoxide (CO), oxides of nitrogen (NO_x), and non-methane hydrocarbon (NMHC) needs to be diagnosed in the same way a three-way catalyst (TWC) is diagnosed today. Even with a monitor for catalyst activity, the filtration efficiency must be diagnosed as mentioned previously. On a DPF application, during a regeneration event, the exhaust lambda ratio may be controlled, and if the incorrect lambda is delivered, it could cause an incomplete regeneration due to the lack of adequate temperatures for proper soot oxidation, thus necessitating a feed gas and feedback control monitoring requirement. Again, on a GPF application, the conditions for regeneration may occur during normal driving without need for active regeneration; therefore the feed gas monitor and feedback control monitor are examples of an OBD requirement that may not apply to a typical GPF application. CARB requires the same level of information storing, fault setting, clearing, and in-use performance ratios as all other emissions-related monitors.

In Europe, the governing regulation is "United Nations Addendum 82: Regulation No. 83" and is managed by the European Union [3]. The revision of this document that includes GPF diagnostics is Euro 6.2. The OBD emissions threshold for activating an MIL is set at a PM of 12 mg/km. The PM level for any specific package is determined with a failed GPF on the Worldwide Harmonized Light Vehicles Test Procedure (WLTP) and the New European Driving Cycle (NEDC). The EU does not have any written requirements for regeneration or regeneration frequency for GPF diagnostics.

The Chinese regulation, "Limits and Measurement Methods for Emissions from Light-Duty Vehicles," is commonly referred to as China 6 [4]. China 6 is published by two regulating bodies, the "Ministry of Environmental Protection" and the "General Administration of Quality Supervision, Inspection and Quarantine of the People's Republic of China (AQSIQ)." For GPF diagnostics, the China 6 regulation has a combination of requirements that are similar to both CARB and Euro 6.2.

Though the PM OBD emissions threshold is closer to Euro 6.2, it is applied to all spark-ignited engines like the CARB regulation. The OBD emissions threshold for activating a MIL is set at a PM of 12 mg/km. The OBD emissions threshold of a specific package is determined with a failed GPF on the WLTP and NEDC drive cycles. China 6 does not require diagnostics for GPF regeneration or regeneration frequency.

14.3 GPF Failure Modes

An OBD-specific failure mode can be defined as the way in which an aftertreatment or emissions control component might fail, having a negative impact on the customer and tailpipe emissions. To ensure emissions systems are operating correctly throughout the life of the vehicle, it is necessary to understand the potential failure modes of a GPF and to determine how they correspond to regulatory and consumer requirements. Throughout this section, several common GPF failure modes will be addressed. It should be noted that these are a few common GPF failure modes and this discussion does not cover all GPF systems and architectures.

14.3.1 Missing GPF Substrate

A GPF can be packaged/canned by itself or with another catalyst such as a TWC. A "missing GPF substrate" should be detected when the GPF is completely removed from the aftertreatment system or severely damaged to the point of extremely low filtration efficiency. Common causes of a "missing GPF substrate" are, but are not limited to, removal by the customer for reduced backpressure and increased engine performance, incorrect build at the canning/assembly plant, and catastrophic failure due to an extreme thermal event.

14.3.2 GPF Cracking and Channel Damage

Due to the thin, porous nature of the GPF channel walls, it is possible for these channels to become cracked resulting in a degradation of GPF filtration efficiency. GPF cracking is most commonly associated with extreme thermal gradients that may occur during the life of the filter as a result of rapid heating and cooling during regeneration or fuel shutoff events. Cracking or channel damage may also occur if the GPF is exposed to localized soot deposits creating areas of excessive heat from regeneration. These localized areas become too hot during the regeneration process, potentially causing cracks or holes in the channel walls. In investigating cracking as a GPF failure mode, it is important to determine what level of cracking constitutes a failed component. This can be determined through regulatory tailpipe emissions standards. Some regulatory requirements may drive the necessity for an analysis of the extent to which the GPF was damaged to determine when the filter should be diagnosed as "failed." It is important to understand that although a GPF is cracked, it may still meet the applicable regulatory requirements. When a failed component exceeds the OBD emissions threshold, the MIL must be illuminated.

14.4 GPF Sensing

This section will discuss all relevant technologies for on-board evaluation of GPFs. Not all of the technologies presented may be useful for OBD compliance of GPFs, and the application of the technology is to be determined by the OBD strategy team within the OEM. The following filter parameters are of concern for the use of sensing technologies: soot loading, ash loading, and filter damage that would result in loss of filtration efficiency.

DPFs have historically used differential pressure sensors (DPS). Though the use of this technology continues into today's market, there has been some adoption of resistive PM sensors. This adoption is driven by a reduction in the allowable tailpipe particulate emissions resulting in a decrease for the filter crack size that must be detected by OBD systems. This reduced crack size often results in a smaller change in differential pressure (DP) as would have been noticed with a more heavily damaged filter. As this reduction in DP reaches a point where the DPS may no longer reliably detect the damaged part, resistive PM sensors may be used given their direct measure of the exhaust soot level.

Though still in the development phase, there are other promising sensor technologies for GPFs: radio frequency, ion charge, and escaping current. As emissions limits become tighter, and as the tailpipe PM value for exceeding the OBD emissions threshold becomes smaller, even the resistive PM sensor may not be capable of properly detecting at these lower limits due to high collection times needed for a sensor response at low soot levels. Given this, the ion charge and escaping current sensing technologies become promising future options as they provide feedback at high frequencies approaching 10 Hz.

Additional parameters of interest in the evaluation of particulate filters are the loading states of ash and soot. These values are used to control regeneration events and determine soot loading dependent filtration efficiency. Though the combination of DPS and virtual models have been used in DPF applications for determining both soot and ash loading, it can be very difficult to resolve these loadings on a GPF for a few reasons. For one, lower engine-out soot levels with respect to diesel enable filter designs to have lower filtration efficiencies and higher permeability. Though this does enhance fuel economy given less restriction of the exhaust gases, it also reduces the pressure drop as measured by the DPS, making detection more difficult. Furthermore, the reduction in filter porosity often results in a loss to substrate integrity, and the filter may not withstand thermal events from soot regeneration that were considered acceptable for a DPF. This effect may demand a lower level of soot be maintained on the GPF with respect to a DPF, furthering difficulties regarding soot level detection. This challenge of resolving soot loading across a filter only continues to grow as the constantly accumulating layer of ash from noncombustible lubrication additives and engine wear must be considered. One potential solution is the use of a radio frequency sensor as it is capable of measuring the soot and ash loading on the GPF independently and at low levels. This sensing technology and those mentioned before are discussed in detail in the subsequent paragraphs.

FIGURE 14.1 Diagram of a differential pressure sensor (DPS) system.

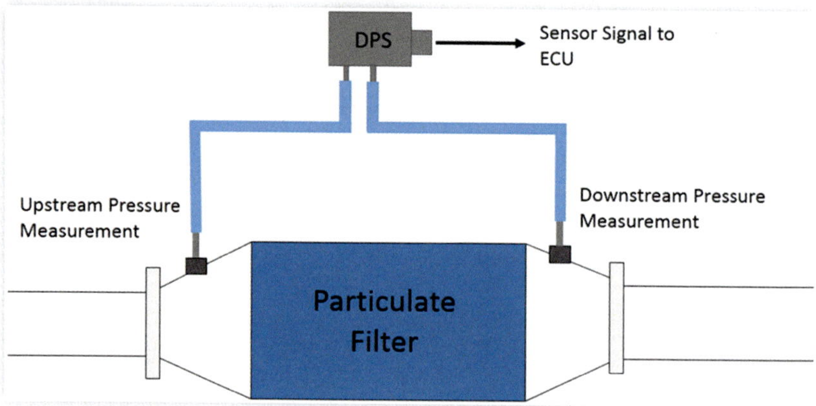

14.4.1 DPS

The DPS is widely used for current production DPF-equipped vehicles. This sensor measures the physical drop in exhaust pressure across the filter substrate as shown in **Figure 14.1** with some sensors providing the option to measure the upstream/downstream pressure in addition to the pressure drop across the filter. DPS are often based on a piezoresistive technology where exhaust gases are exposed to a flexible silicon membrane with an embedded strain gauge connected to a Wheatstone bridge. This silicon membrane will flex when pressure is applied. The voltage output resulting from the pressure-induced flex of the silicon membrane is amplified and sent to the engine control unit (ECU) on the vehicle.

14.4.2 Resistive PM Sensor

The resistive PM sensor has recently been introduced to the market for DPF monitoring and is currently used in some applications. It is a direct measurement of the soot level in the exhaust and can be used to measure soot breakthrough from a damaged DPF. This sensor is similar in size to an oxygen sensor and is mounted in a similar fashion in the exhaust pipe, typically downstream from the particulate filter. The resistive PM sensor uses a ceramic sensor element with two interdigitated comblike electrodes with infinite electrical resistance, an internal heater, and a measuring element. Soot particles are deposited on the sensor element, and as the soot layer on the element builds, the measured resistance across the comblike electrodes drops. After a given amount of soot exposure, the measured resistance will drop below a set threshold and the sensor has completed a measurement event. The more quickly the resistance drops, the higher the soot concentration should be in the exhaust flow. The internal heater is used to burn the soot off the element, thus providing a clean start at the next measurement period. A schematic of the sensor element loading and measured resistance is given in **Figure 14.2**.

FIGURE 14.2 Schematic of resistive PM sensor.

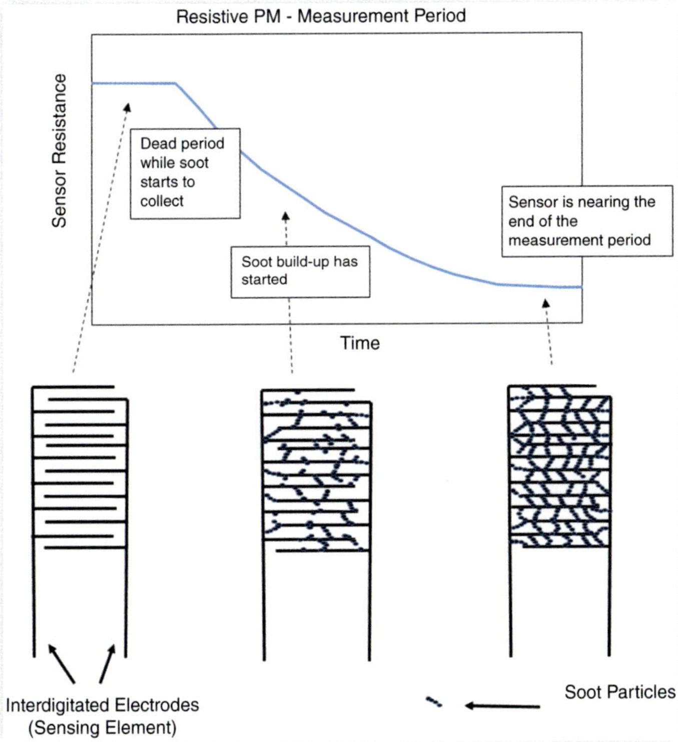

14.4.3 Ion Charge Sensor

The ion charge sensor is another direct measurement technique for exhaust soot levels, though unlike the resistive PM sensor, this sensor operates at a much faster rate, reaching 10 Hz. The ion charge sensor is similar in size to a standard oxygen sensor and is installed into the exhaust in a similar fashion. This technology utilizes two electrically isolated metal electrodes that extend into the exhaust stream. One electrode is given a high voltage bias near 1000 V whereas the other electrode is left as a sensing electrode and connected to a charge amplifier [5]. The high voltage electrode generates a strong electric field between the two electrodes. As portions of particles are naturally charged during combustion [6], these particles drift to the sensing electrode. The sensor output is a measure of the PM level in the exhaust as determined by the flux of charged particles to the sensing electrode [7].

14.4.4 Escaping Current PM Sensor

Another particle measurement technique that allows for fast response times is the "escaping current" method. This sensor type uses electrical detection of aerosols to measure particulate number and mass [8, 9]. During sensor operation, a portion of the

FIGURE 14.3 Radio frequency sensor.

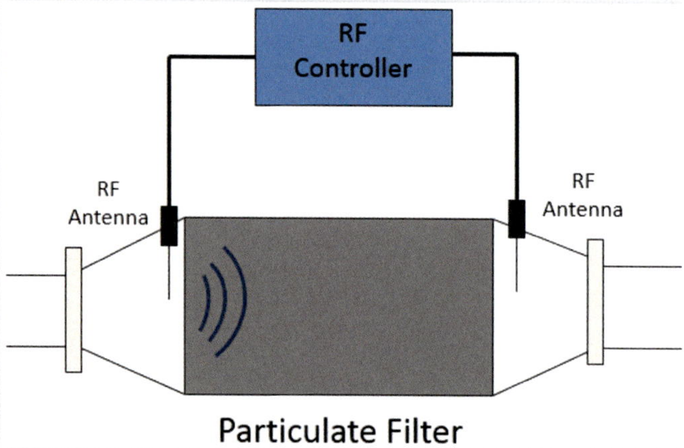

exhaust flow is directed into the sensor where soot particles are charged by a corona ionized flow. This corona ionized flow is created by a corona ionizer and an ejector diluter providing clean air to carry the ions. Most of the corona ions return to the grounded sensor body; however, a portion of these ions find their way to an exhaust particle and escape with the exhaust flow exiting the sensor. These escaping ions are the cause of the name "escaping current." As the level of particulates increases, more current will "escape." This current produced by the charged particles as they leave the sensor is measured through a Faraday cup within the sensor.

14.4.5 Radio Frequency Sensor

The radio frequency sensor differs largely from the previously introduced sensors. The sensor output is not a measure of the soot in the exhaust but of the soot or ash loading state on the filter [10, 11]. Some work has been done to show that damaged filters and soot breakthrough can be detected as well with this sensing technology [11]. The radio frequency sensor may have one or two antennae installed directly into the filter housing. The example given by **Figure 14.3** depicts a two-antenna setup. A series of frequencies are sent through the ceramic filter from one antenna and picked up by the receiving antenna. As a signal moves through the filter, the dielectric properties of the soot or ash change the signal's resonance characteristics, resulting in a received signal that looks different from the one that was sent. Furthermore, an interpretation of this signal shift can be used to determine the state of soot or ash loading on the GPF.

14.5 Algorithm Development

In this section, a basic introduction of different sensor-based algorithms for GPF monitoring will be discussed. Subsequently, the DP-based algorithm is chosen as an example to demonstrate how an algorithm is developed to detect a failed GPF.

One of the most widely accepted techniques for GPF monitoring is DP-based detection. Although the DP-based technique has proven to be a robust detector of complete failures and/or missing substrates, the effectiveness of this technique for partial failures needs to be evaluated on a case-by-case basis. One other method which has been evaluated to detect a missing filter is using an absolute pressure sensor in the front of the GPF. This is a cost-effective solution; however, it requires complex modeling and may not be a viable method for partial failures.

Although there are numerous other methods using information such as inlet/outlet gas temperatures, oxygen storage, and soot sensor values which can be used, DP-based failure detection is used in this work as an example to illustrate how an OBD algorithm is developed. Regional OBD regulations, separation between healthy and failed GPFs, cost, and robustness of the strategy are some of the considerations that can be used to select a detection technique for an application.

14.5.1 Flow Restriction

The most common style of GPF, wall flow, uses an alternating plug pattern with plugs present on both ends, and these plugs, with the porous walls, present a restriction to flow of exhaust gases from the engine. Please reference the illustration in **Figure 14.4**. This inherent characteristic can be used to diagnose the health and filtering capability of the GPF. Flow restriction (R) is defined as the ratio of the difference in measured pressures before and after the GPF (ΔP) to volumetric flow (f).

$$R = \frac{\Delta P}{f} \tag{14.1}$$

At a given volumetric flow, the flow restriction is expected to be higher for a healthy GPF and lower for a cracked GPF, with the lowest value occurring when the GPF is missing entirely. One must also take into consideration any soot or ash inside the filter.

FIGURE 14.4 GPF flow restriction.

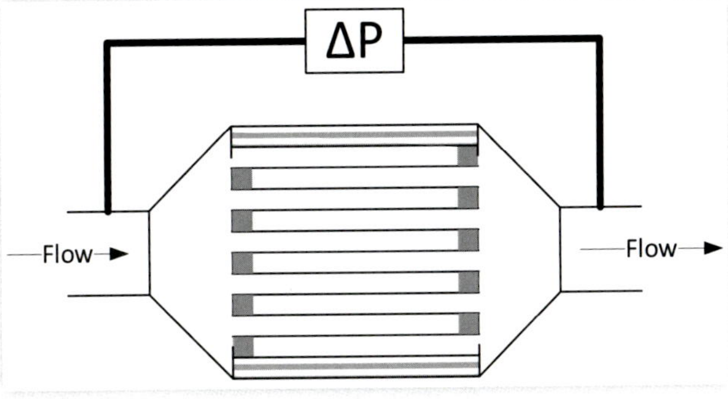

14.5.2 Differential Pressure Signal

The DPS has two probes, one installed in the front and the other installed at the rear of the GPF. When the exhaust gases pass through the GPF, they create a pressure differential between the front (high) and the rear (low) faces of the GPF due to flow restriction. The front probe of the DPS measures the pressure exerted by the exhaust gases on the GPF. The frequency of exhaust pressure depends on firing events or power pulses of the engine resulting in a high frequency DPS signal (100–300 Hz). The raw DP trace for a healthy GPF is shown in **Figure 14.5**. The filtered signal is a moving average of the raw DP trace that can be used for further calculations in place of the high frequency raw signal. Utilizing the filtered signal in place of the raw signal could help improve the consistency of the calculated decision variable, which in turn, could improve the robustness of the diagnostic.

14.5.3 Fault Criteria

One of the important tasks during algorithm development is to define a failed component. In the case of a GPF, a failed component is one that is unable to effectively filter particulates at the level required for OBD compliance. As mentioned in the previous subsection, the resistance to flow is expected to decrease with increasing level of component failures as the filtering capability of the GPF progressively decreases with increasing levels of failure. This is shown in **Figure 14.6**. A new filter is expected to have the least flow restriction, as ash deposits that occur during aging not only contribute to a slight increase in filtration efficiency but also to a marginal increase in the flow restriction. A missing or a completely damaged filter is considered the most severe level of GPF failure. This type of failure also has the maximum separation from the healthy green filter flow restriction line. A crack in the filter that is sufficient

FIGURE 14.5 GPF DP trace.

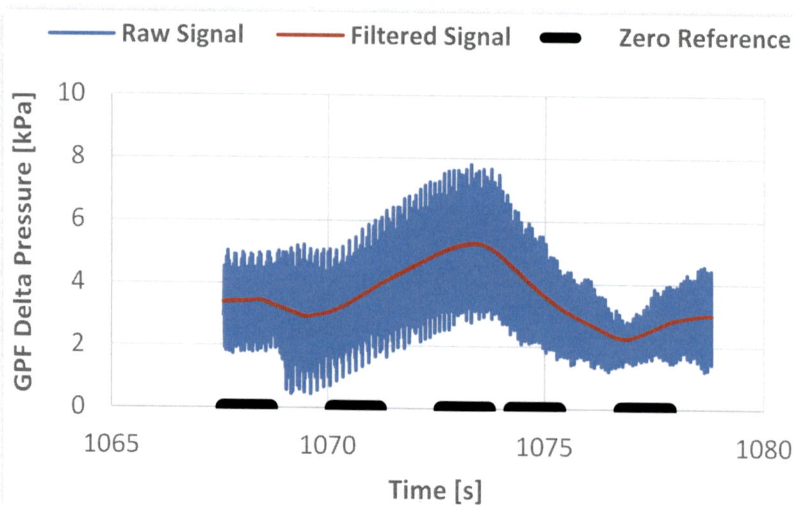

FIGURE 14.6 GPF DP-based fault criteria.

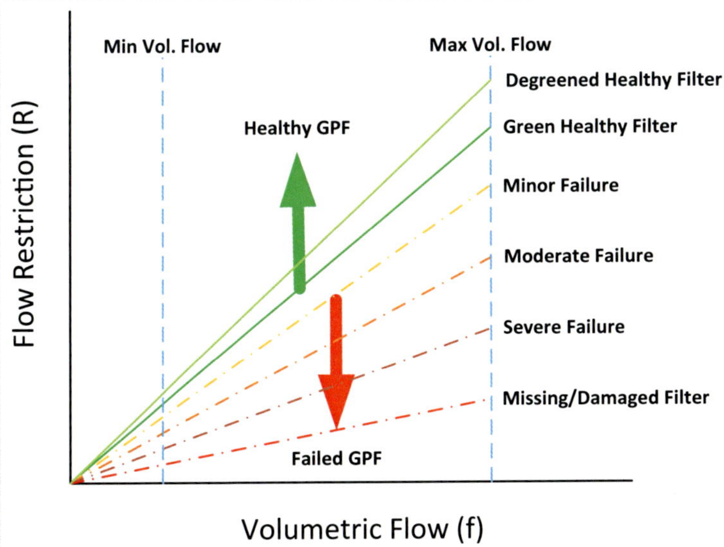

to cause the particulate to exceed the emission limit, but not the OBD limit, has the minimum separation from the healthy green filter flow restriction line. It is important to note that due to several noise factors and measurement uncertainties, the actual flow restriction is always a distribution around the lines indicated in Figure 14.6. Hence, extensive data collection and analysis for both healthy and failed components are essential to determine if the DP-based failure detection can be used to diagnose the GPF failures.

14.5.4 System Identification

The diagnostic algorithm should be able to reliably and consistently make a "fail decision" when the GPF is missing or damaged such that the OBD emissions threshold is breached. The algorithm should also be able to reliably and consistently make a "pass decision" when the GPF is healthy and the tailpipe particulate value is less than the regulatory OBD limits.

The actual response from the algorithm is determined by the noise factors which affect the decision-making process. Control factors are adjustments available to the engineer to help suppress noise factors as much as possible, thus enabling robust algorithm results. One way to design robust algorithms is to develop a P-diagram using the following system components: inputs, outputs, noise, and control factors. Preliminary identification of the relevant parameters helps in designing an algorithm to yield accurate and consistent results. An example of a P-diagram for GPF filtration efficiency monitor and GPF missing monitor is shown in **Figure 14.7**.

Some common noise and control factors have been identified in the P-diagram. These noise factors have been categorized into environmental, manufacturing, and deterioration. Noise factors can be consciously introduced to understand their effect

FIGURE 14.7 GPF filtration efficiency and GPF missing P-diagram.

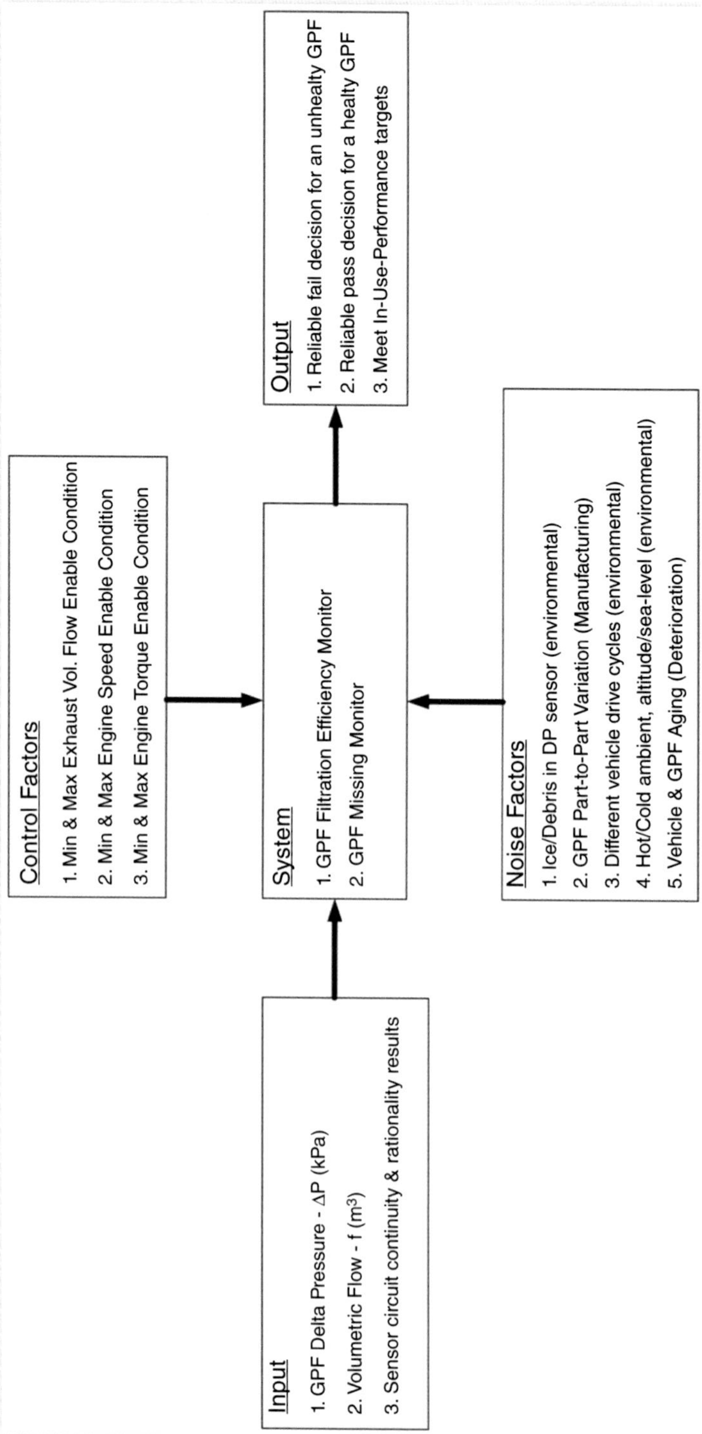

on algorithm results. Control factors not only determine the output of the algorithm but also specify the conditions to execute the algorithm. Many combinations of the control factors are possible, and one of these combinations should give the least sensitivity to undesirable variations in response.

14.5.5 Monitor Enable Conditions, Calculations, and Thresholds

Control factors such as min and max exhaust flow thresholds, exhaust temperature thresholds, engine speed thresholds, and torque thresholds are common monitor enable conditions that may be included to execute the algorithm under known conditions. The algorithm executes only when the enable conditions are true, as the enable conditions should be set to provide separation between healthy and failed GPFs. When the enable conditions are met, the flow restriction could be calculated as shown in Equation 14.1.

The calculated flow restriction should be compared against an error threshold to make a pass or a fail decision. The error thresholds for GPF failures may be established through either constant error thresholds with normalization of the decision variable or dynamic error thresholds without normalization of the decision variable. The former method uses a constant failure threshold value but includes a multiplication or a normalization factor for the calculated decision parameter (flow restriction). The latter method uses a dynamic failure threshold but does not need a multiplier to the calculated decision variable (flow restriction). For either error threshold type, when the flow restriction is greater than the error threshold, a "pass decision" is made, and when the flow restriction is less than the error threshold, a "fail decision" is made. A flowchart of an example algorithm is shown in **Figure 14.8**.

14.6 Algorithm Validation

Validation is the process of ensuring that the system under test meets predefined requirements and fulfills its intended purpose. Existing validation standards and methodologies include IEEE 1012, DO-178C, ISO 26262 Part 6, and MISRA.

The range of validation tests performed will vary depending on applicable regulations. These tests typically include, but are not limited to, the general categories summarized in **Table 14.1** below.

14.6.1 Validation Strategies

Various tools are available to aid validation testing including Model-in-the-Loop, Software-in-the-Loop, and Hardware-in-the-Loop systems, flow benches, dynamometers, and surrogate vehicles. Care should be taken to verify that the selected tests provide full requirements coverage and traceability.

GPF-specific testing, such as tests with damaged GPF substrates, should be performed on appropriate regulatory drive cycles, and diagnostic capability should be validated against the damaged parts to determine what level of damage constitutes a failed part. In addition, robust detection of emissions failures, on and off cycle, with

FIGURE 14.8 GPF health monitor.

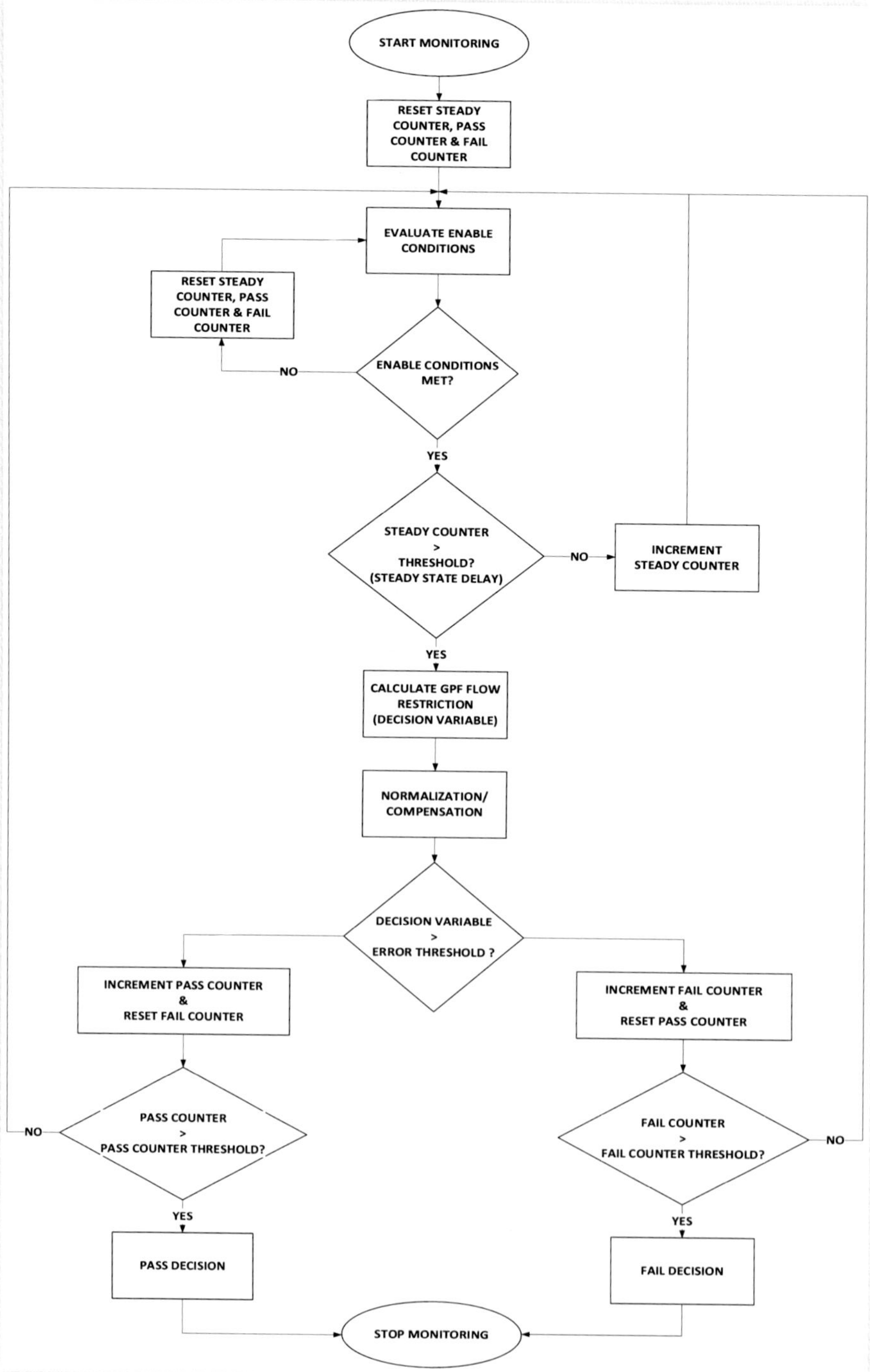

TABLE 14.1 GPF diagnostic validation examples

Validation category	Testing criteria
Algorithm requirements	Validation of enable conditions, thresholds
Sensor failures	Circuit continuity, input rationalities
GPF failures	Filter damage including cracks, missing plugs, melting
System level failures	Partially or fully blocked exhaust system, partially or completely pinched pressure sensor lines
System robustness	Diagnostic variations due to environmental factors, exhaust resonance and turbulence, varying GPF and exhaust geometry, sensor placement
Regulatory compliance	Compliance over applicable drive cycles including regulated emission constituent limits, monitoring conditions, In-Use Monitoring Performance Ratio (IUMPR)

good monitor separation should be validated. It should be noted again that testing with damaged GPF substrates is only required when tailpipe particulate emissions may exceed the relevant OBD emissions threshold at some level of measurable GPF damage. If the tailpipe particulate emissions are below the OBD emissions threshold when the GPF is removed entirely, testing of damaged GPF substrates is not required.

As an example of a tool that may aid validation, six sigma can be used as the target pass/fail separation threshold for evaluating the robustness of GPF missing diagnostics. Regardless of the threshold selected, the system should be tested for robustness against varying factors including GPF system hardware, software implementation, and environmental variations. For example, exhaust and GPF substrate geometry and sensor placement may create system variability over different engine operating conditions. Figure 14.9 below shows an example GPF missing diagnostic output plot, with six-sigma separation. It should be noted that this data is in no way taken from a GPF system and has been included in the plot as a visual aid.

FIGURE 14.9 GPF missing monitor separation.

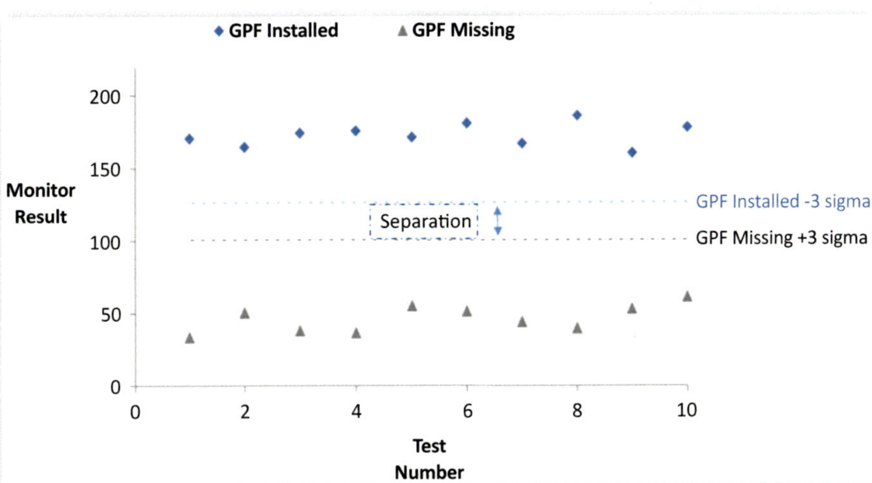

14.6.2 Regulatory Requirements

Lastly, an important part of diagnostic validation is confirming that the system meets applicable location specific regulatory requirements, including relevant use conditions and drive cycles.

References

1. OBD_1. Title 13, California Code Regulations, Section 1968.2, Malfunction and Diagnostic System Requirements for 2004 and Subsequent Model Year Passenger Cars, Light-Duty Trucks, and Medium-Duty Vehicles and Engines (OBD 11).
2. Regenfuss, M., *SAE OBD Symposium*, (Anaheim: SAE International, 2014).
3. 83, United Nations Addendum 82: Regulation No. Agreement Concerning the Adoption of Uniform Conditions of Approval and Reciprocal Recognition of Approval for Motor Vehicle Equipment and Parts.
4. Supervision, Ministry of Environmental Protection and General Administration of Quality and (AQSIQ), Inspection and Quarantine of the People's Republic of China. *Limits and Measurement Methods for Emissions from Light-Duty Vehicles*, (China: National Standard of the People's Republic of China).
5. Diller, T., Osara, J., Hall, M., and Matthews, R., "Electronic Particulate Matter Sensor - Mechanisms and Applications in a Modern Light-Duty Diesel Vehicle," SAE Technical Paper 2009-01-0647, 2009, doi:10.4271/2009-01-0647.
6. Moon, K., "Charging Mechanisms of Submicron Diesel Particulates," Ph.D. thesis, University of Minnesota, Minneapolis, 1984.
7. Steppan, J., Henderson, B., Johnson, K., Yusuf Khan, M. et al., "Comparison of an On-Board, Real-Time Electronic PM Sensor with Laboratory Instruments Using a 2009 Heavy-Duty Diesel Vehicle," SAE Technical Paper 2011-01-0627, 2011, doi:10.4271/2011-01-0627.
8. Amanatidis, L., Ntziachristos, L., Samaras, Z., Janka, K. et al., "Applicability of the Pegasor Particle Sensor to Measure Particulate Number, Mass and PM Emissions," SAE Technical Paper 2013-24-0167, 2013, doi:10.4271/2013-24-0167.
9. Ntziachristos, L., Fragkiadoulakis, P., Samaras, Z., Janka, K. et al., "Exhaust Particle Sensor for OBD Application," SAE Technical Paper 2011-01-0626, 2011, doi:10.4271/2011-01-0626.
10. Sappok, A. and Bromberg, L., "Development of Radio Frequency Sensing for In-Situ Diesel Particulate Filter State Monitoring and After-Treatment System," *ASME International Combustion Engine Division 2013 Fall Technical Conference*, 2013.
11. Sappok, A. and Bromberg, L., "Radio Frequency Diesel Particulate Filter Soot and Ash Level Sensors: Enabling Adaptive Controls for Heavy-Duty Diesel Applications," *SAE Int. J. Commer. Veh.* 72, no. 2 (2014): 468-477.

About the Authors

Jim Daley, Advanced OBD Systems Manager, FCA US LLC received his Masters of Science in Mechanical Engineering from West Virginia University (WVU). While at WVU, Jim was a co-captain of the SAE Formula team and vice president of the WVU chapter of SAE. Jim's research work was on heavy-duty diesel engine emissions and development of a chassis dynamometer-based heavy-duty class 8 emissions test. After graduating college, Jim started working for DaimlerChrysler Corporation and was lead emissions calibrator for the 3.7L V6 and 4.0L I6 engine families. Jim then transferred to OBD calibration where he was responsible for all aftertreatment system calibrations and later became the Center of Excellence leader for Aftertreatment Systems Diagnostics. After Fiat and Chrysler Group LLC merged, Jim was a feature team leader on the Fiat 500 Federalization project that launched the Fiat 500 in the USA.

Jim moved on to become the leader of the Advanced OBD Systems team where he has built a team comprised of outstanding engineers from around the globe. His team is dedicated and focused on designing compliant OBD control systems for all of FCA US future product portfolio.

Joseph Dekar, Advanced OBD Systems Engineer, FCA US LLC received his bachelor's degree in Mechanical Engineering from Binghamton University. He later joined FCA US as an Advanced OBD Systems Engineer working to develop diagnostics for future engine and aftertreatment technologies. He began his career working on aftertreatment sensor diagnostics specifically related to oxygen sensors and their associated in-use aging and failure modes. Most recently, Joseph supported the global implementation of gasoline particulate filters within FCA US, focusing on development of sensor and filter diagnostics. During his time at FCA US, he has been awarded one patent and is listed on three more pending patents.

Jordan Easter, Advanced Emissions Systems Engineer, FCA US LLC completed her Bachelors of Science in Mechanical Engineering at the University of Alabama–Tuscaloosa and her Masters of Science and Doctorate of Philosophy degree in Mechanical Engineering at the University of Michigan–Ann Arbor. Her research has resulted in several publications and has focused on aftertreatment for advanced combustion modes, advanced particulate sensors, and gasoline soot oxidation characteristics. She has worked for FCA US for the past four years in advanced powertrain emissions and diagnostic organizations. During this time, she has focused on advanced monitor development and future compliance regarding particulate emissions of gasoline direct injection engines.

Nithin Baradwaj, Advanced OBD Systems Engineer, FCA US LLC holds a bachelor's degree in Mechanical Engineering from Birla Institute of Technology and Science. After graduation, he went on to work for John Deere India Pvt Ltd as a test and analysis engineer in the product verification and validation group. He also holds a master's degree in Mechanical Engineering from The Ohio State University. During his graduate studies, his research focused on determining the uncertainties involved in soot measurements using on-board resistive PM sensors. After earning his master's degree, he worked at Cummins Inc. as an aftertreatment OBD calibration engineer. In his current role at FCA US, his focus is on developing OBD-compliant technologies that improve fuel economy and/or reduce evaporative and tailpipe emissions.

Brian Terwedo, Advanced OBD Systems Engineer, FCA US LLC holds degrees in Electrical Engineering from Michigan Technological University and Computer Science from Western Michigan University. He started his career at General Electric Aviation (formerly Smiths Industries Aerospace), where he developed aircraft operational flight software for data recorders, loads monitors, and Integrated Vehicle Health Monitoring (IVHM) systems. Later, he joined the General Motors/Delphi Forward Engine Management Systems Group and developed gasoline engine control software and systems including idle and electronic throttle control. Currently, he is a member of the FCA US Advanced OBD Systems group and is responsible for development of advanced engine, OBD, and aftertreatment systems and software. He has over 20 years embedded systems design experience and is an IEEE member.

CHAPTER 15

Modeling of Gasoline Particulate Emission Control Systems and Components

Michail A. Mitsouridis and Dimitrios Karamitros
Exothermia SA

Grigorios C. Koltsakis
Aristotle University of Thessaloniki

Julie Le Louvetel-Poilly, Kotaro Maeda, and Francois-Alexandre Lafossas
Toyota Motor Europe

To support the reduction of the particle emission in gasoline engines, in parallel with the improvement of the combustion process, the particulate filter technology can also be used. Previous experience in the area of diesel particulate filters (DPFs) has shown that mathematical modeling of the filtration, transport, and reaction processes facilitates in-depth understanding and supports system design optimization and control. Although the Gasoline Particulate Filter (GPF) is based on the same filtration and reaction engineering fundamentals, it is subjected to substantially different operating conditions due to the large differences in gasoline engine exhaust gas emission compared to the diesel engine exhaust. In this chapter, the phenomena of interest and the equations used to describe them in mathematical models are presented in a systematic manner. Starting from wall-scale phenomena, the filtration processes occurring at clean and loaded wall state are described. Due to the much lower engine-out particulate matter (PM) emissions of the gasoline engine compared to the diesel engines, depth-filtration processes in the wall are of particular interest to maintain the desired filtration efficiency under all operating modes. The impact of soot accumulation in the wall, and on the wall surface, on pressure drop is presented with the help of measurements and respective simulations. In the case of coated filters, the reaction phenomena in the wall interplay with diffusion processes and can be described by respective equations to account for the conversion of gaseous pollutants, as well as their impact on the accumulated soot oxidation. In a next step, the channel-scale phenomena are presented with emphasis on the flow distribution along the inlet/outlet channels and through the wall. These can be well described by the solution of the mass and momentum balance equations. Special emphasis will be given for the case of zone-coated filters that may produce significant impact on the flow in the channels and affect the preferred soot accumulation in certain areas of the filter. The solution of the energy and species balance equations in the gas phase are important to calculate the temperature and species concentration and provide the proper boundary conditions for the solution of the wall-scale phenomena. The impact of soot accumulation on the diffusion processes in the channel and the resulting effects on the pollutant conversion are examined by respective simulations at steady and transient operating conditions. The oxidation of soot accumulated on the filter is a rather complex phenomenon affected by the nature of the emitted particles as well as the non-carbonaceous components emitted by the engine and filtered on the soot cake. The impact of the non-carbonaceous elements on the soot oxidation chemistry is more important comparatively to diesel exhaust due to their higher concentration in the soot cake. It is therefore important to account for that in predictive mathematical models used to study filter regeneration events. As a next step, the equations describing the flow and heat transfer phenomena at filter scale are presented along with results highlighting the importance of such phenomena under conditions of non-uniform inlet flow and temperature distribution. Last, the effect of ash accumulation and the respective deposition pattern (layer, plug type) on filter performance aspects will be presented with the help of experimental and simulation results.

15.1 Introduction

To reduce carbon dioxide emissions in the road transport sector, the European Commission has adopted mandatory CO_2 emission standards, defined as fleet average values, for all new passenger cars and light commercial vehicles [1]. Regarding gasoline-propelled vehicles, the current trend for most manufacturers is to produce gasoline direct injection (GDI) engines, aiming to improve fuel consumption, and thus lower CO_2 emissions.

However, GDI engines come with the downside of increased particulate emissions, both in terms of particulate mass (PM) and particulate number (PN). According to a European Commission report [2], GDI engines emit an order of magnitude more PN than similar-sized DPF-equipped diesel engines, and at least one order of magnitude more PM than gasoline engines with port fuel injection. Legislative limits have recently been imposed for both PM and PN emissions from gasoline engines while major changes have been introduced to the conformity procedures as well. The introduction of the real driving emission (RDE) cycle with the Euro6d amendment is a typical paradigm. The public road is now regarded as a new certification environment, in addition to the laboratory scale, for the regulation of both gaseous and particulate emissions under real-world driving conditions. Within this framework, additional measures need to be taken for the reduction of the particle emissions in gasoline engines. Passive measures can be adopted to tackle the source of particulates and include improvements in the injection system and the combustion process. In parallel, active measures can also be taken with the introduction of the particulate filter technology [3] in the gasoline exhaust line.

Previous experience in the area of DPFs (diesel particulate filters) has shown that mathematical modeling of the filtration, transport, and reaction processes facilitates in-depth understanding and supports system design optimization and control [4,5]. Although the GPF is based on the same fundamental filtration and reaction mechanisms, it is subjected to substantially different operating conditions. Indicatively, the emitted PM and the oxygen concentration are generally much lower, while the engine-out exhaust gas temperature is much higher in the case of a GDI engine compared to a diesel engine [6].

In this chapter, the phenomena of interest and the equations used to describe them in mathematical models are presented in a systematic manner. Starting from wall-scale phenomena, the filtration processes occurring at clean and loaded wall state are described. Due to the much lower engine-out PM emissions of gasoline compared to diesel engines [7], depth-filtration processes in the wall are of particular interest to maintain the desired filtration efficiency under all operating modes. The impact of soot accumulation in the wall and on the wall surface on the pressure drop are presented with the help of measurements and respective simulations. In the case of coated filters, the reaction phenomena in the wall interplay with diffusion processes and can be described by respective equations to account for the conversion of gaseous pollutants, as well as their impact on the accumulated soot oxidation.

In a next step, the channel-scale phenomena are presented with emphasis on the flow distribution along the inlet/outlet channels and through the wall. These can be well described by the solution of the mass and momentum balance equations. The solution of the energy and species balance equations in the gas phase are important to calculate the temperature and species concentration. They also provide the proper boundary conditions for the solution of the wall-scale phenomena. Then, the solution of the species transfer equation is essential to model coated GPFs with respect to pollutant conversion and their interactions with the accumulated soot. As a coated GPF can perform all three-way catalyst (TWC)-related reactions, TWC reactions are listed as well. Special emphasis is given for the case of zone-coated filters that may produce significant impact on the flow in the channels and affect the preferred soot accumulation in certain areas of the filter [8]. The impact of soot accumulation on the diffusion processes in the channel and the resulting effects on the pollutant conversion are examined by respective simulations at steady and transient operating conditions. The oxidation of soot accumulated on the filter is a rather complex phenomenon affected by the nature of the emitted particles as well as the non-carbonaceous components emitted by the engine and filtered on the soot cake. The impact of the non-carbonaceous elements on the soot oxidation chemistry is more important comparatively to diesel exhaust due to their higher concentration in the soot cake [9]. It is therefore important to account for that in predictive mathematical models used to study filter regeneration events.

As a next step, the equations describing the flow and heat transfer phenomena at filter scale are presented along with results highlighting the importance of such phenomena under conditions of non-uniform inlet flow and temperature distribution.

Last, the effect of ash accumulation and the respective deposition pattern (layer, plug type) on filter performance aspects will be presented with the help of experimental and simulation results.

15.2 Wall-Scale Modeling

This section summarizes the phenomena that take place in the wall scale and presents the main mathematical equations that are used to describe them. The processes of particulate filtration and deposition, both in the porous wall and the soot cake, together with the induced pressure drop are of critical importance and are discussed with the aid of experimental and simulation results.

15.2.1 Wall Filtration

The separation of the particles from the exhaust stream is achieved via a porous barrier (collective surface) which retains them as the flow passes through it. Starting from clean filter state, soot particles are collected via depth filtration inside the empty pores of the filter wall, as schematically illustrated in **Figure 15.1**.

For the length scales of particle sizes and pore size diameters in automotive exhaust conditions, depth filtration relies primarily on two mechanisms [10], shown in **Figure 15.2**:

1. Diffusional deposition, or simply diffusion, which depends on the Brownian movement of smaller size particulates. Those particles do not move uniformly along the gas streamlines but rather diffuse onto the porous surface where they are collected.
2. Inertial deposition, or interception, which becomes more important with increasing particle size. According to this mechanism, particles of significant mass tend to follow the streamlines but may strike onto the porous surface because of their inertia.

FIGURE 15.1 Wall-flow filtration principle.

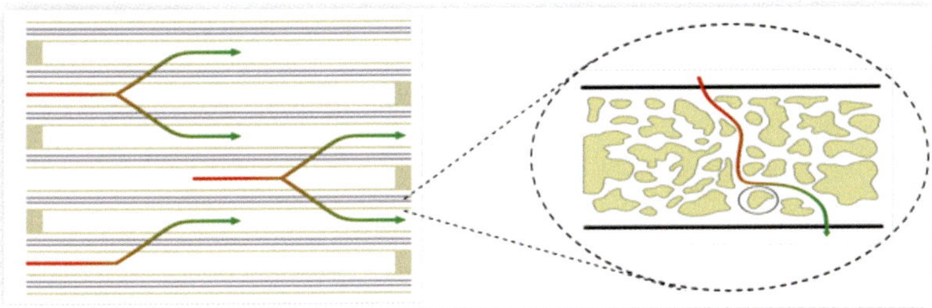

FIGURE 15.2 Filtration mechanisms considered in wall filtration.

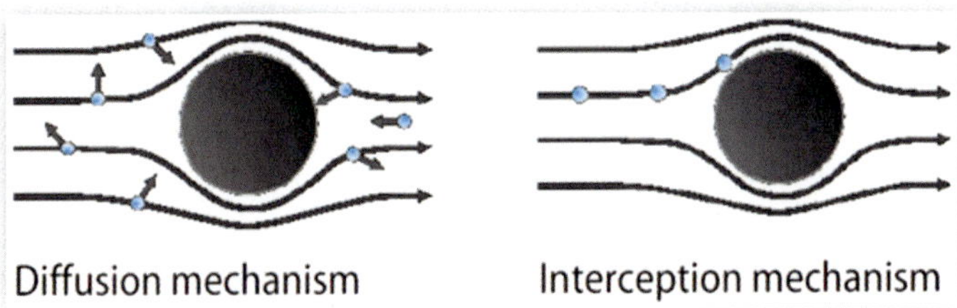

Since particle mass emissions from gasoline engines are generally very low, compared to diesel engines [7], depth filtration is comparatively more important. The mathematical model presented below is based on the standard theory of diffusion and interception for granular media [11] to calculate the filtration efficiency in the wall. Assuming that the filter microstructure can be represented as a structure of objects with a known shape (cylinder or sphere), an analytical solution of the flow field can then be described for the elementary object which is defined as a unit collector.

The clean grain diameter, $d_{grain,0}$ for spherical unit collectors, is computed as a function of the pore diameter d_{pore} and the substrate porosity $\varepsilon_{pore,0}$ at clean state:

$$d_{grain,0} = \frac{3}{2}\left(\frac{1-\varepsilon_{pore,0}}{\varepsilon_{pore,0}}\right)d_{pore} \quad \text{(Eq. 15.1)}$$

The single collector filtration efficiency by diffusion, n_D, is calculated as:

$$n_D = n_{clean,D} \cdot 3.5\left(\frac{\varepsilon_{pore,0}}{Ku}\right)^{\frac{1}{3}} Pe^{n_{Pe}} \quad \text{(Eq. 15.2)}$$

The Peclet number is:

$$Pe = \frac{u_w \cdot d_{grain,0}}{D_{part}} \quad \text{(Eq. 15.3)}$$

The particle diffusivity D_{part} is computed as a function of particle mobility diameter d_{part} as defined in reference [12]. The Kuwabara hydrodynamic factor is calculated for sphere collectors:

$$Ku = 1 - \frac{9}{5}\left(1 - \varepsilon_{pore,0}\right)^{\frac{1}{3}} + \left(1 - \varepsilon_{pore,0}\right) - \frac{1}{5}\left(1 - \varepsilon_{pore,0}\right)^2 \qquad \text{(Eq. 15.4)}$$

The single collector filtration efficiency by interception, n_R, is calculated as:

$$n_R = n_{clean,R} \cdot 1.5\left(\frac{\varepsilon_{pore,0}}{Ku}\right)\frac{R^2}{\left(1+R\right)^m} \qquad \text{(Eq. 15.5)}$$

The m and R terms are calculated by the following equations:

$$m = \frac{3 - 2 \cdot \varepsilon_{pore,0}}{3 \cdot \varepsilon_{pore,0}} \qquad \text{(Eq. 15.6)}$$

$$R = \frac{d_{part}}{d_{grain,0}} \qquad \text{(Eq. 15.7)}$$

Equations (15.2) and (15.5) contain three tunable parameters ($n_{clean,D}$, n_{Pe}, $n_{clean,R}$) which can be tuned in order to capture the behavior of different substrate microstructures and materials.

The total single collector filtration efficiency of the clean wall with both mechanisms is given as a sum:

$$n_{g,0} = n_D + n_R \qquad \text{(Eq. 15.8)}$$

The total bed filtration efficiency of the clean wall is calculated by the following relationship:

$$E_{wall,0} = 1 - \exp\left(-n_{g,0}\frac{3}{2}\frac{\left(1-\varepsilon_{pore,0}\right)}{\varepsilon_{pore,0}}\frac{\Delta w}{d_{grain,0}}\right) \qquad \text{(Eq. 15.9)}$$

With the aid of the mathematical model, a sensitivity analysis was performed in order to illustrate the effect of certain substrate and flow parameters on the particulate filtration efficiency. The basic geometrical and microstructural properties of the GPF that were used in the model are presented in **Table 15.1**.

Filtration efficiency results for clean filters with same porosity and different mean pore sizes, but similar pore size distributions, as function of the mean particle diameter, are given in **Figure 15.3**. For all curves filtration efficiency shows a local minimum for particle sizes around 100 nm. This well-known behavior is attributed to the opposing size effects of the two governing filtration mechanisms.

TABLE 15.1 Properties of the GPF used in the model

Geometry	Length [mm]	122
	Diameter [mm]	117
	Volume [l]	1.3
Wall structure	Cell density [cpsi]	300
	Wall thickness [mils]	10
Substrate	Material	Cordierite
	Density [kg/m^3]	1100
	Pore diameter [μm]	20
	Porosity [%]	63

FIGURE 15.3 Simulated clean filter filtration efficiency as a function of particle size for different mean pore diameters (same substrate porosity of 63%). Inlet conditions: Mass flow rate = 235 kg/h, T = 630°C.

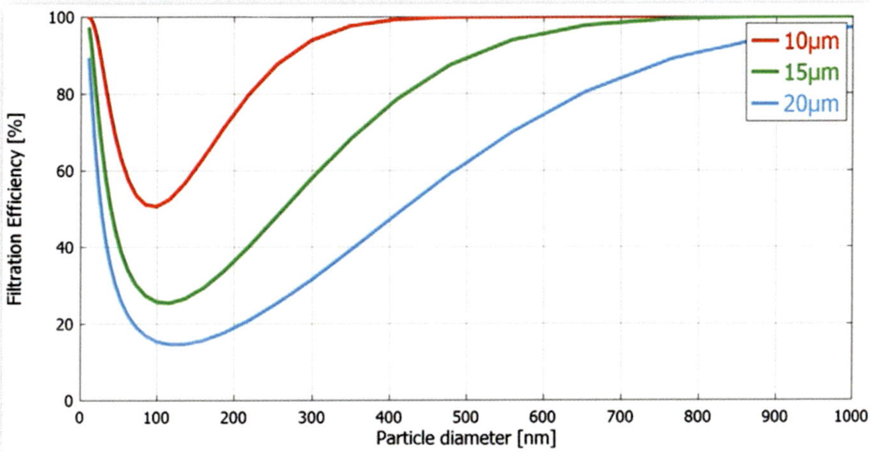

Diffusional deposition is predominant for smaller particles whereas interception is higher for larger particles [1, 2]. As a result, a lower efficiency is achieved for intermediate particle diameters, where neither diffusion nor interception mechanisms are strong enough. Focusing now on the impact of the substrate mean pore diameter, it is observed that a lower pore diameter results in a higher filtration efficiency for the whole range of particle sizes due to the enhanced depth-filtration characteristics.

The impact of wall porosity on filtration efficiency is presented in **Figure 15.4**. Higher substrate porosity, keeping the substrate pore diameter constant (20 μm), leads to a higher filtration efficiency.

In a similar manner the impact of exhaust flow rate on the filtration efficiency was evaluated in **Figure 15.5**. Interestingly, increasing flow rate leads to a lower filtration efficiency only for particle sizes below 500 nm. Since interception mechanism is independent of flow velocity, a change in exhaust flow rate will only affect the diffusion

FIGURE 15.4 Simulated clean filter filtration efficiency as a function of particle size for different wall porosity values. Inlet conditions: Mass flow rate = 235 kg/h, T = 630°C.

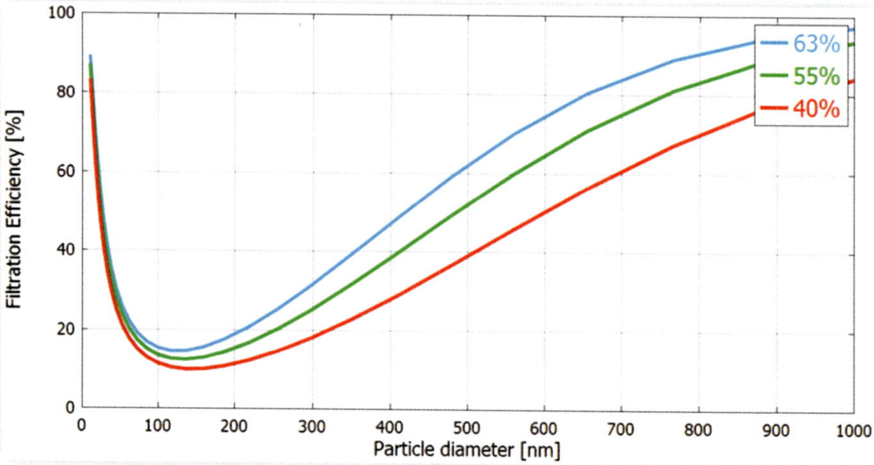

FIGURE 15.5 Simulated clean filter filtration efficiency as a function of particle size for different flow rates. Inlet conditions: T = 630°C.

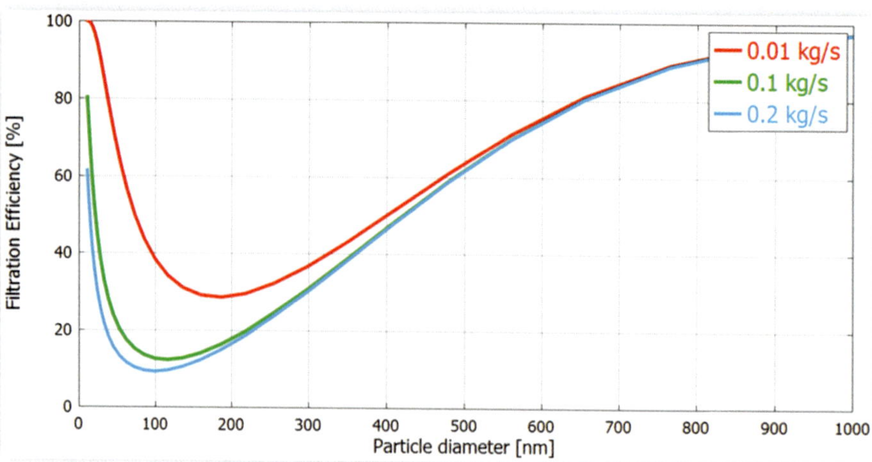

mechanism and thus the lower particle diameters. As the typical range of particle diameters in the gasoline exhaust is between 10 and 200 nm, the influence of the flow rate on the filtration performance can be particularly important.

15.2.2 Effect of Accumulated Soot on Filtration

In the general case of a wall loaded with soot, the filtration efficiency n_g is different than that of the clean wall $n_{g,0}$. Using an assumed density of the deposited soot on the collector, it is possible to calculate the change in the collector diameter and calculate the filtration efficiency of the loaded filter [13]. According to this approach, the substrate wall is discretized in an *i* number of layers, called slabs. The size evolution of the unit collector is calculated at each slab by:

$$d_c(i,t) = 2\left[\frac{3}{4\pi}\frac{m_w(i,t)}{\rho_{soot,w}} + \left(\frac{d_{c,0}}{2}\right)^3\right]^{\frac{1}{3}} \quad \text{(Eq. 15.10)}$$

The local porosity $\varepsilon(i, t)$ and permeability $k(i, t)$ also change with time, as soot is deposited inside the filter, according to Equations (15.11) and (15.12), respectively:

$$\varepsilon(i,t) = 1 - \left(\frac{d_c(i,t)}{d_{c,0}}\right)^3 (1-\varepsilon_0) \quad \text{(Eq. 15.11)}$$

$$\frac{k(i,t)}{k_0} = \left(\frac{d_c(i,t)}{d_{c,0}}\right)^2 \frac{f(\varepsilon(i,t))}{f(\varepsilon_0)} \quad \text{(Eq. 15.12)}$$

where $f(\varepsilon)$ is a geometric function for the Kuwabara unit cell model [14]. The local filtration efficiency is calculated as:

$$E(i,t) = 1 - exp\left[\frac{3n_{DR}(i,t)(1-\varepsilon(i,t))(x_{i+1}-x_i)}{2\varepsilon(i,t)d_c(i,t)}\right] \quad \text{(Eq. 15.13)}$$

In the above equation, n_{DR} is the combined filtration efficiency due to diffusion and interception:

$$n_{DR} = n_D + n_R - n_D \cdot n_R \quad \text{(Eq. 15.14)}$$

In practice, this approach typically underpredicts the effect of accumulated soot on filtration efficiency increase, and it is therefore customary to use semi-empirical correlations that allow a closer correlation with measured trends. The semi-empirical correlations calculate the filtration efficiency as function of the clean filter efficiency $E_{wall,0}$ and the local soot loading m_{wall}.

$$E_{wall} = f(m_{wall}) \cdot E_{wall,0} \quad \text{(Eq. 15.15)}$$

As the porous wall capacity becomes gradually saturated, a layer of soot starts to form on top of the inlet channel wall. This layer, commonly referred to as "cake," becomes itself the principal filtration medium. This process is called "cake" or "surface" filtration. In order to model the transition period from wall filtration towards cake filtration, when the soot layer has only started to form, one has to resort to semi-empirical methods using percolation factors [15]. Based on this approach the fraction of the inlet soot mass that will be deposited as particulate layer on the surface of the filter wall is calculated by a partition coefficient Φ:

$$\Phi(t) = \frac{(d_c(1,t))^2 - d_{c,0}^2}{(\psi \cdot b)^2 - d_{c,0}^2} \quad \text{(Eq. 15.16)}$$

In the above equation ψ is a dimensionless constant $(0 < \psi < 1)$ which controls the onset of the unit cell blocking and is estimated based on experimental data. The quantity $(1 - \Phi)$ corresponds to the fraction of the inlet soot which will be deposited inside the filter wall. When the unit collector of the first discretization node $d_c(1,t)$ reaches the value $\psi \cdot b$, the particulate filtration is only due to the cake filtration regime.

To avoid mesh-dependent issues that may arise from the use of such approaches depending on the result of the first calculation node, alternative methods may be used as described below.

During the transition period from wall filtration to cake filtration, two mechanisms are assumed to take place simultaneously:

1. The first mechanism represents the initial formation of the first cake layer due to the saturation of the wall. This is mathematically expressed by a linear relationship of the cake filtration efficiency as a function of wall porosity. Cake filtration starts after wall porosity has been reduced below a critical limit and reaches 100% when wall porosity approaches zero.
2. The second mechanism represents the final steps of cake formation from a sparse collection of particles to a dense particle structure. This is mathematically expressed by a linear relationship of cake filtration as a function of cake mass. When no cake exists, this mechanism does not contribute to soot filtration. In the opposite, when cake mass has reached a critical value, cake filtration becomes 100%.

During the transition period, the total cake filtration efficiency is calculated as:

$$E_{cake} = E_{mech1} + E_{mech2} - E_{mech1} \cdot E_{mech2} \tag{Eq. 15.17}$$

A typical example of a steady-state particle filtration experiment for a wall-flow ceramic filter is presented in **Figure 15.6**. The experiment was performed on an engine dyno equipped with a 2 liter Euro6 gasoline engine (Toyota 8AR). The engine is turbocharged with both port- and direct-injection capability. The engine-out PM emissions were artificially increased by operating at rich conditions to minimize the duration of the experiments. Soot mass emissions were measured, upstream and downstream of the filter, by an AVL micro soot sensor. Since the filter is initially

FIGURE 15.6 Experimental and simulated GPF filtration efficiency during soot loading. Inlet conditions: Mass flow rate = 160 kg/h, T = 400°C, λ = 0.9.

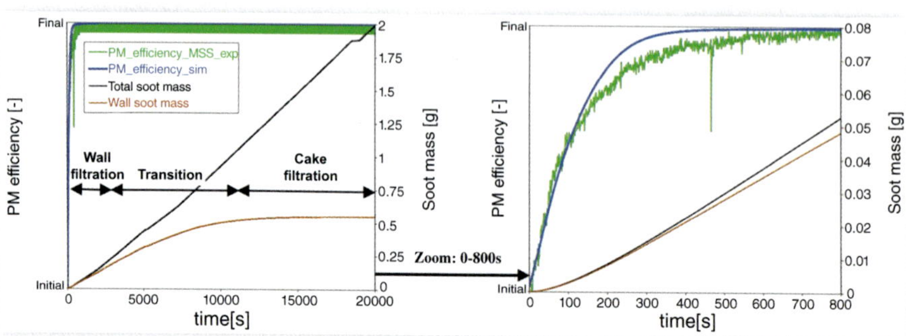

clean, particles are depth-filtered into the porous wall. As the wall is filled with soot, filtration efficiency gradually increases until it reaches ~100%. After the soot cake is formed, incoming particles are primarily filtered through cake filtration. Soot particles may still penetrate the soot cake and be filtered by the wall, but the respective rate steadily decreases over time as the soot cake becomes thicker and the wall pores saturated.

15.2.3 Pressure Drop

To solve the velocity and pressure distribution in the inlet and outlet channels, it is necessary to calculate the pressure drop through the soot layer and the porous filter wall. A schematic of the front view of a filter channel is given in **Figure 15.7**.

With the definitions shown above, the channel width available to the flow is given by the following relationship:

$$b(w) = d - 2(w_p - w) \quad \text{(Eq. 15.18)}$$

The particulate layer thickness w_p can be calculated from the following expression:

$$w_p = \frac{1}{2}\left(d - \sqrt{d^2 - \frac{2m_p}{\rho_p \cdot V_{trap} \cdot cpsm}}\right) \quad \text{(Eq. 15.19)}$$

where V_{trap} is the "effective" volume of the filter, excluding any cement and the plug length at the inlet and outlet of the filter.

Starting from the elementary Darcy law, the differential pressure drop over dw will be as follows:

$$\frac{dP}{dw} = \frac{\mu \cdot v(w)}{k_p} \quad \text{(Eq. 15.20)}$$

FIGURE 15.7 Cross section of a filter inlet channel.

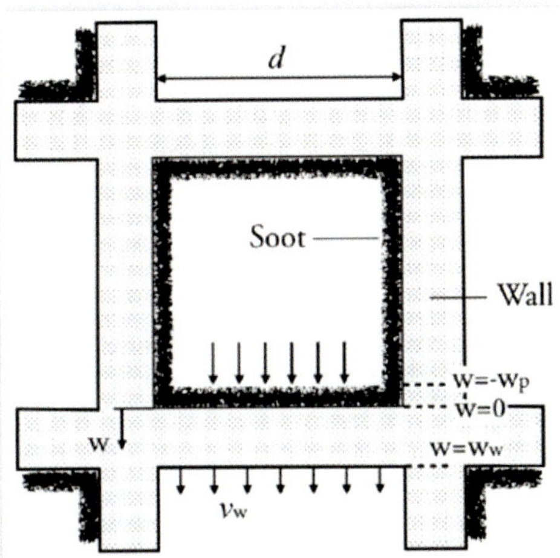

Although mass flow rate is steady throughout the soot layer, the gas velocity varies due to changes in gas density and flow area, as shown by the continuity equation:

$$v(w) = \frac{\rho_w \cdot v_w \cdot d}{\rho(w) \cdot b(w)} \quad \text{(Eq. 15.21)}$$

The gas density across the soot layer is a function of the local pressure. With the assumption of ideal gas the density can be expressed as follows:

$$\rho(w) = \frac{p(w) \cdot M_g}{RT} \quad \text{(Eq. 15.22)}$$

The permeability of the particulate deposit depends on the gas mean free path due to slip phenomena, which may be described by the following expression [16]:

$$k_p = k_{p,0}\left(1 + C_4 \frac{p_0}{p} \mu \sqrt{\frac{T}{M_g}}\right) \quad \text{(Eq. 15.23)}$$

The above formula expresses the dependency of soot permeability on local temperature and pressure for a constant soot density. The parameters $k_{p,0}$ and C_4 are estimated based on experimental data. Integration of Equation (15.20) provides the pressure drop over the soot layer thickness, taking into account Equations (15.21)–(15.23):

$$\Delta P_{soot} = \frac{R \cdot T \cdot \mu \cdot d \cdot \rho_w \cdot v_w}{M_g \cdot \bar{p} \cdot 2 \cdot k_p(\bar{p})} \ln\left(\frac{d}{d - 2w_p}\right) \quad \text{(Eq. 15.24)}$$

where $\bar{p}$ is the mean value of inlet and outlet pressures. The pressure drop through the porous filter wall, assuming constant flow velocity across the wall, can be described by the Darcy law:

$$\Delta P_{wall} = \frac{\mu \cdot v_w}{k_w} w_w \quad \text{(Eq. 15.25)}$$

The dependence of wall permeability k_w on temperature T and pressure p is taken into account, similarly to Equation (15.23). Furthermore, an extra correction is included to account for the sudden decrease of wall permeability at the initial stage of soot loading, as soot particles block a portion of the wall pores:

$$k_w = \frac{1}{\frac{1}{k_{w,0}} + C_1\rho_p + C_2\rho_p^2}\left(1 + C_4 \frac{p_0}{p} \mu \sqrt{\frac{T}{M_g}}\right) \quad \text{(Eq. 15.26)}$$

The clean wall permeability $k_{w,0}$ and parameters C_1 and C_2 define a trinomial function of ρ_p which is defined as total soot accumulated in the wall per wall volume. In order to calculate the extra parameters, the following values are estimated based on experimental data:

- Clean wall permeability $k_{w,0}$
- Loaded wall permeability $k_{w,1}$ at reference soot loading $\rho_{p,1}$

FIGURE 15.8 Experimental and simulated impact of wall- and cake-soot accumulation on pressure drop. Inlet conditions: Mass flow rate = 235 kg/h, T = 500°C, λ = 0.9, PM = 1 g/h.

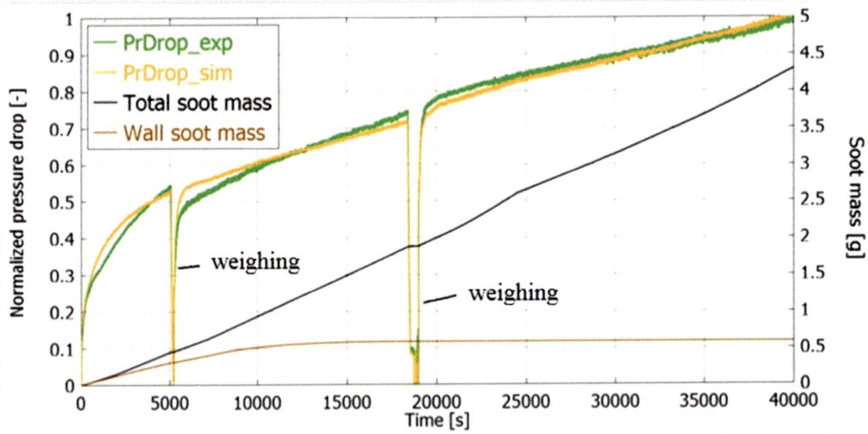

Based on the above analysis for the wall and deposit layer, the pressure drop between the inlet and the outlet channel can be evaluated as below:

$$p_1 - p_2 = \Delta P_{soot} + \Delta P_{wall} \tag{Eq. 15.27}$$

Experimental and simulation normalized pressure drop results during a steady-state soot loading are presented in **Figure 15.8**. Additionally, the model prediction for wall and total soot mass is also reported. The experiment was divided in three parts with intermediate weight measurements to verify the accumulated soot mass. The filter is initially in a clean state and the engine operates at rich conditions throughout the measurement to artificially increase soot emissions and to avoid any oxidation of the accumulated soot. Since the filter is initially clean, particles are depth-filtered into the porous wall. The depth-filtration region can be distinguished by the rapid increase of the pressure drop curve (0 to 3000 s). As the wall is being filled with soot, a less steep increase of the backpressure is observed, and gradually the soot cake layer starts to form. After a certain transition period (3000 to 12,000 s), wall-soot mass stabilizes and cake filtration becomes the predominant particle collection mechanism. In this case, pressure drop increases almost linearly with increasing soot loading.

15.3 Channel-Scale Modeling

Assuming a spatially uniform inlet flow and temperature field, the mathematical solution of the complete particulate filter can become equivalent to the solution of a representative pair of inlet and outlet channels (see **Figure 15.9**). In this section, the formulation of mass-momentum, energy, and species balance equations is discussed.

FIGURE 15.9 Side view of a filter, inlet (1) and outlet (2) channels.

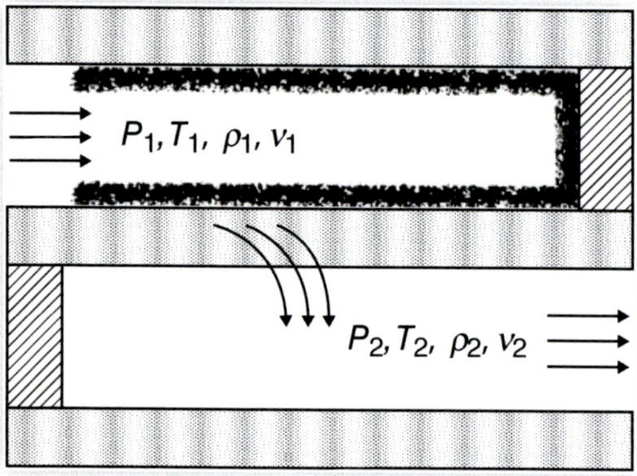

15.3.1 Mass-Momentum Balance

Mass conservation of channel gas: The mass balance equation for the gas flowing in the inlet and outlet channels is as follows:

$$\frac{\partial}{\partial z}\left(d_i^2 \rho_i v_i\right) = (-1)^i \cdot 4d\rho_w v_w \quad \text{(Eq. 15.28)}$$

where the subscript i identifies regions 1 (inlet channel) and 2 (outlet channel).

Conservation of axial momentum of channel gas: Considering the mass gain/loss through the porous wall and the friction in the axial direction z, the momentum balance of exhaust gas can be written as follows:

$$\frac{\partial p_i}{\partial z} + \frac{\partial}{\partial z}\left(\rho_i v_i^2\right) = -\alpha_1 \mu v_i / d_i^2 \quad \text{(Eq. 15.29)}$$

15.3.2 Pressure Drop

Filter-induced backpressure, attributed to pressure losses along the channels and through the substrate wall, was discussed in the previous section. Moving to filter scale, additional pressure drop losses due to flow entrance/exit effects are considered. More specifically, the checkered face of the filter causes a sudden contraction of the flow at the inlet and a sudden expansion at the outlet. The associated pressure losses can be estimated by using the following equations:

$$\Delta P_{contraction} = \left[1.1 - 0.4\frac{\left(d - 2w_p\right)^2}{2\left(d + w_w\right)^2}\right]\frac{\rho_1 v_1^2\big|_{z=0}}{2} \quad \text{(Eq. 15.30)}$$

$$\Delta P_{expansion} = \left[1 - \frac{d^2}{2\left(d + w_w\right)^2}\right]^2 \frac{\rho_2 v_2^2\big|_{z=L}}{2} \quad \text{(Eq. 15.31)}$$

FIGURE 15.10 Effect of temperature and mass flow rate on GPF pressure drop: experimental (markers) and simulation (lines) results.

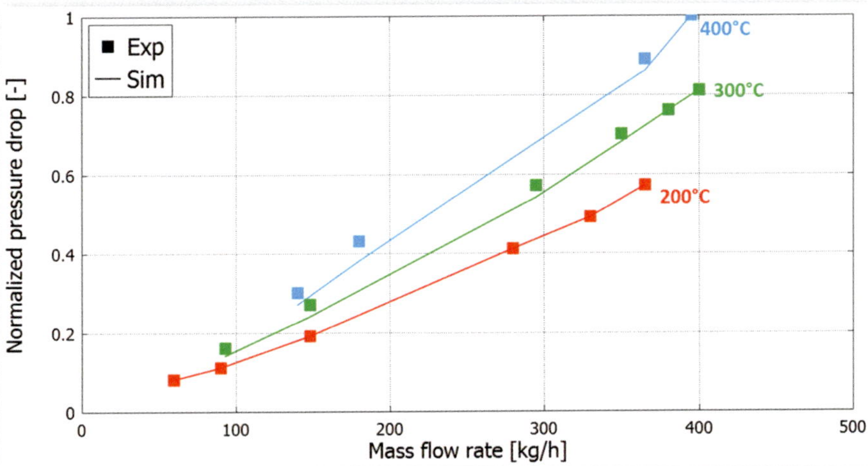

Experimental and simulation pressure drop results are presented for the case of the clean filter. **Figure 15.10** shows a typical dependency of the pressure drop on the exhaust mass flow rate $\dot{m}$ and temperature T, when considering a soot-free GPF. The presented results are normalized with respect to the maximum measured pressure drop (at the highest flow rate and temperature).

Figure 15.11 presents the pressure drop breakdown and the temperature impact on the contribution of the different pressure drop terms as well. We may identify four pressure drop terms, that is, contraction (Equation 15.30), channel flow (derived by Equation 15.29), wall flow (Equation 15.25), and expansion (Equation 15.31). In the general case where a soot cake or an ash layer exists, the related pressure drop terms obey to (Equation 15.25) as well. The contribution of each term may vary with respect

FIGURE 15.11 Pressure drop breakdown.

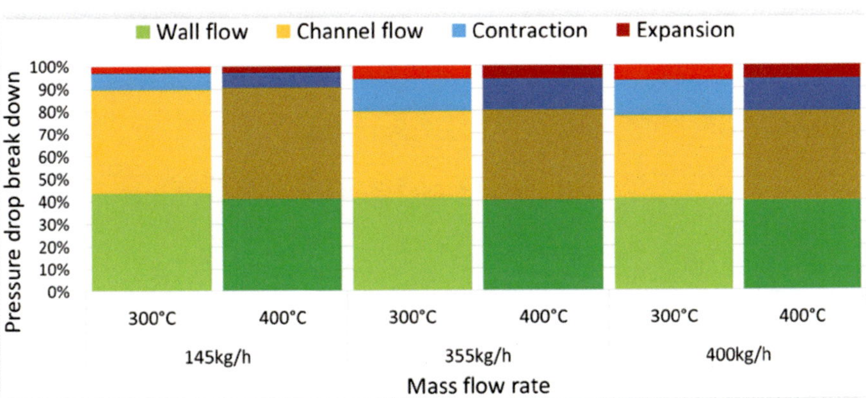

CHAPTER 15

to the operating conditions. In the examined configuration, the channel flow term has the highest contribution on filter pressure drop at low mass flow rates, while its contribution is significantly decreased at higher mass flow rates. This can be explained as the contraction and expansion terms are parabolically scaled with velocity, hence their significance increases at higher mass flow rates. For this particular case, the wall flow term contribution appears to have practically the same influence irrespective of the mass flow rate or temperature.

In addition to pressure drop itself, the consideration of the axial flow non-uniformities in the model, based on the mass-momentum balance equations, is also critical, since the flow field greatly affects local soot deposition and filtration behavior. Again, for the case of the clean filter, steady-state simulations were performed for three different exhaust flow rates. The filter properties are the same as in Table 15.1 while the clean wall permeability of the substrate was set to 4e–13 m^2. A comparison of the flow-through-wall distribution is given in **Figure 15.12**. For low flow rates, the wall profiles follow a U-shaped curve, while for higher flow rates, local velocities are increasing in the rear part of the filter. The actual shape of the wall-flow distribution depends on the different pressure drop losses between the inlet and outlet channels and across the wall as it has been reported in earlier DPF modeling studies [16, 17].

Substrate permeability may vary along the axial direction for certain filter configurations. A typical example is the zone-coated filters which may contain two or more zones with different wall properties. A model-based investigation of the axial coating zoning and its effect on the GPF flow field is presented in this section. Three different exemplary configurations were selected, as illustrated in **Figure 15.13**.

In addition to the reference GPF, two zoned filters were considered, which are referred to as "Zoned High:Low" and "Zoned Low:High" GPF. In the first case, more washcoat is placed at the front zone, whereas in the second case more washcoat is placed at the rear. The impact of washcoat amount variation in each zone is taken

FIGURE 15.12 Simulated effect of mass flow rate on wall-flow distribution, at clean state. Inlet conditions: Mass flow rate = 180 kg/h, T = 630°C.

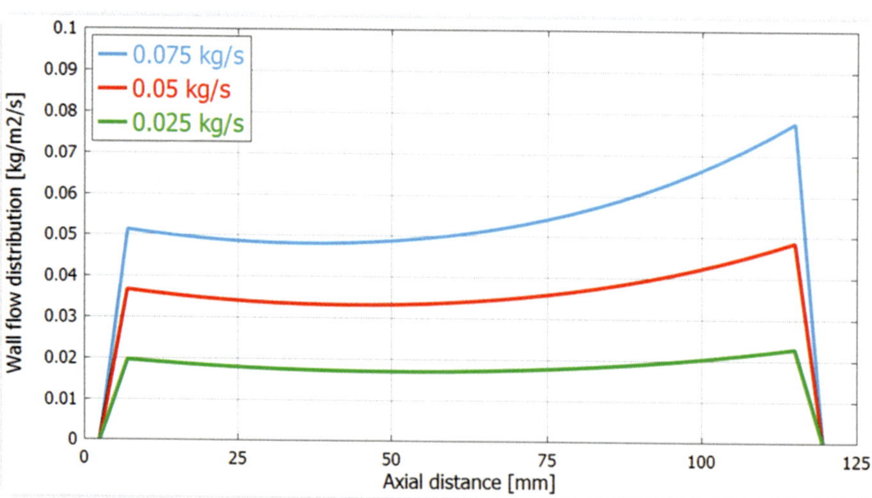

FIGURE 15.13 Coated GPF configurations used to study the effect of non-uniform wall permeability.

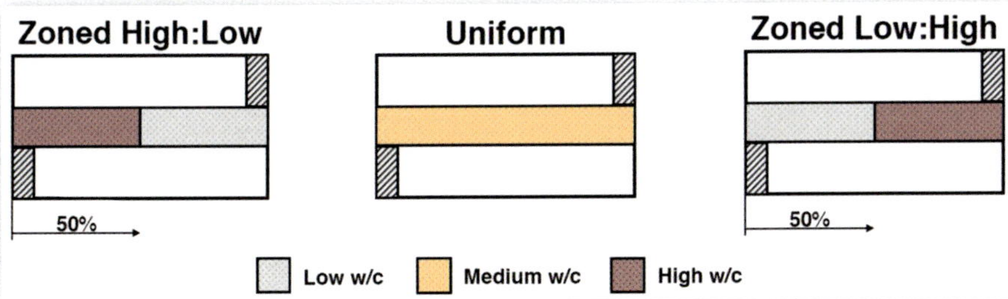

TABLE 15.2 Clean wall permeabilities used in the model-based investigation

Zone	Clean wall permeability [m^2]
Low w/c	1e−13
Medium w/c	0.75e−13
High w/c	0.5e−13

into account by considering a different clean wall permeability. The assumed clean wall permeabilities for each zone are given in Table 15.2.

Figure 15.14 shows a comparison of the flow distribution patterns for the three different wall configurations. The presence of zones with different washcoat loadings leads to a significant alteration of the wall-flow distribution along the filter length. Flow is preferably directed through the zones of lower washcoat loading, rear for the

FIGURE 15.14 Comparison of simulated wall-flow distribution patterns, at clean state for uniform zone-coated filters. Inlet conditions: Mass flow rate = 180 kg/h, T = 630°C.

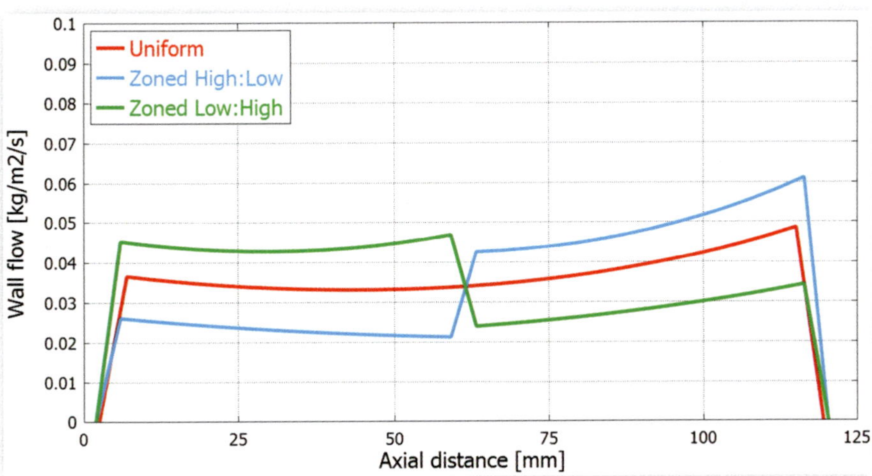

FIGURE 15.15 Comparison of simulated axial soot distribution for uniform and zone-coated filters at 2 g/L of soot. Inlet conditions: Mass flow rate = 180 kg/h, T = 630°C, PM = 1 g/h.

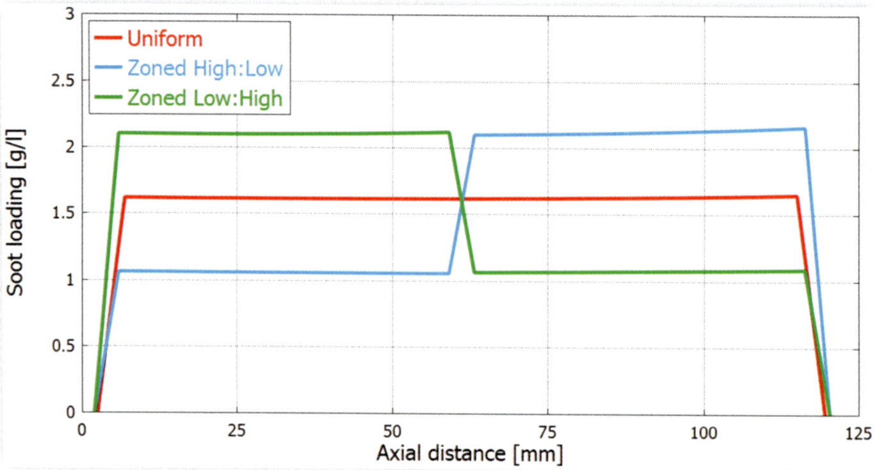

"Zoned High:Low" and front for the "Zoned Low:High" filters. Indicatively at clean state, the flow rate passing through the low washcoat zone is almost two times higher than the flow through the high washcoat one. The reason is associated with the lower wall permeability and increased resistance of the high washcoat areas.

Next, the behavior of the three filter configurations was compared under soot-loaded state by simulating a soot loading at 2 g/L. The predicted axial soot loading profiles at 2 g/L of soot are presented in **Figure 15.15**. Since soot particles follow the gas streamlines, they tend to accumulate in the areas of higher permeability. For this reason, particulates are preferably deposited in the zones of lower washcoat amount, rear for the "Zoned High:Low" and front for the "Zoned Low:High" filter, generating a highly non-uniform soot distribution along the inlet channel.

15.3.3 Energy Balance

Gas phase: The conservation of energy of the channel gas for the inlet (Equation 15.32) and the outlet channels is given below (Equation 15.33).

$$C_{p,g}\rho_1 v_1\Big|_z \frac{\partial T_1}{\partial z} = h_1 \frac{4}{d_1}(T_s - T_1) \qquad \text{(Eq. 15.32)}$$

$$C_{p,g}\rho_2 v_2\Big|_z \frac{\partial T_2}{\partial z} = \left(h_2 + C_{p,g}\rho_w v_w\right)\frac{4}{d}(T_s - T_2) \qquad \text{(Eq. 15.33)}$$

Solid phase: The one-dimensional temperature field in the filter is described by the transient heat conduction equation with heat sources:

$$\rho_s \cdot C_{p,s} \frac{\partial T_s}{\partial t} = \lambda_{s,z} \frac{\partial^2 T_s}{\partial z^2} + S \qquad \text{(Eq. 15.34)}$$

If the non-uniformities in the inlet temperature and flow distribution need to be considered, a 2D or a 3D model is required. In this case, the transient energy balance equation is extended to two or three dimensions. The heat conduction equation is formulated in polar coordinates for 2D simulations and in Cartesian coordinates for 3D simulations.

$$\rho_s C_{p,s} \frac{\partial T_s}{\partial t} = \lambda_{s,z} \frac{\partial^2 T_s}{\partial z^2} + \lambda_{s,r} \frac{1}{r} \frac{\partial}{\partial r}\left(r \frac{\partial T_s}{\partial r} \right) + S \quad \text{(Eq. 15.35)}$$

$$\rho_s C_{p,s} \frac{\partial T_s}{\partial t} = \lambda_{s,x} \frac{\partial^2 T_s}{\partial x^2} + \lambda_{s,y} \frac{\partial^2 T_s}{\partial y^2} + \lambda_{s,z} \frac{\partial^2 T_s}{\partial z^2} + S \quad \text{(Eq. 15.36)}$$

The source term S includes the contribution of the convective heat transfer of the gas flow in the channel H_{conv} and through the wall H_{wall}, as well as the exothermic heat release H_{react}.

$$S = H_{conv} + H_{wall} + H_{react} \quad \text{(Eq. 15.37)}$$

$$H_{conv} = h_1 \cdot S_F \cdot (T_1 - T_s) + h_2 \cdot S_F \cdot (T_2 - T_s) \quad \text{(Eq. 15.38)}$$

$$H_{wall} = \rho_w \cdot v_w \cdot S_F \cdot C_{p,g} \cdot (T_1 - T_s) \quad \text{(Eq. 15.39)}$$

$$H_{react} = S_F \sum_k \left(\int_{-w_p}^{w_w} f_w R_k dw \right) \cdot \Delta H_k \quad \text{(Eq. 15.40)}$$

The definition of the geometrical parameter f_w is as follows:

$$f_w = \frac{b(w)}{d} \quad \text{(Eq. 15.41)}$$

15.3.4 Species Balance

The importance of species transfer to account for oxygen convection to the reacting soot layer was initially recognized for the case of uncontrolled regeneration modeling of DPF [15]. In addition, the solution of the species transfer equation is essential to model the coated filters with respect to pollutant conversion and their interactions with the accumulated soot.

The governing advection-reaction-diffusion equation for the conservation of mass of any species within the soot layer and wall is as follows:

$$v_w \frac{\partial y_j}{\partial w} - D_{w,j} \frac{\partial}{\partial w}\left(f_w \frac{\partial y_j}{\partial w} \right) = \frac{f_w}{c_m} \sum_k n_{j,k} R_k \quad \text{(Eq. 15.42)}$$

The calculation of the effective diffusivity $D_{w,j}$ is based on the mean transport pore model which uses the following expression:

$$\frac{1}{D_{w,j}} = \frac{\tau}{\varepsilon_{pore}} \left(\frac{1}{D_{mol,j}} + \frac{1}{D_{knud,j}} \right) \quad \text{(Eq. 15.43)}$$

with the Knudsen diffusivity:

$$D_{knud,j} = \frac{d_{pore}}{3}\sqrt{\frac{8RT}{\pi M_j}} \quad \text{(Eq. 15.44)}$$

The porosity ε_{pore} and the mean pore size d_{pore} can be extracted from the microstructural properties of the washcoat while tortuosity τ is an empirical parameter which can be tuned by dedicated experiments.

The boundary conditions couple the wall with the gas concentrations in the channels. The "film" approach is used to compute the convective mass transfer from the bulk gas to the wall surface. The mass transfer coefficients $k_{i,j}$ correspond to laminar flow for both inlet and outlet channels:

$$\frac{\partial(v_1 y_{1,j})}{\partial z} = -\frac{4}{d \cdot f_w^2} v_w y_{1,j} + \frac{4}{d \cdot f_w} k_{1,j}\left(y_{1s,j} - y_{1,j}\right) \quad \text{(Eq. 15.45)}$$

$$\frac{\partial(v_2 y_{2,j})}{\partial z} = \frac{4}{d \cdot f_w^2} v_w y_{2s,j} + \frac{4}{d \cdot f_w} k_{2,j}\left(y_{2s,j} - y_{2,j}\right) \quad \text{(Eq. 15.46)}$$

The GPFs shown in Figure 15.13, previously evaluated in terms of pressure drop and soot deposition, are now compared in terms of species conversion. TWC intrinsic reactivity is proportional to the amount of active washcoat that is impregnated inside the filter. To take this into account in the model, a linear relationship between TWC reactivity and washcoat loading was assumed. The specific washcoat loadings that were considered for each zone are listed in **Table 15.3**. To ensure direct comparability, all three filters are characterized by the same global washcoat loading (45 g/L).

The three configurations were simulated under transient warm-up conditions, where an inlet temperature ramp (2°C/s) from 20 to 500°C was imposed. The gas mixture was stoichiometric, containing 0.75% CO, 0.2% NO, and 0.15% HC. The CO conversion curves with respect to the inlet temperature are given in **Figure 15.16a** for all three filters. Placing more washcoat in the frontal part of the filter appears to have a beneficial effect under transient conditions, with the temperature required to reach 50% CO conversion being 10°C lower, compared to the uniform filter. Placing more washcoat at the rear has an overall negative effect, especially at temperatures greater than 250°C, while the benefits at lower temperatures are very small. To understand the phenomena behind the beneficial effect of placing more washcoat in the frontal part of the filter, wall temperature and inlet/outlet CO concentration profiles are also presented in Figure 15.16. These graphs correspond to the time instance of 250°C inlet temperature. The "Zoned High:Low" filter is benefited from the fast heat-up of the frontal part, as the majority of the washcoat is exposed to higher temperatures. In addition, the exotherm generated by the TWC reactions self-enhances this effect and leads to a further increase of the temperature along the whole filter

TABLE 15.3 Washcoat loadings used in the model-based investigation

Zone	Washcoat loading [g/L]
Low w/c	30
Medium w/c	45
High w/c	60

FIGURE 15.16 Comparison of light-off behavior between uniform and zone-coated GPF: (a) CO conversion, (b) wall temperature profiles, (c) inlet and (d) outlet channel CO concentrations at 250°C. Inlet conditions: Mass flow rate = 0.1 kg/s, λ = 1.006.

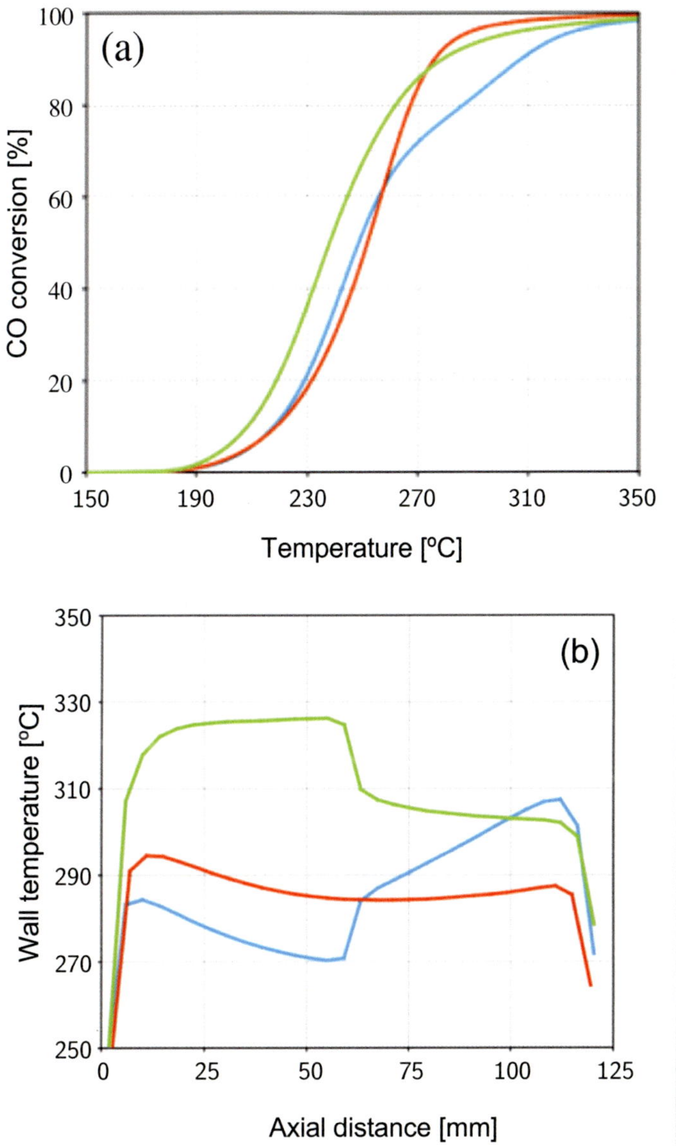

FIGURE 15.16 *(Continued)* Comparison of light-off behavior between uniform and zone-coated GPF: (a) CO conversion, (b) wall temperature profiles, (c) inlet and (d) outlet channel CO concentrations at 250°C. Inlet conditions: Mass flow rate = 0.1 kg/s, λ = 1.006.

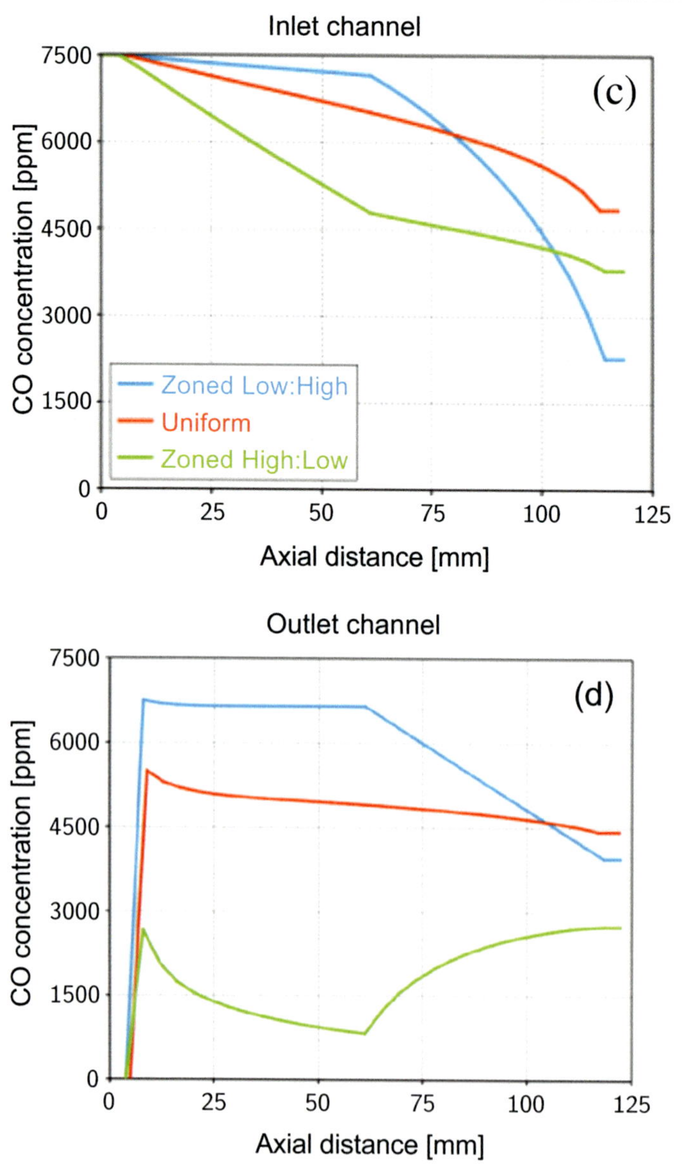

length, as shown in Figure 15.16b. The axial CO concentration profiles for the inlet and outlet channels are plotted in Figure 15.16c and d, respectively. As explained above, the front zone is more active for the case of "Zoned High:Low" filter, indicated by the steep decrease of the outlet CO concentration. Conversely for the "Zoned Low:High" filter, the front part has almost no contribution with the majority of CO oxidation taking place in the rear part, as expected due to its higher washcoat loading. Still, at 250°C, the efficiency of the "Zoned Low:High" filter is better than the uniform one, a behavior related to the advantage of the high-washcoat zone at lower temperatures. At temperatures higher than 270°C though, the performance of the "Zoned Low:High" filter is significantly lower than both the other configurations. Since more washcoat is placed at the rear of the filter, the exotherm generated by the TWC reactions cannot be exploited in order to self-enhance the heat-up of the low-washcoat zone.

15.3.5 TWC Reactions

The abatement of the gaseous emissions (CO, HC, NOx) from gasoline engines has been successfully carried out for more than 30 years with the use of three-way catalytic converters. With the introduction of the GPF for the reduction of particulate emissions, catalyst manufacturers focused also on the development of a combined TWC+GPF device (known as TWC/GPF, catalyzed GPF (cGPF) or as a four-way catalyst (FWC) [18, 19, 20]) in which the porous walls of the GPF are impregnated with TWC catalytic material. By using a TWC/GPF instead of a bare GPF, the required volume of the TWC can be reduced or alternatively more TWC catalytic material can be accommodated in the same packaging volume, increasing the total efficiency of the system. The main catalytic reactions that take place in a TWC/GPF are the same as for the TWC:

Oxidation reactions

$$CO + \frac{1}{2}O_2 \rightarrow CO_2 \qquad \text{(reac. 15.1)}$$

$$H_2 + \frac{1}{2}O_2 \rightarrow H_2O \qquad \text{(reac. 15.2)}$$

$$C_xH_y + \left(x + \frac{y}{4}\right)O_2 \rightarrow xCO_2 + \frac{y}{2}H_2O \qquad \text{(reac. 15.3)}$$

NO reduction

$$CO + NO \rightarrow CO_2 + \frac{1}{2}N_2 \qquad \text{(reac. 15.4)}$$

$$C_xH_y + \left(2x + \frac{y}{2}\right)NO \rightarrow xCO_2 + \frac{y}{2}H_2O + \left(x + \frac{y}{4}\right)N_2 \qquad \text{(reac. 15.5)}$$

$$H_2 + NO \rightarrow H_2O + \frac{1}{2}N_2 \qquad \text{(reac. 15.6)}$$

$$CO + 2NO \rightarrow CO_2 + N_2O \qquad \text{(reac. 15.7)}$$

$$H_2 + 2NO \rightarrow H_2O + N_2O \quad \text{(reac. 15.8)}$$

$$NO + \frac{5}{2}H_2 \rightarrow NH_3 + H_2O \quad \text{(reac. 15.9)}$$

Steam reforming

$$C_xH_y + xH_2O \rightarrow xCO + \left(x + \frac{y}{2}\right)H_2 \quad \text{(reac. 15.10)}$$

Water-gas shift

$$CO + H_2O \leftrightarrow CO_2 + H_2 \quad \text{(reac. 15.11)}$$

Cerium (reduced state)

$$Ce_2O_3 + \frac{1}{2}O_2 \rightarrow 2CeO_2 \quad \text{(reac. 15.12)}$$

$$Ce_2O_3 + NO \rightarrow 2CeO_2 + \frac{1}{2}N_2 \quad \text{(reac. 15.13)}$$

$$Ce_2O_3 + H_2O \rightarrow 2CeO_2 + H_2 \quad \text{(reac. 15.14)}$$

$$Ce_2O_3 + CO_2 \rightarrow 2CeO_2 + CO \quad \text{(reac. 15.15)}$$

Cerium (oxidized state)

$$2CeO_2 + CO \rightarrow Ce_2O_3 + CO_2 \quad \text{(reac. 15.16)}$$

$$2CeO_2 + H_2 \rightarrow Ce_2O_3 + H_2O \quad \text{(reac. 15.17)}$$

$$C_xH_y + 2\left(x + \frac{y}{2}\right)CeO_2 \rightarrow \left(x + \frac{y}{2}\right)Ce_2O_3 + xCO + \frac{y}{2}H_2O \quad \text{(reac. 15.18)}$$

Light-off performance is an important aspect in the design of a catalyst since a significant part of CO and HC emissions are generated during cold-start. Opitz et al. [21] performed an experimental and modeling comparison of the light-off behavior between a flow-through TWC and a wall-flow GPF with integrated TWC reactivity. Typical wall geometries were selected, 600 cpsi/4.3 mils for the TWC and 300 cpsi/12 mils for the cGPF. In addition, the washcoat and precious metal loadings were identical while the thermal mass was kept the same in both devices to ensure direct comparability. Temperature step experiments were performed in the temperature range of 250°C to 325°C in order to evaluate the fast light-off performance. The respective experimental and modeling results are given in **Figure 15.17**.

A significantly delayed light-off was observed for the wall-flow filter compared to the conventional flow-through monolith. The numerical model was able to accurately capture this experimental trend while it provided insight into the physical effects behind it. The authors concluded that a significant part of the observed performance difference can be attributed to the lower cell density of the filter which is associated

FIGURE 15.17 Comparison of the fast light-off performance between a conventional monolith and a coated filter. Experiment versus model prediction. Mole fraction of CO at the outlet of the catalyst plotted against time at four gas inlet temperatures in the range from 250°C to 325°C (5000 ppm CO, 3000 ppm O_2, GHSV: 20,000 h^{-1}). Reprinted from Opitz et al. [21] with permission from Elsevier.

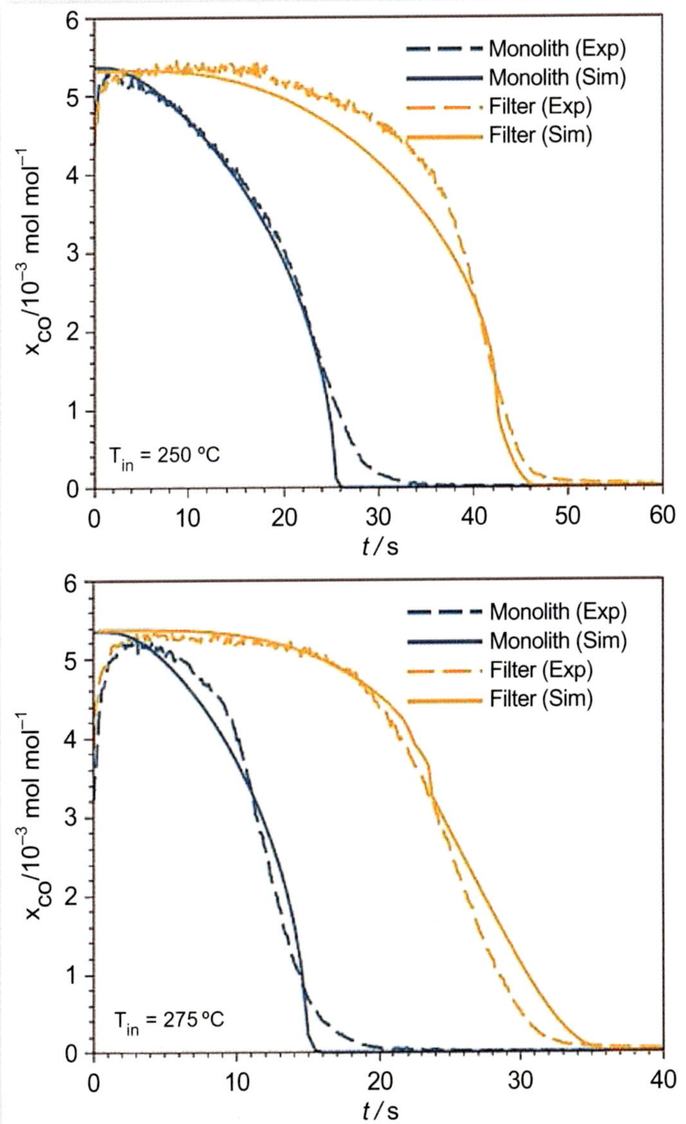

FIGURE 15.17 (*Continued*) Comparison of the fast light-off performance between a conventional monolith and a coated filter. Experiment versus model prediction. Mole fraction of CO at the outlet of the catalyst plotted against time at four gas inlet temperatures in the range from 250°C to 325°C (5000 ppm CO, 3000 ppm O_2, GHSV: 20,000 h^{-1}). Reprinted from Opitz et al. [21] with permission from Elsevier.

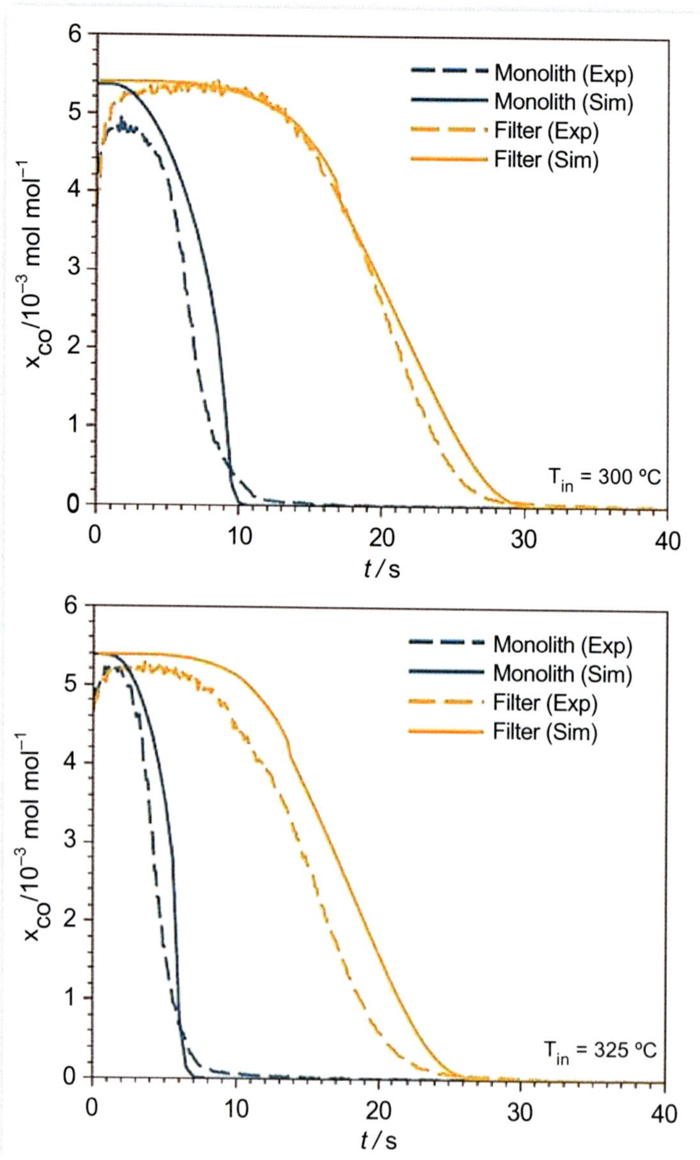

with a lower surface area. In addition, the thermal capacity of the outlet channel plugs has a negative effect as it increases the thermal mass at the frontal part of the filter, a part which has a higher contribution during light-off. Finally, the intrinsic differences between the wall-flow filter and the flow-through monolith were also considered. More specifically, in the filter case, only half of the total channels (inlet channels) contribute to the heat and mass transfer from the gas towards the filter wall. As a result, the heat and mass transfer per reactor volume is only half of that of the open monolith.

15.3.6 Soot Reactions

More than 90% of the engine-emitted particles consist of soot, that is, elemental carbon and small hydrocarbon fractions derived by lube oil and fuel. In the case of gasoline engines, the emitted PM is only a fraction of similar-sized diesel engine emissions (about 30 times lower for MPI and 5 times for GDI engines [2]). Moreover, the soot oxidation rate of gasoline soot has been reported to be 2-5 times higher compared to diesel soot oxidation rate [22], whereas the exhaust gas temperatures of gasoline engines are typically higher compared to diesel engines. An important difference for the case of gasoline engine exhaust compared to diesel engines is related to oxygen concentration. In the case of a stoichiometric engine, the concentration of oxygen downstream an operational three-way catalyst is virtually zero. Therefore, despite the higher reactivity of gasoline particles and the higher temperatures, the lack of oxygen may pose challenges with respect to soot oxidation in real-world operation.

Passive regeneration in GPFs is nevertheless possible during fuel cut-off events which occur rather frequently in practice during a vehicle deceleration. Under such events, the GPF is purged with oxygen-rich air. If the GPF temperature during this event is adequately high (more than 600°C), the accumulated soot will be at least partially oxidized. If these passive regeneration events are frequent and the incoming PM emissions low, then the soot loading in the GPF will stay very low. Under such conditions, one would expect that this small amount is insufficient to form a cake, therefore most of the accumulated soot may remain inside the wall pores. This is in fact contrary to the operation mode of a typical DPF in passenger car applications, where most of the soot amount is found as cake.

As mentioned in the previous sections, the location of accumulated soot is of high importance for the filtration and flow resistance aspects of the filter. Therefore, it is important to develop a good understanding of how soot accumulates in the filter as function of design and operational parameters. Even though the passive regeneration conditions are significantly more likely to occur in a gasoline engine compared to a respective passenger car diesel engine, the possibility of continuous soot accumulation in the filter cannot be excluded, for example, after prolonged low load operation. In this case, an active regeneration strategy needs to be implemented. The active regeneration strategy should respect the permissible pressure drop, the thermal durability of the filter, and possible emission penalty associated with forced regeneration triggering. It is therefore important to have a good understanding of the underlying physical and chemical processes and use high fidelity mathematical models for design and control.

A GPF thermal regeneration involves the oxidation of solid particles to gaseous products. This process can be described by the following reactions:

$$C + O_2 \rightarrow CO_2 \quad \text{(reac. 15.19)}$$

$$C + \frac{1}{2}O_2 \rightarrow CO \quad \text{(reac. 15.20)}$$

The respective rate expressions of the above reactions can be formulated as:

$$R'_{19} = A_{19} \cdot exp^{\frac{-E_{19}}{RT}} \cdot P_{O_2}^{\alpha_{19}} \quad \text{(Eq. 15.47)}$$

$$R'_{20} = A_{20} \cdot exp^{\frac{-E_{20}}{RT}} \cdot P_{O_2}^{\alpha_{20}} \quad \text{(Eq. 15.48)}$$

where the pre-exponential factor A, the activation energy E, and the dependency from oxygen concentration α are tunable parameters which are identified based on dedicated experiments.

Soot particles in contact with catalytically active surface are known to oxidize faster. This process is usually referred to as direct soot catalysis as opposed to indirect soot catalysis occurring via NO_2 formation in precious metal-coated surfaces. The direct soot catalysis is applicable to soot that is in physical contact with the active components that are present in some of the commercial washcoats. To account for this effect in the model, it is customary to assume an additional soot oxidation pathway that is applicable only to the soot trapped in the wall pores, which are assumed to be in contact with the active catalyst. The following two reactions are therefore applicable to soot in the wall and are characterized by lower activation energy compared to the soot on the cake.

$$C + O_2 \rightarrow CO_2\,(\text{catalyzed}) \quad \text{(reac. 15.21)}$$

$$C + \frac{1}{2}O_2 \rightarrow CO\,(\text{catalyzed}) \quad \text{(reac. 15.22)}$$

The respective rate expressions of the above reactions can be formulated as:

$$R'_{21} = A_{21} \cdot exp^{\frac{-E_{21}}{RT}} \cdot P_{O_2}^{\alpha_{21}} \quad \text{(Eq. 15.49)}$$

$$R'_{22} = A_{22} \cdot exp^{\frac{-E_{22}}{RT}} \cdot P_{O_2}^{\alpha_{22}} \quad \text{(Eq. 15.50)}$$

In contrast to diesel exhaust where NO_2 chemistry is important for soot oxidation, in the case of a TWC + GPF exhaust line NO_2 is not expected to reach the GPF at significant levels. However, this may not be the case for single filter-coated GPF systems without a TWC upstream. For such designs the reactions of soot with NO_2 will have to be considered as well in the reaction models.

It has been reported that gasoline soot oxidation is enhanced in the presence of ash particles [9]. Ash particles are naturally produced mainly due to lube oil consumption both in gasoline and diesel engines. However, since gasoline soot emissions are lower, the resulting ash-to-soot ratio is much higher in the case of GDI engines [9].

The effect of lube oil-derived metallic ash additives (primarily Ca, Na, and Zn) on soot reactivity was investigated by Choi et al. [9]. The authors sampled soot from various engine operating points of a commercial 2.4 L GDI engine. Depending on the particular engine conditions, a soot with different ash fractions was obtained. Further details are listed in **Table 15.4** below.

A thermogravimetric analysis was performed on all samples at 600°C under 8% of oxygen. The comparative results of weight loss due to soot oxidation are summarized in **Figure 15.18**. These results indicate that soot reactivity can significantly vary depending on the engine operating condition and that the ash fraction is a major factor which enhances soot oxidation. Indeed, increasing ash fraction (especially over 0.50%) leads to significantly shorter times for 90% soot oxidation.

TABLE 15.4 GDI soot sampling conditions by Choi et al. [9]

Soot	Ash [%]	Engine condition description
Printex U	0.00	—
GDI-1	0.10	High soot concentration
GDI-2	0.26	Sequential fuel cuts
GDI-3	0.50	Cold idle
GDI-4	0.58	Low soot concentration
GDI-5	1.35	Low soot concentration (TWC out)
GDI-6	17.27	2% oil-dosed fuel

FIGURE 15.18 Ash fraction effect on soot oxidation under isothermal (600°C) conditions. Reprinted from Choi et al. [9].

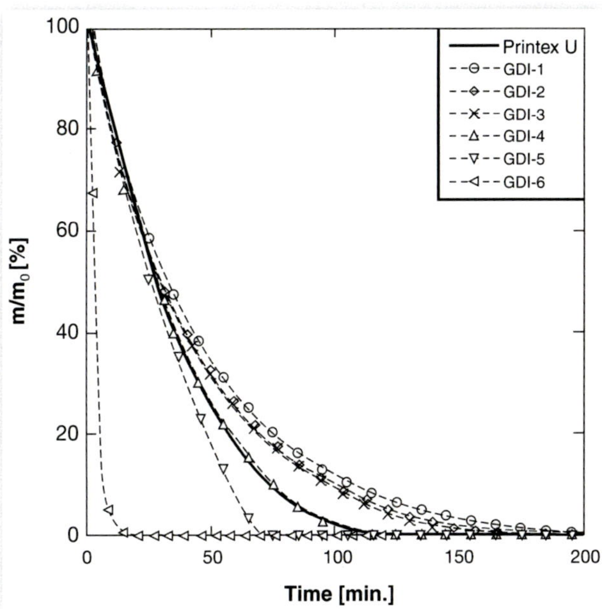

A similar study was conducted by the authors of this paper with a coated GPF. Instead of thermogravimetric analysis, the soot oxidation rate characterization was performed using full-length GPF cores, which were preloaded with soot on an engine bench equipped with a 2 liter Euro6 gasoline engine. The soot-loaded cores were subsequently placed in a synthetic gas bench (SGB) and were exposed to controlled temperature ramps under constant flow rate and gas composition (20% O_2 in nitrogen). Two cases were compared: The "high" ash/soot ratio, where the engine fuel was doped with oil, and the "low" ash/soot ratio case without doping. Steady-state soot loadings were performed on an engine bench. The instantaneous soot mass was calculated based on the carbon balance from the outlet CO and CO_2 concentrations. The respective results for the low and the high ash/soot ratio are given in **Figure 15.19** and verify the before-mentioned enhancing effect of ash on soot reactivity.

To obtain these results, in addition to the previously presented soot oxidation reaction mechanisms including non-catalyzed soot oxidation (reac. 15.19 and 15.20) and direct soot catalysis (reac. 15.21 and 15.22), the additive-promoted soot oxidation is introduced. The respective reaction form is described below:

$$C + AdditiveO_2 \rightarrow AdditiveO_3 + CO_2 + CO \qquad \text{(reac. 15.23)}$$

In the low ash/soot ratio case, two separate mechanisms are identified, observing the soot oxidation slope changes. The first one (first turning point at approximately 300°C) is attributed to direct soot catalysis, while the second one (second turning point) to non-catalyzed soot oxidation. In this case the additive-promoted reactions are weak and their influence cannot be identified.

In the high ash/soot ratio case though, the additive-promoted soot oxidation reaction mechanism is the predominant one, masking the direct soot catalysis and non-catalyzed soot oxidation reactions effect.

FIGURE 15.19 Measured impact of non-carbonaceous elements on soot oxidation chemistry of a coated GPF. The tests were performed on a synthetic gas bench (SGB) facility. Inlet conditions: Mass flow rate = 1.1 kg/m^2 s, O_2 = 20% experimental (markers) and simulation (lines) results.

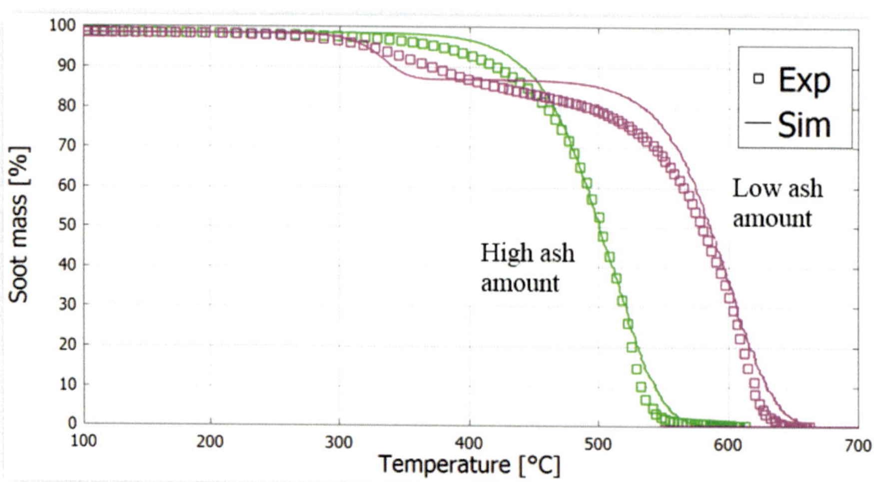

15.4 Filter Scale Modeling

15.4.1 Filter Regeneration in Fuel Cut-Off Events

As stated before, soot oxidation can occur during fuel cut-off phases where the conditions of high oxygen availability and temperature are met at the same time. The results of such a soot oxidation scenario during a fuel cut-off event were published by Nicolin et al. [23] and are shown in **Figure 15.20**.

An initially soot-loaded coated GPF placed in a close-coupled position was subjected to high exhaust temperatures followed by a long fuel cut-off event (0-40 s). High exotherms of more than 200°C are generated due to the oxidation of the accumulated soot with the highest temperatures appearing at the rear part of the filter. The reaction of soot with oxygen is very fast during the fuel-cut phase as verified by the measured O_2 concentrations before and after the GPF. Indeed, inlet O_2 is almost completely consumed during the first 20 s of the experiment. During 30-40 s, outlet O_2 reaches the inlet levels while in-filter temperatures drop. The above-described trends were successfully predicted by the authors as the axial temperature field and the oxygen consumption are properly captured by the model.

Soot oxidation rate can be significantly enhanced in the case of a catalytically active GPF, which leads to a higher exotherm production. This is a result of the highly exothermic oxidation of CO produced by the soot oxidation in the catalytic surface. The additional CO oxidation exotherm is responsible for the acceleration of the soot

FIGURE 15.20 Experimental and simulated fuel cut-off effect on soot oxidation exotherm and oxygen consumption. Reprinted from Nicolin et al. [23].

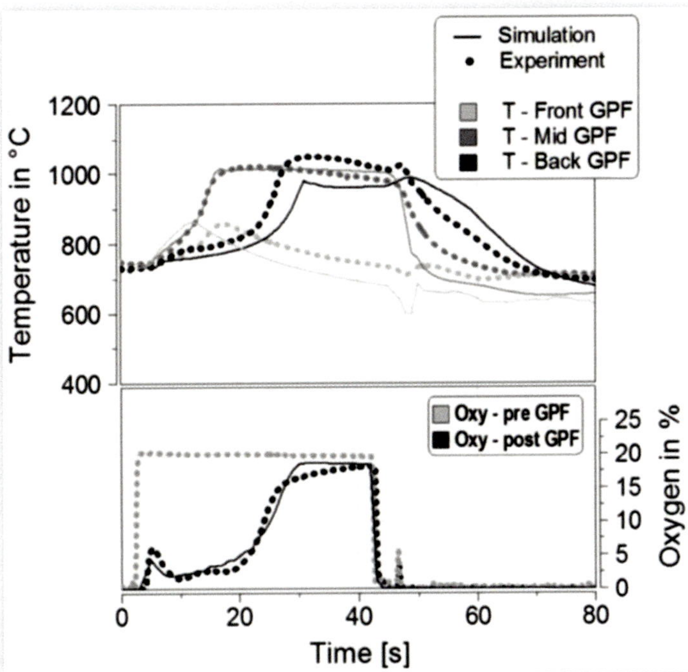

FIGURE 15.21 Comparison of the peak temperature response of TWC-coated and uncoated filters. Reprinted from Boger et al. [24] with permission from Springer.

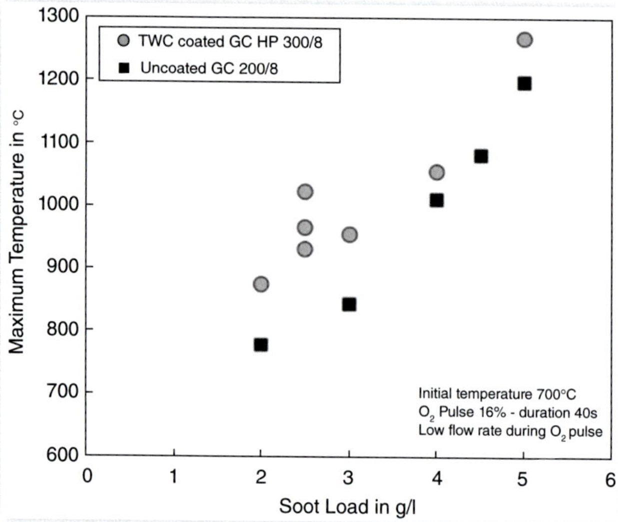

oxidation process compared to the case of an uncoated GPF. The effect of TWC coating existence on soot oxidation was investigated by Boger et al. [24]. The authors compared a 200 cpsi/8 mils uncoated and a 300 cpsi/8 mils catalytically active GPF under fuel-cut conditions (40 s of 20% O_2 at an inlet temperature of 700°C) with different initial soot loadings (2-5 g/L) while they monitored the temperature field inside the filters. The maximum temperatures produced during the fuel cut-off event are summarized in **Figure 15.21** for both filters.

In all cases, the TWC-coated filters exhibited a higher exotherm (50-100°C) compared to the uncoated filters of same initial soot loading. A direct comparison with respect to the axial filter temperatures and CO, CO_2 concentrations in the case of a coated and an uncoated GPF was performed by the same authors (**Figure 15.22**).

For the case of the uncoated filter, both CO_2 and CO are produced by the oxidation of soot. Interestingly, the selectivity of carbon oxidation is favorable for CO with its concentration being almost double than CO_2. Moving to the case of the coated filter, only CO_2 is produced. This is expected since a complete oxidation of CO to CO_2 takes place due to precious metal reactions. This results in a significantly higher amount of energy released per unit of carbon oxidized, leading to higher filter temperatures, which in turn favors the soot oxidation rate.

15.4.2 Ash Accumulation

According to literature studies [25] approximately half of the formed ash emissions is derived by lube-oil consumption while the other half comes from material corrosion and the catalytic washcoat. During normal engine operation, these ash particles are accumulated in the GPF substrate pores and on the substrate surface affecting the GPF performance with respect to filtration efficiency and pressure drop.

FIGURE 15.22 Comparison of the temperature, CO, and CO2 evolution during a "simulated" fuel cut experiment with a) an uncoated and b) a TWC-coated GPF. Conditions: Initial temperature 700°C, soot load ~0.52 g, and duration of oxygen pulse 40 s. Filter size: Uncoated GPF ⌀2 × 5.5 in., coated GPF ⌀2 × 4.5 in. Reprinted from Boger et al. [24] with permission from Springer.

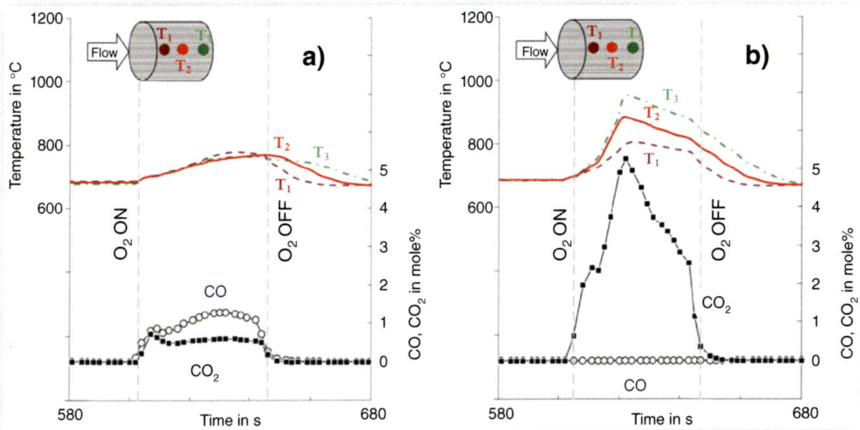

The impact of ash loading on the filter's filtration efficiency was investigated by Lambert et al. [26]. Two cordierite-coated GPF samples with 300 cpsi/12 mils wall structure were aged under different conditions (on the engine dyno and on-road) while an identical fresh (green) filter was used as reference. The respective aging conditions and the estimated ash mass are given in **Table 15.5**.

The above samples were evaluated in terms of pressure drop and filtration efficiency with the respective results shown in **Figure 15.23**. The impact of small ash accumulation (1-3 g) on the GPFs filtration efficiency can be quite considerable, increasing from ~60% to ~80%. Similarly, an increase of ~50 mbar on pressure drop (at 600 kg/h and 750°C) was induced by the accumulation of ash particles, compared to the ash-free ("green") filter.

The impact of ash loading on pressure drop was also investigated by Custer et al. [27]. Five identical cordierite wall flow monolith GPFs with a 220 cpsi/6 mils wall structure were loaded with ash using an accelerated procedure. More specifically, engine oil was introduced into a gasoline burner combustion chamber to generate elevated levels of oil-derived ash constituents in the exhaust stream. Each filter was loaded with different ash amounts, as summarized in **Table 15.6**.

The ash loading experiments imitate an underfloor GPF placement. The GPF is exposed to a maximum temperature of 650°C and the average loading temperature

TABLE 15.5 Ash loading of the five samples tested by Lambert et al. [27]

Filter	Ash loading method	Estimated ash mass [g]
Green	N/A	N/A
3k1	3000 km on road (2.3l GTDI)	0.95
FM50h	50 h on engine dynamometer	3.27

FIGURE 15.23 Clean backpressure and mass filtration efficiencies of green-, 3k1-, and FM50h-coated GPFs. (This figure is adapted based on the respective reference [26] figure.)

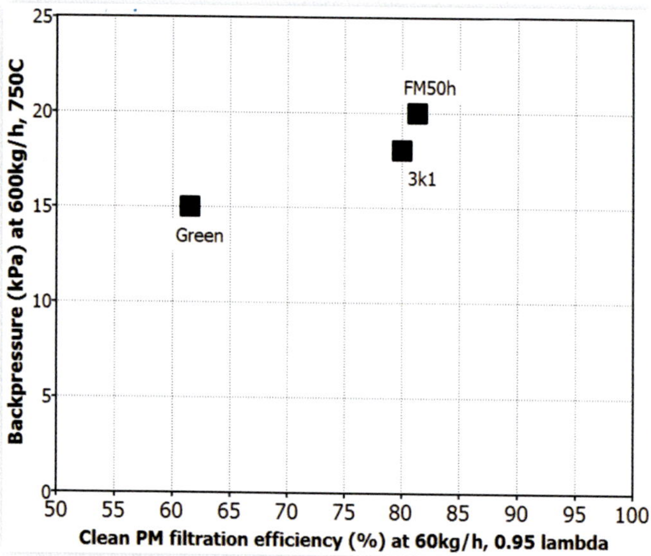

is 400°C. The loading cycle utilizes a periodic regeneration strategy (3.5 h loading – 0.5 h regeneration) to oxidize any intermediately accumulated soot. At the end of each loading cycle, a weight measurement was performed to determine the added ash mass. A compilation of pressure drop versus ash loading is shown in Figure 15.24 for all filters. The authors demonstrated that during an ash loading process, two separate mechanisms can be identified. The first mechanism is related to the depth filtration of ash particles inside the wall ("wall accumulation" as noted on Figure 15.24) which restrict the pores and lead to a lower porosity. This is evident by the initial steep pressure drop increase during the first 4 g/L of ash loading. The second mechanism is linked to the ash layer formation for ash loads greater than 8 g/L. This "layer formation" region as noted on Figure 15.24 is characterized by a linear increase of the pressure drop which is proportional to the ash layer thickness.

As a further step Custer et al. [27] investigated the contribution of ash accumulation (blue lines) on pressure drop compared to the contribution of soot accumulation (red lines). In this respect, they loaded Filter 5 with ash, and at specific ash levels, they performed soot loading tests (to 3 g/L) before regenerating and continuing the ash loading. Figure 15.25 illustrates the effect of soot on pressure drop at specific ash loadings. As the ash amount inside the filter increases, the pressure drop evolution during the soot loading becomes more

TABLE 15.6 Ash loading of the five samples tested by Cluster et al. [27]

Filter	Ash loading [g/L]
1	3
2	6.1
3	12.2
4	20.3
5	30.5

FIGURE 15.24 Pressure drop vs. ash load at 45,000 h^{-1}. Reprinted from Custer et al. [27].

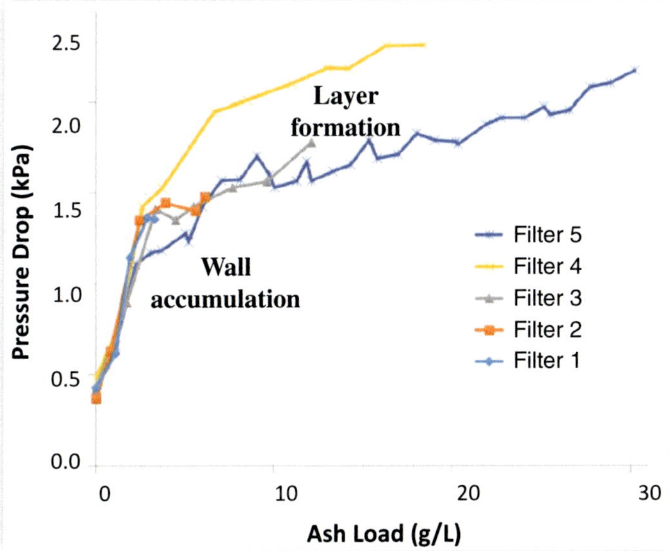

FIGURE 15.25 Pressure drop vs. combined ash and soot load at 45,000 h^{-1} in filter 5. This plot shows the pressure drop due to ash (diamonds) and due to ash+soot (squares). The dotted line represents ash + 3 g/l soot. Reprinted from Custer et al. [27].

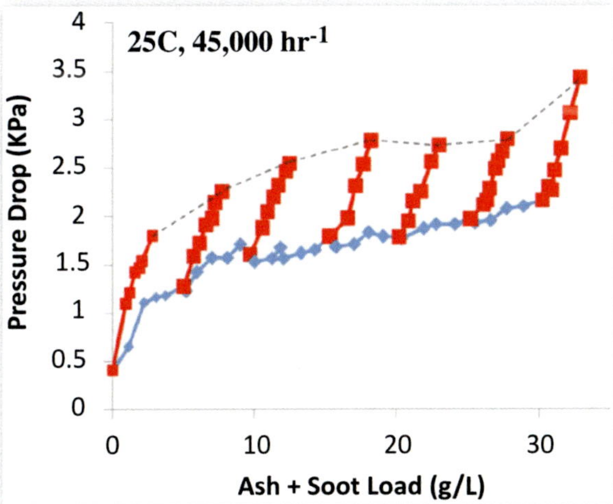

linear. According to Custer et al., the reason is predominantly associated with the generated ash layer which prevents soot particles from penetrating the wall. In this respect, the pressure drop increase during soot loading is attributed only to the cake formation on top of the ash layer.

A simulation study demonstrating the effect of plug ash and soot on pressure drop was evaluated in Figure 15.26. Applying steady-state flow and temperature conditions,

FIGURE 15.26 Simulated plug ash and soot effect on pressure drop. Inlet conditions: Mass flow rate = 215 kg/h, *T* = 500°C.

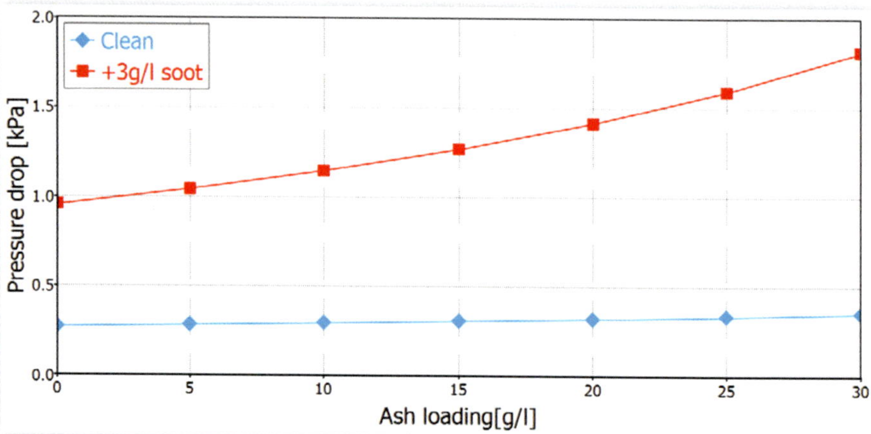

the case of a bare (uncoated) GPF in seven different states, that is, clean, 5 g/L, 10 g/L, 15 g/L, 20 g/L, 25 g/L, and 30 g/L initial plug ash loading, was investigated. The cases where a 3 g/L initial soot loading coexists with the above indicated initial plug ash loadings were examined as well. The filter properties are the same as in Table 15.1 while the clean wall permeability of the substrate was set to 2e−13 m^2. Plug ash significantly affects the filter flow distribution as it blocks a part of the inlet channels forcing the flow to pass through the front part of the filter. The impact of soot loading on pressure drop is enhanced by the plug ash existence, as a higher plug ash loading leads to a decrease of the effective length where soot can be deposited. Therefore, the soot cake thickness is increased creating a more important flow barrier and hence a higher pressure drop as the plug ash loading increases.

Experimental data concerning the effect of ash loading on a filter's reactivity were provided by Seong et al. [28]. The CO and NOx conversion efficiency was evaluated at different ash loadings for two coated GPFs with different washcoat loadings. The respective tests were conducted at laboratory conditions, using an engine dynamometer. To achieve the desired ash loading levels, accelerated protocols were performed by mixing engine fuel with 2% conventional oil. Coated cordierite GPF samples of same wall structure (300 cpsi/8 mils), same precious metal loading (0.5 g/L), and different ash loading were tested. The respective investigations were performed both on filters of 25 g/L and 50 g/L washcoat loading. The respective results are given in Figure 15.27. In all the different ash loading cases, a slight performance decrease was measured between 250 and 350°C while no effect was observed for temperatures above 400°C. This was validated in both the 25 g/L washcoated samples and the 50 g/L washcoated samples. The authors additionally concluded that the GPF performance was degraded at low temperatures as the TWC washcoat loading increased.

Mathematical modeling is of particular interest when it comes to studying effects of ash on GPF performance given the high complexity and long durations involved

FIGURE 15.27 Experimental effect of ash loading on the conversion efficiency of CO and NOx [28].

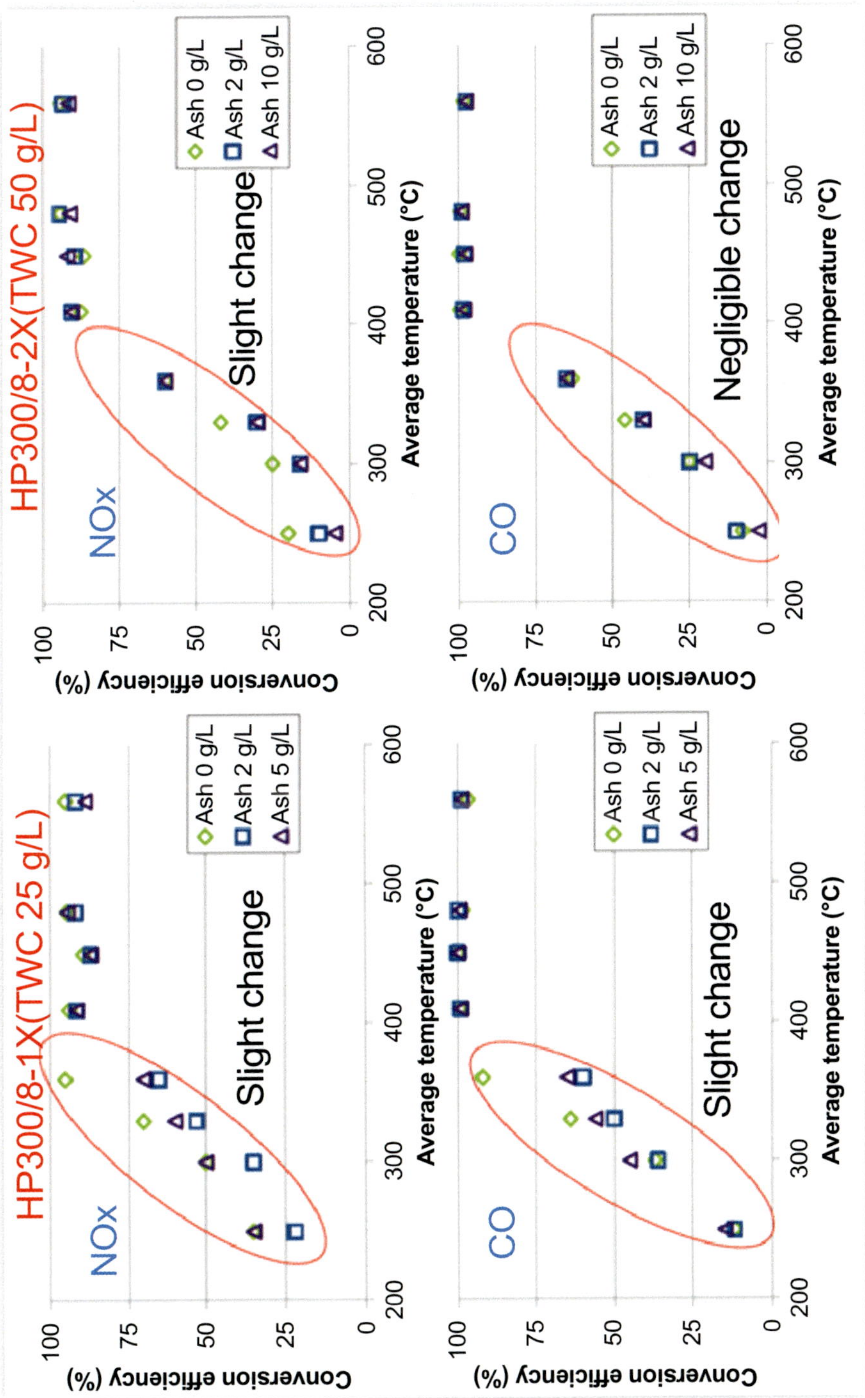

FIGURE 15.28 Predicted effect of (a) layer and (b) plug ash loading on CO conversion efficiency. Inlet conditions: Mass flow rate = 43 kg/h, T = 230°C, λ = 1.04.

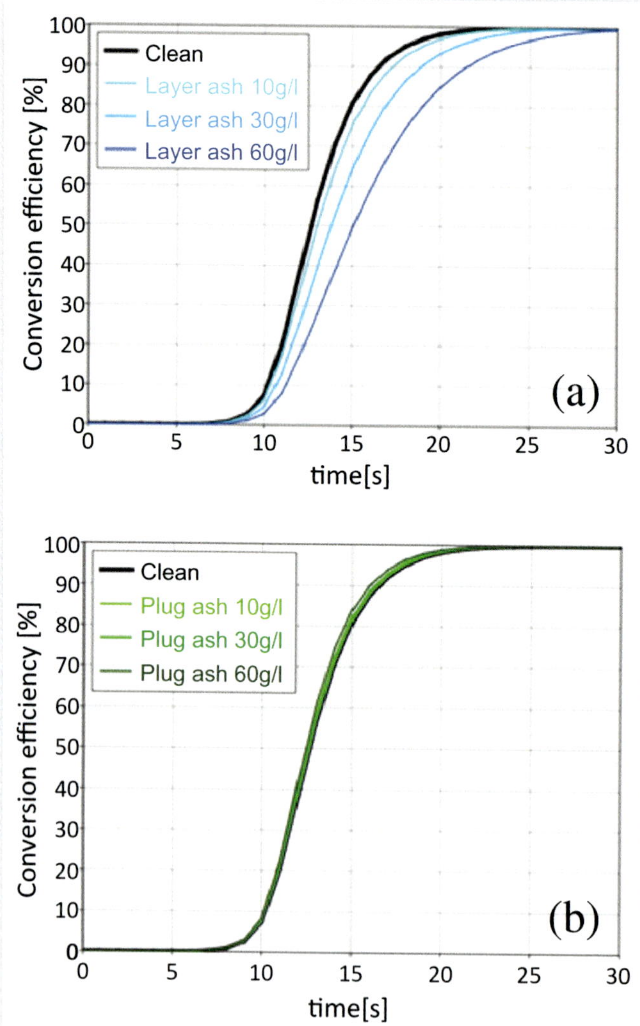

in purely experimental studies. A simulation study regarding the effect of layer and plug ash with respect to a uniformly coated GPF's TWC reactivity is presented below. Three different ash loading levels were investigated, namely, 10, 30, and 60 g/L, examining the effect of ash both as layer type and as plug type. The investigated scenario is replicating a cold-start by imposing a constant inlet temperature of 230°C while the initial filter temperature is 20°C.

Focusing first on the effect of the ash layer, CO conversion curves for various ash loadings are presented in **Figure 15.28a**. The increasing ash amount leads to a delayed light-off, especially for ash loadings higher than 30 g/L. According to the model prediction, this trend is attributed to both thermal and mass-transfer effects.

Ash layer acts as a barrier to the diffusion of species from the gas phase towards the TWC catalytic wall, leading to a decreased conversion compared to the ash-free filter. On the other hand, the non-negligible mass of the ash layer increases the total thermal capacity of the filter, which delays its warm-up and therefore the light-off occurrence.

Furthermore, similar investigations were performed for plug ash. The respective CO conversion curves, shown in **Figure 15.28b**, suggest an insignificant effect of plug ash on conversion efficiency. The presence of plug ash reduces the effective length of the inlet channels, forcing the flow to pass through the front part of the filter. Since this part exhibits higher temperatures than the rear one under cold-start, the predicted effect is slightly beneficial for conversion efficiency. However, it is expected that plug ash may have a more important impact in the case of zone-coated filters where the plugging will block the rear part of the channel and force the flow to pass preferentially through the front zone.

15.5 Conclusions

Even though the wall-flow filter technology has already proven its potential in the area of diesel engines, its introduction in the gasoline exhaust line can pose significant challenges and require extensive studies and in-depth understanding of highly complex physical-chemical processes. This understanding can be effectively supported by mathematical models and respective simulation tools. Carefully calibrated, high fidelity models are nowadays applicable during the development process as a cost- and time-efficient solution for timely introduction of advanced technologies and controls. While the mathematical principles to describe the filtration, transport, and reaction processes are similar to diesel applications, the phenomena of interest may vary due to the inherent differences of gasoline exhaust conditions.

Wall-scale modeling is important to predict the effects of wall properties on soot filtration and resulting pressure drop. Transferring the modeling techniques already known from DPF, it is possible to describe the main trends observed experimentally with reasonable accuracy. More predictive modeling approaches that would account for detailed microstructure geometry effects on transient filtration and pressure drop may further enhance the application area of modeling. Nevertheless, models based on apparent wall permeability properties at clean and loaded state are a good compromise between simplicity and accuracy and can be used in the development of new substrate designs.

Modeling of the coated GPF follows similar principles as the case of coated DPF using reaction schemes and rate expressions that have been established by previous TWC chemistry research. As the majority of gaseous emissions are generated during cold-start, the thermal characteristics of the coated GPF is a crucial design aspect, especially when considered as a replacement of the traditional flow-through TWC. Additionally, new coating techniques can be introduced to enhance light-off behavior, leading to optimum coating utilization by selectively concentrating the washcoat along the filter length. Modeling examples, describing the flow, thermal, and reaction behavior of zone-coated GPFs, were presented while the complex underlying phenomena were analytically discussed. Accurate modeling of the processes in the abovementioned technologies rely on deep understanding and characterization of wall permeabilities, effective diffusivities, and proper level of modeling detail.

Current research efforts are focusing on deeper understanding of the soot oxidation processes. Accurate modeling of transient and multi-dimensional phenomena such as fuel cut-off events is of major interest.

Lastly, the effect of ash accumulation on filtration efficiency, pressure drop, and species conversion is a topic of particular interest, due to the increased ash/soot ratio expected in gasoline engines compared to diesel engine. Intensive work among researchers is in progress to obtain a good understanding of the phenomena leading to ash layer and plug ash formation as well as to quantify their impact on pressure drop, soot oxidation, and TWC reactions.

The presented GPF model will be able to support exhaust line design. It could be used, coupled with three-way catalyst models and/or engine models, to support RDE and worst cases model-based design investigation through a reduction of needed tests. It can also be used for a wider scope of investigated conditions.

Acknowledgments

Special thanks are accorded to Toyota Motor Corporation members, N. Ohashi, S. Nakagoshi, and Y. Namura, for their financial support and their assistance during the project.

We also want to thank M. Guingouain for his advices during the projects of GPF modeling.

Nomenclature

A. Latin Letters

a_1	constant in channel pressure drop correlation	–
A	pre-exponential factor	units depend on rate expression
A_{ze}	pore size distribution constant	$(mol/J)^2$
b	unit cell diameter	m
c	concentration	mol/m^3
C_P	specific heat capacity	$J/(kg \cdot K)$
$cpsm$	channels per square meter	$cells/m^2$
d	hydraulic diameter of clean channel	m
d_c	unit collector diameter	m
$d_{c,0}$	unit collector diameter of clean filter	m
d_h	hydraulic diameter of a channel	m
D_{knud}	Knudsen diffusivity	m^2/s

D_{mol}	molecular diffusivity	m^2/s
D_{part}	particle diffusion coefficient	m^2/s
d_{pore}	mean pore size	m
D_w	effective diffusivity	m^2/s
E	activation energy	J/mol
f_w	geometric parameter	–
h	heat transfer coefficient	$W/(m^2 \cdot K)$
H	heat source component	W/m^3
k	permeability	m^2
k_p	apparent permeability of the particulate layer	m^2
k_0	permeability of the clean filter	m^2
M_g	gas molecular mass	kg/mol
M_j	molecular mass of each species	kg/mol
m_p	cake soot mass	kg
m_w	deposited particulate mass per unit collector	kg
n	filtration efficiency	–
$n_{clean, D}$	diffusion mechanism correction factor	–
$n_{clean, R}$	interception mechanism correction factor	–
p_j	partial pressure	Pa
$p_{sat, j}$	saturation pressure	Pa
R	universal gas constant	$J/(mol \cdot K)$
r	radial coordinate	–
R_k	species reaction rate	$mol/(m^3 \cdot s)$
R'_k	soot reaction rate	$mol/(molC \cdot s)$
S	heat source term	W/m^3
S_F	monolith specific surface area	m^2/m^3
T	temperature	K
t	time	s
u_w	velocity perpendicular to wall	m/s
v	velocity	m/s
w	dimension perpendicular to wall surface	–
W_0	total volume of micropores per reactor volume	m^3/m^3
w_c	washcoat layer thickness	m
w_p	soot layer thickness	m
w_w	wall thickness	m
x_i	wall thickness coordinate	m
y_j	molar fraction	–
z	axial coordinate along monolith	m

B. Greek Letters

a	dependency from oxygen concentration	–
a_1	constant in channel pressure drop correlation	–
ΔH_k	reaction heat	J/mol
Δw	elementary width in the wall direction	m
ε	macroscopic void fraction	–
ε_0	porosity of the clean filter	–
λ	thermal conductivity	$W/(m \cdot K)$
μ	dynamic viscosity	$Pa \cdot s$
ρ	density	kg/m^3
ρ_p	soot packing density	kg/m^3
$\rho_{soot, w}$	soot packing density in wall	kg/m^3

C. Subscripts and Superscripts

g - exhaust gas

j - species index

k - reaction index

s - solid

w - substrate wall

D. Acronyms

CC - close-coupled

DPF - diesel particulate filter

FWC - four-way catalyst

GDI - gasoline direct injection

GPF - gasoline particulate filter

PM - particulate matter

RDE - real driving emissions

TWC - three-way catalyst

UF - underfloor

References

1. European Environmental Agency, "Monitoring CO2 Emissions from Passenger Cars and Vans in 2015," European Environmental Agency Report, No 27, 2016, accessed December 22, 2017, Available: https://www.eea.europa.eu/publications/monitoring-co-2-emissions-from.

2. Andersson, J., Giechaskiel, B., Muñoz-Bueno, R., Sandbach, E. et al., "Particle Measurement Programme (PMP) Light-Duty Inter-Laboratory Correlation Exercise (ILCE_LD) Final Report," European Commission Joint Research Centre, Report EUR 22775 EN, 2007, accessed December 22, 2017, Available: http://publications.jrc.ec.europa.eu/repository/bitstream/111111111/429/2/7386%20-%20PMP_LD_final.pdf.

3. Spiess, S., Wong, K.F., Richter, J.M., and Klingmann, R., "Investigations of Emission Control Systems for Gasoline Direct Injection Engines with a Focus on Removal of Particulate Emissions," *Topics in Catalysis* 56 (2013): 434-439, Springer.

4. Koltsakis, G., Haralampous, O., Depcik, C., and Ragone, J.C., "Catalyzed Diesel Particulate Filter Modeling," *Reviews in Chemical Engineering* 29 (2013): 1-61, De Gruyter.

5. Haralampous, O.A. and Koltsakis, G.C., "Back-Diffusion Modeling of NO2 in Catalyzed Diesel Particulate Filters," *Industrial & Engineering Chemistry* 43, no. 4 (2004): 875-883.

6. Saito, C., Nakatani, T., Miyairi, Y., Yuuki, K. et al., "New Particulate Filter Concept to Reduce Particle Number Emissions," SAE Technical Paper 2011-01-0814, 2011, doi:10.4271/2011-01-0814.

7. Myung, C.L. and Park, S., "Exhaust Nanoparticle Emissions from Internal Combustion Engines: A Review," *International Journal of Automotive Technology* 13, no. 1 (2012): 9-22, Springer.

8. Karamitros, D. and Koltsakis, G., "Model-Based Optimization of Catalyst Zoning on SCR-Coated Particulate Filters," *Chemical Engineering Science* 173 (2017): 514-524, Elsevier.

9. Choi, S. and Seong, H., "Oxidation Characteristics of Gasoline Direct-Injection (GDI) Engine Soot: Catalytic Effects of Ash and Modified Kinetic Correlation," *Combustion and Flame* 162, no. 6 (2015): 2371-2389, Elsevier.

10. Ohara, E., Mizuno, Y., Miyairi, Y., and Mizutani, T., "Filtration Behavior of Diesel Particulate Filters," SAE Technical Paper 2007-01-0921, 2007, doi:10.4271/2007-01-0921.

11. Gutfinger, C. and Tardos, G., "Theoretical and Experimental Investigation on Granular Bed Dust Filters," *Atmospheric Environment* 13 (1978): 853-867, Elsevier.

12. Dullien, F., *Porous Media: Fluid Transport and Pore Structure*, (New York: Elsevier, Academic Press, 1979).

13. Konstandopoulos, A.G., Kostoglou, M., Skaperdas, E., Papaioannou, E. et al., "Fundamental Studies of Diesel Particulate Filters: Transient Loading,

Regeneration and Aging", SAE Technical Paper 2000-01-1016, 2000, doi:10.4271/2000-01-1016.

14. Lee, K.W. and Gieseke, J.A., "Collection of Aerosol Particles by Packed Beds," *Environmental Science & Technology* 13, no. 4 (1979): 466-470.

15. Kladopoulou, E., Yang, S., Johnson, J., Perker, G. et al., "A Study Describing the Performance of Diesel Particulate Filters during Loading and Regeneration—A Lumped Parameter Model for Control Applications," SAE Technical Paper 2003-01-0842, 2003, doi:10.4271/2003-01-0842.

16. Haralampous, O., Kandylas, I., Koltsakis, G., and Samaras, Z., "Diesel Particulate Filter Pressure Drop. Part I: Modeling and Experimental Validation," *International Journal of Engine Research* 5 (2004): 149-162.

17. Konstandopoulos, A.G., Skaperdas, E., Warren, J., and Allansson, R., "Optimized Filter Design and Selection Criteria for Continuously Regenerating Diesel Particulate Traps," SAE Technical Paper 1999-01-0468, 1999, doi:10.4271/1999-01-0468.

18. Green Car Congress, "BASF's New Four-Way Conversion Catalyst: TWC Plus Particulate Filter," Green Car Congress, 2014, accessed December 22, 2017, Available: http://www.greencarcongress.com/2014/12/20141210-basf.html.

19. Richter, J.M., Klingmann, R., Spiess, S., and Wong, K.F., "Application of Catalyzed Gasoline Particulate Filters to GDI Vehicles", *SAE Int. J. Engines* 5, no. 3 (2012): 1361-1370, doi:10.4271/2012-01-1244.

20. Boger, T., Gunasekaran, H., Bhargava, R., and Bischof, C., "Particulate Filers for DI Gasoline Engines", *MTZ* 74, no. 06 (2013).

21. Opitz, B., Drochner, A., Vogel, H., and Votsmeier, M., "An Experimental and Simulation Study on the Cold Start Behavior of Particulate Filters with Wall Integrated Three Way Catalyst," *Applied Catalysis B: Environmental* 144 (2014): 203-215, Elsevier.

22. Konieczny, R., Zaldua Moreno, N., Brück, R., "The Challenge of Future Particle Number Legislation; Particle Formation and Particle Reduction," *6th International Exhaust Gas and Particulate Emissions Forum*, Ludwigsburg, Germany, 2010.

23. Nicolin, P., Rose, D., Kunath, F., and Boger, T., "Modeling of the Soot Oxidation in Gasoline Particulate Filters," *SAE Int. J. Engines* 8, no. 3 (2015): 1253-1260, doi:10.4271/2015-01-1048.

24. Boger, T., Rose, D., Nicolin, P., Gunasekaran, N. et al., "Oxidation of Soot (Printex® U) in Particulate Filters Operated on Gasoline Engines," *Emission Control Science and Technology* 1 (2015): 49–63, Springer Paper.

25. Lambert, C., Bumbaroska, M., Dobson, D., Hangas, J. et al., "Analysis of High Mileage Gasoline Exhaust Particle Filters," *SAE Int. J. Engines* 9, no. 2 (2016): 1296-1304, doi:10.4271/2016-01-0941.

26. Lambert, C., Chanko, T., Jagner, M., Hangas, J. et al., "Analysis of Ash in Low Mileage, Rapid Aged, and High Mileage Gasoline Exhaust Particle Filters," *SAE Int. J. Engines* 10, no. 4 (2017): 1595-1603, doi:10.4271/2017-01-0930.

27. Custer, N., Kamp, C., Sappok, A., Pakko, J. et al., "Lubricant-Derived Ash Impact on Gasoline Particulate Filter Performance," *SAE Int. J. Engines* 9, no. 3 (2016): 1604-1614, doi:10.4271/2016-01-0942.

28. Seong, H.J., Choi, S., Howden, K., and Singh, G., "Ash-Durable Catalyzed Filters for Gasoline Direct Injection (GDI) Engines," Argonne National Laboratory, (AOP lab call), 2016, accessed December 22, 2017, Available: https://energy.gov/sites/prod/files/2016/06/f32/ace024_seong_2016_o_web.pdf.

About the Authors

Michail Mitsouridis followed a double degree program, receiving his Mechanical Engineering Diploma from Aristotle University of Thessaloniki and his Diplôme d'Ingénieur from École Centrale Paris. He is currently a PhD candidate at Aristotle University of Thessaloniki, under the supervision of Prof. Koltsakis. In his PhD thesis he studies the performance of catalyzed gasoline particulate filters with the support of mathematical models. He is also employed as project manager at Exothermia SA.

Dimitrios Karamitros is a mechanical engineer. He holds an MSc in powertrain engineering from IFP School in France and a PhD in exhaust aftertreatment modeling from Aristotle University of Thessaloniki in Greece. He is currently employed as project engineer at Exothermia SA.

Dr. Grigorios C. Koltsakis received his diploma and PhD from the Department of Mechanical Engineering of Aristotle University of Thessaloniki. He is currently a Professor at the Laboratory of Applied Thermodynamics of Aristotle University, specializing in internal combustion engines and exhaust aftertreatment technology. He has authored more than 100 papers in scientific Journals and international conferences and holds 2 patents. Dr. Koltsakis has received the Arch T. Colwell Merit Award, two Oral presentation awards from SAE International as well as the first Innovation prize from the Research Committee of Aristotle University. Dr. Koltsakis leads a research group active in the simulation and evaluation of after-treatment technologies in close collaboration with the automotive industry. He is a co-founder and scientific director of Exothermia SA, a university spin-off company developing exhaust system simulation software.

Julie Le Louvetel-Poilly did a PhD on fluid mechanics and turbulent flow from 2005 to 2008. From 2008 to 2014, she did fundamental research in Keio University (Japan) and Beihang University (China) on fluid mechanism modeling. She joined Toyota Motor Europe in 2015 as engineer in Powertrain Model Based Development group, where she is in charge to develop the exhaust aftertreatment model for gasoline engines.

Kotaro Maeda studied mechanical engineering at Tokushima University in Japan. He joined Toyota Motor Corporation diesel engine system development and mass production calibration department from 2007 to 2014. He joined Toyota Motor Europe in 2015 as manager in Powertrain Model Based Development group, where he is in charge of the development of engine and exhaust modeling.

Francois-Alexandre Lafossas graduated from MATMECA engineering school in 2000. He started his activity in 2000 in IFP Energies Nouvelles as combustion research engineer to support the development of gasoline and diesel combustion models in 3D and 0D applications. He joined Toyota Motor Europe in 2009 as senior engineer in the Powertrain Pre-development group where he was in charge to develop the integration of a modeling tool in the pre-development process. He became manager in Model Based Development group in 2011 with a focus on expanding the modeling area from combustion engines to drivetrain and hybrid systems. In 2015, he became senior manager of the Model Based Development group. He also contributed to more than 30 international publications in the field of powertrain and engine systems modeling.

SAE book category descriptions

An **SAE Core Title** delivers core content covering all the essentials of a specific area of engineering.

An **SAE Deep Dive** provides advanced, expert-level coverage of an engineering subject.

An **SAE Handbook** is a practical guide or reference for a particular subject of applied engineering.

An **SAE Technical Paper Compilation** gathers judiciously selected papers important to a topic or theme, often with extensive editorial input and additional content.

An **SAE Compendium** contains an update on the critical developments on particular topic, such as important research, events, legislation, and standards.

index

A

B

C

H

I

U

V

Z

Printed in the United States
By Bookmasters